PROCESS MODELING, SIMULATION, and ENVIRONMENTAL APPLICATIONS in CHEMICAL ENGINEERING

PROCESS MODELING, SIMULATION, and ENVIRONMENTAL APPLICATIONS in CHEMICAL ENGINEERING

Edited by

Bharat A. Bhanvase, PhD

Rajendra P. Ugwekar, PhD

Apple Academic Press Inc.
3333 Mistwell Crescent
Oakville, ON L6L 0A2
Canada

Apple Academic Press Inc.
9 Spinnaker Way
Waretown, NJ 08758
USA

First issued in paperback 2021

Exclusive worldwide distribution by CRC Press, a member of Taylor & Francis Group
No claim to original U.S. Government works

ISBN 13: 978-1-77463-604-6 (pbk)
ISBN 13: 978-1-77188-324-5 (hbk)

Library and Archives Canada Cataloguing in Publication

Process modeling, simulation, and environmental applications in chemical engineering / edited by Bharat A. Bhanvase, PhD, Rajendra P. Ugwekar, PhD.

Includes bibliographical references and index.
Issued in print and electronic formats.
Process modeling, simulation, and environmental applications in chemical engineering.
ISBN 978-1-77188-324-5 (hardcover).--ISBN 978-1-77188-325-2 (pdf)

1. Chemical engineering--Mathematical models. 2. Chemical engineering--Computer simulation. 3. Chemical engineering--Environmental aspects. I. Bhanvase, Bharat A., author, editor II. Ugwekar, Rajendra P., author, editor
TP155.2.M35P76 2016 660'.281 C2016-904681-8 C2016-904682-6

Library of Congress Cataloging-in-Publication Data

Names: Bhanvase, Bharat A., editor. | Ugwekar, Rajendra P., editor.
Title: Process modeling, simulation, and environmental applications in chemical engineering / editors, Bharat A. Bhanvase, PhD, Rajendra P. Ugwekar, PhD.
Description: Toronto ; New Jersey : Apple Academic Press, 2017. | Includes bibliographical references and index.
Identifiers: LCCN 2016030184 (print) | LCCN 2016031088 (ebook) | ISBN 9781771883245 (hardcover : alk. paper) | ISBN 9781771883252 (ebook) | ISBN 9781771883252 ()
Subjects: LCSH: Chemical engineering--Mathematical models. | Chemical engineering--Computer simulation. | Chemical engineering--Environmental aspects.
Classification: LCC TP155.2.M35 P76 2017 (print) | LCC TP155.2.M35 (ebook) | DDC 660/.281--dc23
LC record available at https://lccn.loc.gov/2016030184

Apple Academic Press also publishes its books in a variety of electronic formats. Some content that appears in print may not be available in electronic format. For information about Apple Academic Press products, visit our website at **www.appleacademicpress.com** and the CRC Press website at **www.crcpress.com**

CONTENTS

List of Contributors .... *ix*

List of Abbreviations .... *xiii*

List of Symbols .... *xv*

Preface .... *xvii*

About the Editors .... *xix*

Introduction .... *xxi*

PART I: MODELING AND SIMULATION **1**

1. Discharge and Pressure Loss Coefficient Analysis of Non-Newtonian Fluid Flow Through Orifice Meter Using CFD **3**

A. Tamrakar and S. A. Yadav

2. Numerical Simulation of Fixed Bed Liquid Chromatography in Multiple Columns **19**

A. Nag and B. C. Bag

3. Statistical Modeling for Adsorption of Congo Red onto Modified Bentonite **35**

T. Mohan Rao and V. V. Basava Rao

PART II: ENVIRONMENTAL APPLICATIONS **53**

4. Surface Altered Alumino-Silicate Resin (Zeolite-Y) for Remediation of Oil Spillage **55**

S. U. Meshram, C. M. Shah, and H. J. Balani

5. Removal of Cr(VI) by Using Sweetlime Peel Powder in a Fixed Bed Column **71**

N. M. Rane, S. P. Shewale, A. V. Kulkarni, and R. S. Sapkal

6. Removal of Cr(VI) from Wastewater Using Red Gram Husk as Adsorbent **87**

V. S. Wadgaonkar and R. P. Ugwekar

7. **Low Cost Adsorbents in the Removal of Cr(VI), Cd and Pb(II) from Aqueous Solution** 97
P. Semil and A. Awasthi

8. **Adsorption of Anionic Dye onto TBAC-Modified Halloysite Nanotubes** 119
P. V. Mankar, S. A. Ghodke, S. H. Sonawane, and S. Mishra

9. **Kinetic Study of Adsorption of Nickel on GNS/δ-MnO_2** 141
M. P. Deosarkar, S. Varma, D. Sarode, S. Wakale, and B. A. Bhanvase

10. **Ultrasound Assisted Synthesis of Hydrogels and Its Effects on Water/Dye Intake** 161
R. S. Chandekar, K. Pushparaj, G. K. Pillai, M. Zhou, S. H. Sonawane, M. P. Deosarkar, B. A. Bhanvase, and M. Ashokkumar

PART III: MATERIALS AND APPLICATIONS 179

11. **Ultrasonically Created Rectangular Shaped Zinc Phosphate Nanopigment** 181
S. E. Karekar, A. J. Jadhav, C. R. Holkar, N. L. Jadhav, D. V. Pinjari, A. B. Pandit, B. A. Bhanvase, and S. H. Sonawane

12. **Biosynthesis of Silver Nanoparticles Using *Raphanus sativus* Extract** 199
P. D. Jolhe, B. A. Bhanvase, V. S. Patil, and S. H. Sonawane

13. **Activated Carbon from Karanja (*Pongamia pinnata*) Seed Shell By Chemical Activation with Phosphoric Acid** 217
M. L. Meshram and D. H. Lataye

14. **Rice Husk Based Co-Firing Plants in India: A Green Perspective** 227
S. U. Meshram, A. Mohan, and P. S. Dronkar

15. **Foamability of Foam Generated by Use of Surf Excel and Sodium Lauryl Sulfate** 247
P. Chattopadhyay, R. A. Karthick, and P. Kishore

16. **Production of Zinc Sulphide Nanoparticles Using Continuous Flow Microreactor** 257
K. Ansari, S. H. Sonawane, B. A. Bhanvase, M. L. Bari, K. Ramisetty, L. Shaikh, Y. Pydi Setty, and M. Ashokumar

PART IV: PROCESSES AND APPLICATIONS **279**

17. Hydrogenation with Respect to Rancidity of Foods **281**
D. C. Kothari, P. V. Thorat, and R. P. Ugwekar

18. Experimental Studies on a Plate Type Heat Exchanger for Various Applications **297**
V. D. Pakhale and V. A. Arwari

19. Ultrasound Assisted Extraction of Betulinic Acid from Leaves of *Syzygium cumini* (Jamun) **307**
S. V. Admane, S. M. Chavan, and S. G. Gaikwad

PART V: ANALYTICAL METHODS **317**

20. Separation, Analysis and Quantitation of Hesperidin in Citrus Fruits Peels Using RPHPLC-UV **319**
S. Kulkarni and B. A. Bhanvase

21. Quantification of Aluminum Metal in Cosmetic Products by Novel Spectrophotometric Method **329**
S. B. Gurubaxani and T. B. Deshmukh

Index *337*

LIST OF CONTRIBUTORS

S. V. Admane
Chemical Engineering Department, Sinhgad College of Engineering, Vadagaon, Pune, Maharashtra, India

K. Ansari
Chemical Engineering Department, National Institute of Technology, Warangal, Telangana, India

V.A. Arwari
Chemical Engineering Department, MIT Academy of Engineering, Alandi (D), Pune–412105, Maharashtra, India

M. Ashokumar
School of Chemistry, University of Melbourne, Parkville, VIC 3010, Australia

A. Awasthi
Department of Chemical Engineering, Harcourt Butler Technological Institute, Kanpur, Uttar Pradesh, India

B. C. Bag
Defence Research and Development Establishment, Ministry of Defence, Government of India Mahanagar Palika Marg, Nagpur–440001, India

H. J. Balani
Research Assistant, Department of Oil, Fats and Surfactants Technology, Laxminarayan Institute of Technology, RTM Nagpur University, Amravati Road, Nagpur–33, India

M. L. Bari
Institute of Chemical Technology, North Maharashtra University Jalgaon, Maharashtra–425001, India

B. A. Bhanvase
Chemical Engineering Department, Laxminarayan Institute of Technology, Rashtrasant Tukadoji Maharaj Nagpur University, Nagpur, Maharashtra, India

R. S. Chandekar
Department of Chemical Engineering, Vishwakarma Institute of Technology, Pune–411037, Maharashtra, India

S. M. Chavan
Chemical Engineering Department, Sinhgad College of Engineering, Vadagaon, Pune, Maharashtra, India

M. P. Deosarkar
Department of Chemical Engineering, Vishwakarma Institute of Technology, Pune–411037, Maharashtra, India

T. B. Deshmukh
Department of Chemistry, Institute of Science, Civil Lines, Nagpur–440001, India. E-mail: sevakgurubaxani@gmail.com

P. S. Dronkar
Laxminarayan Institute of Technology, RTM Nagpur University, Nagpur, Maharashtra, India

S. G. Gaikwad
Chemical Engineering Department, National Chemical Laboratory (NCL), Pune, Maharashtra, India

S. A. Ghodke
Chemical Engineering Department, Sinhgad College of Engineering, Pune, Maharashtra, India

S. B. Gurubaxani
Department of Chemistry, Institute of Science, Civil Lines, Nagpur–440001, India. E-mail: sevakgurubaxani@gmail.com

C. R. Holkar
Chemical Engineering Department, Institute of Chemical Technology, Matunga, Mumbai 400019, India

A. J. Jadhav
Chemical Engineering Department, Institute of Chemical Technology, Matunga, Mumbai 400019, India

P. D. Jolhe
University Institute of Chemical Technology, North Maharashtra University, Jalgaon, Maharashtra, India

S. E. Karekar
Chemical Engineering Department, Institute of Chemical Technology, Matunga, Mumbai 400019, India

D. C. Kothari
Chemical Engineering and Polymer Technology Department, Shri Shivaji Education Society, Amravati's College of Engineering and Technology, Babulgaon (Jh.), Akola, Maharashtra, India

A. V. Kulkarni
Department of Chemical Engineering, MIT Academy of Engineering, Alandi, Pune, India

S. Kulkarni
Chemical Engineering Department, Vishwakarma Institute of Technology, Pune, Maharashtra, India

D. H. Lataye
Department of Civil Engineering, V.N.I.T., Nagpur, India

P. V. Mankar
Chemical Engineering Department, Sinhgad College of Engineering, Pune, Maharashtra, India

M. L. Meshram
Department of Civil Engineering, Laxminarayan Institute of Technology, Nagpur, India. E-mail: manojlmesh@gmail.com

S. U. Meshram
Laxminarayan Institute of Technology, RTM Nagpur University, Nagpur, Maharashtra, India

S. Mishra
University Institute of Chemical Technology, North Maharashtra University, Jalgaon, Maharashtra, India

A. Mohan
Laxminarayan Institute of Technology, RTM Nagpur University, Nagpur, Maharashtra, India

A. Nag
Defence Research and Development Establishment, Ministry of Defence, Government of India Mahanagar Palika Marg, Nagpur–440001, India

V. D. Pakhale
Chemical Engineering Department, MIT Academy of Engineering, Alandi (D), Pune–412105, Maharashtra, India

A. B. Pandit
Chemical Engineering Department, Institute of Chemical Technology, Matunga, Mumbai 400019, India

V. S. Patil
University Institute of Chemical Technology, North Maharashtra University, Jalgaon, Maharashtra, India

G. K. Pillai
Department of Chemical Engineering, Vishwakarma Institute of Technology, Pune–411037, Maharashtra, India

D. V. Pinjari
Chemical Engineering Department, Institute of Chemical Technology, Matunga, Mumbai 400019, India, E-mail: dv.pinjari@ictmumbai.edu.in, dpinjari@gmail.com

K. Pushparaj
Department of Chemical Engineering, Vishwakarma Institute of Technology, Pune–411037, Maharashtra, India

V. V. Basava Rao
Faculty of Technology, University College of Technology, Osmania University, Hyderabad, India

T. Mohan Rao
Department of Chemical Engineering, Bapatla Engineering College, Bapatla, India

Y. Pydi Setty
Chemical Engineering Department, National Institute of Technology, Warangal, Telangana, India

K. Ramisetty
Chemical Engineering Department, Institute of Chemical Technology, Mumbai, Maharashtra, India

N. M. Rane
Department of Chemical Engineering, MIT Academy of Engineering, Alandi, Pune, India

R. S. Sapkal
UDCT, Sant Gadge Baba Amravati University, Amravati, India. E-mail: nmrane@chem.maepune.ac.in

D. Sarode
Chemical Engineering Department, Vishwakarma Institute of Technology, Pune, Maharashtra, India

P. Semil
Department of Chemical Engineering, Harcourt Butler Technological Institute, Kanpur, Uttar Pradesh, India

C. M. Shah
Research Assistant, Department of Oil, Fats and Surfactants Technology, Laxminarayan Institute of Technology, RTM Nagpur University, Amravati Road, Nagpur–33, India

L. Shaikh
Chemical Engineering Process Division, National Chemical Laboratory, Pune, Maharashtra, India

S. P. Shewale
Department of Chemical Engineering, MIT Academy of Engineering, Alandi, Pune, India

S. H. Sonawane
Chemical Engineering Department, National Institute of Technology, Warangal, Telangana–506004, India

A. Tamrakar
Chemical Engineering Department, MIT Academy of Engineering, Alandi (D), Pune–412105, Maharashtra, India

P. V. Thorat
Chemical Engineering and Polymer Technology Department, Shri Shivaji Education Society, Amravati's College of Engineering and Technology, Babulgaon (Jh.), Akola, Maharashtra, India

R. P. Ugwekar
Chemical Engineering Department, Laxminarayan Institute of Technology, Rashtrasant Tukadoji Maharaj Nagpur University, Nagpur, Maharashtra, India

S. Varma
Chemical Engineering Department, Vishwakarma Institute of Technology, Pune, Maharashtra, India

V. S. Wadgaonkar
Petrochemical Engineering Department, Maharashtra Institute of Technology, Pune, Maharashtra, India

S. Wakale
Chemical Engineering Department, Vishwakarma Institute of Technology, Pune, Maharashtra, India

S. A. Yadav
Chemical Engineering Department, MIT Academy of Engineering, Alandi (D), Pune–412105, Maharashtra, India

M. Zhou
School of Chemistry, University of Melbourne, VIC 3010, Australia

LIST OF ABBREVIATIONS

AA	acrylic acid
ACGIH	American Conference of Governmental Industrial Hygenists
APS	ammonium persulfate
BA	betulinic acid
BCM	billion cubic meters
BDST	bed depth service time
BET	brunauer emmet teller
BHA	butylated hydroxyanisole
CB	carbon black
CEC	cation exchange capacity
CFA	coal fly ash
CFD	computational fluid dynamics
CMC	critical micelles concentration
CR	congo red
DDW	double distilled water
DLS	dynamic light-scattering
DOE	design of experiments
DST	Department of Science and Technology
EDS	equilibrium degree of swelling
EDX	energy dispersive X-ray
ESP	electro static precipitator
FDM	finite difference method
FEM	finite element method
FTIR	fourier transform infrared ray
FVM	finite volume method
FWHM	full-width at half-maximum height
GESAMP	Group of Experts on Scientific Aspects of Marine Environmental Protection
GNS	graphene nanosheets
HDTMA-Br	hexadecyl trimethyl ammonium bromide

HNTs	halloysite nanotubes
IOCC	iron oxide-coated cement
IR	infrared spectroscopy
KBr	potassium bromide
LDPE	low-density polyethylene
MB	methylene blue
MNRE	Ministry of New and Renewable Energy
MTMS	methyltrimethoxysilane
MTPS	Mini Thermal Power Stations
NCL	National Chemical Laboratory
NOAA	National Oceanic and Atmospheric Administration
NR	number of runs
NRC	National Research Council
OAC	oil adsorption capacity
ODS	octyl decyl silane
OSTM	oil spill trajectory models
PEG	polyethylene glycol
PVP	poly vinyl pyrrolidone
RHA	rice husk ash
RSD	relative standard deviation
RTD	residence time distribution
SDS	sodium dodecyl sulfate
SEM	scanning electron microscope
SLS	sodium lauryl sulfate
SPSS	sigma scan pro software
TBAC	tetrabutylammonium chloride
TBHQ	tertiary butylhydroquinone
TEM	transmission electron microscopy
TGA	thermo-gravimetric analysis
ULPROM	ultra-low pressure reverse osmosis membrane
USGS	U.S. Geological Survey
WEC	World Energy Council
WSH	water sanitation and hygiene
XRD	X-ray diffraction

LIST OF SYMBOLS

A	total heat transfer area (m^2)
a_i	constant in Langmuir isotherm for component *i*
b_i	adsorption equilibrium constant for component *i*
B_i	Biot number of mass transfer
C_{bi}	bulk-fluid phase concentration of component *i*
C_{fi}	feed concentration profile of component *i*, a time dependent variable
C_{oi}	concentration used for nondimensionalization
C_{pi}	concentration of component *i* inside the particle macro-pores
C_{pi}^*	equilibrium concentration of component *i* in the solid phase of particle
D	inner diameter of a column
D_{bi}	axial dispersion coefficient of component *i*
D_{pi}	effective diffusivity of component *i,* porosity not included
k_i	film mass transfer coefficient of component i
L	packed column length
Pe_{Li}	Peclet number of axial dispersion for component *i*
Q	mobile phase volumetric flow rate
$Q_{Average}$	average heat transfer between hot and cold water (W)
R	radial coordinates for particle
R_p	particle radius
T_1	hot water temperature at inlet (°C)
T_2	hot water temperature at outlet (°C)
T_3	cold water temperature at outlet (°C)
T_4	cold water temperature at inlet (°C)
U	overall heat transfer coefficient (W/m^{2-} °C)
V_1	cold water supply valve
V_2	hot water flow control valve
V_3	cold water flow control valve
V_4	drain valve
V_{CW}	cold water flow rate (L/min)

V_{HW}	hot water flow rate (L/min)
Z	axial coordinate
z	dimensionless axial coordinate, Z/L

Greek Letters

ΔT_{CW}	temperature change of cold water (°C)
Δ_{THW}	temperature change of hot water (°C)
ΔT_{ln}	log mean temperature difference (LMTD)
ΔT_{outlet}	temperature difference of hot and cold water
u	interstitial velocity
ε_b	bed void volume fraction
ε_p	particle porosity
η_i	dimensionless constant
$\Theta_{average}$	average thermal length
τ	dimensionless time

Subscripts

b	bulk-fluid phase
i	*i*-th component
L	liquid phase

PREFACE

This book, *Process Modeling, Simulation and Environmental Applications in Chemical Engineering*, is the result of the 2015 national conference REACT-15, organized by the Laxminarayan Institute of Technology, Nagpur, Maharashtra, India. Out of 60 original research articles, we have selected 21 articles to publish in this book. Our goal is to compile the research articles related to applications of chemical processes for environment, materials, and modeling and simulation of chemical processes in a single book that can benefit students, researchers, faculties and industrialists concurrently. This book covers different areas of chemical engineering and technology.

The important and recent topics in the field of chemical engineering and technology include modeling and simulation, material synthesis, wastewater treatment, analytical techniques, and microreactors are presented in this book. We, the editors, are pleased to bring out this special type of book that reports on these areas.

We would also like to acknowledge the team of Apple Academic Press, Mr. Ashish Kumar, President and Publisher, and Mr. Rakesh Kumar, for their prompt and supportive attention to all our queries related to editorial assistance.

With all humbleness, we acknowledge the initial strength derived for this book from Dr. V. S. Sapkal, Former Vice Chancellor, RTM Nagpur University, Nagpur; the Eminent Scientist Dr. B. D. Kulkarni of National Chemical Laboratory, Pune; and Professor A.B. Pandit, Institute of Chemical Technology, Mumbai, for their inspiration and unwavering encouragement. The editors would like to acknowledge RTM Nagpur University authorities, various research laboratories, and advisory and working committee for their support and encouragement from time to time. We would like to thank all the contributors and their respective organizations.

—Dr. Bharat A. Bhanvase
Dr. Rajendra P. Ugwekar

ABOUT THE EDITORS

Bharat A. Bhanvase, PhD

Dr. Bharat A. Bhanvase is currently working as Associate Professor in the Chemical Engineering Department at the Laxminarayan Institute of Technology, Rashtrasant Tukadoji Maharaj Nagpur University, Nagpur, Maharashtra, India. His research interests are focused on conventional and cavitation-based synthesis of nanostructured materials, ultrasound-assisted processes, polymer nanocomposites, heat transfer enhancement using nanofluid, process intensification, and microreactors for nanoparticle and chemical synthesis. He has published 38 and four articles in international and national journals. He has written nine book chapters in internationally renowned books and applied for three Indian patents. He is the recipient of a Summer Research Fellowship from the Indian Academy of Sciences, Bangalore, India, in 2009. He has more than 13 years of teaching experience. He has completed a research project received from the University of Pune. He is a reviewer for various international journals. Dr. Bhanvase completed his BE in Chemical Engineering from the University of Pune, his ME in Chemical Engineering from Bharati Vidyapeeth University Pune, and his PhD in Chemical Engineering from the University of Pune, India.

Rajendra P. Ugwekar, PhD

Rajendra P. Ugwekar, PhD, is currently working as Associate Professor and Head in the Chemical Engineering Department at the Laxminarayan Institute of Technology, Rashtrasant Tukadoji Maharaj Nagpur University, Nagpur, Maharashtra, India. His research interests are focused on hydrogen energy, nanotechnology, wastewater treatment, membrane separation technologies, and heat transfer enhancement using nanofluid. He has published eight articles in international and national journals and has presented papers at international and national conferences. He has more than 23 years of teaching experience and three years of industrial experience as well. He has completed research projects received from the All India Council for Technical Education and he has worked with many MTech and PhD students. He has worked as Head of the Department at Anuradha Engineering College, Chikhali, Buldhana, India, and at the Priyasharshani Institute of Engineering and Technology, Nagpur, India. He is a trainer and motivator for entrepreneurship. Dr. Ugwekar has completed his BTech in Chemical Engineering from Nagpur University, his MTech in Chemical Engineering from Nagpur University, Nagpur, and his PhD in Chemical Technology from Sant Gadge Baba Amravati University, Maharashtra, India.

INTRODUCTION

This book is divided into five parts. The volume includes selected articles from the National Conference on Recent Trends in Chemical Engineering and Technology in India, held at Laxminarayan Institute of Technology, RTM Nagpur University, Nagpur in the year 2015. The five parts are modeling and simulation, environmental applications, materials and applications, processes and applications, and analytical methods.

The first part contains three chapters covering different areas of modeling and simulation of chemical processes. In Part I, we attempted to highlight modeling and simulation of chemical processes such as discharge and pressure loss coefficient analysis of non-Newtonian fluid flow through orifice meter using CFD, numerical simulation of fixed bed liquid chromatography in multiple columns and statistical modeling for adsorption of Congo red onto modified bentonite.

The second part of this book includes seven chapters related to environmental applications such as surface altered alumino-silicate resin (Zeolite-Y) for remediation of oil spillage, removal of Cr(VI) by using sweetlime peel powder in a fixed bed column, removal of Cr(VI) from wastewater using red gram husk as adsorbent, low cost adsorbents in the removal of Cr(vi), Cd and Pb(ii) from aqueous solution, adsorption of anionic dye onto TBAC-modified halloysite nanotubes, kinetic study of adsorption of nickel on GNS/δ-MnO_2, and ultrasound assisted synthesis of hydrogels and it's effects on water/dye intake. All these chapters are related to chemical processes for environmental applications.

The part three of this book consists of six chapters related to materials and applications. This part include the chapters such as ultrasonically created rectangular shaped zinc phosphate nanopigment, biosynthesis of silver nanoparticles using *Raphanus sativus* extract, activated carbon from karanja (*Pongamia pinnata*) seed shell by chemical activation with phosphoric acid, rice husk based co-firing plants in India: a green perspective, foamability of foam generated by use of surf excel and sodium lauryl sulfate, and production of zinc sulfide nanoparticles using continuous flow microreactor. In this section we have included the chemical processes such as ultrasound assisted synthesis, green synthesis, etc.

Part four of this book includes three chapters and that are hydrogenation with respect to rancidity of foods, experimental studies on a plate type heat exchanger for various applications and ultrasound assisted extraction of betulinic acid from leaves of syzygium cumini (Jamun). This section highlights the different methods such as ultrasound assisted extraction, heat transfer enhancement, etc.

Finally, part five of this book consists of analytical methods such as RPHPLC-UV, spectrophotometric method, etc., and the chapters are separation, analysis and quantitation of hesperidin in citrus fruits peels using RPHPLC-UV, and quantification of aluminum metal in cosmetic products by novel spectrophotometric method.

We believe that these varied chapters of this book will stimulate new ideas, methods, and applications in ongoing advances in this growing area of chemical engineering and technology. Also we believe that this book will give new insight in the area of environment, materials, and modeling and simulation of chemical processes to benefit students, researchers, faculties and industrialists concurrently. The book chapters included in this book are written by authors and editors have complied the book chapters and included in this book.

PART I

MODELING AND SIMULATION

CHAPTER 1

DISCHARGE AND PRESSURE LOSS COEFFICIENT ANALYSIS OF NON-NEWTONIAN FLUID FLOW THROUGH ORIFICE METER USING CFD

A. TAMRAKAR and S. A. YADAV

Chemical Engineering Department, MIT Academy of Engineering, Alandi (D), Pune–412105, Maharashtra, India

CONTENTS

1.1 Introduction 4
1.2 Theory 5
1.2.1 Discharge Coefficient 5
1.2.2 Pressure Loss Coefficient 7
1.3 Numerical Methodology 8
1.3.1 Geometry and Grid Details 8
1.3.2 Material Properties 9
1.3.3 Boundary Conditions 10
1.3.4 Solver Details 11
1.4 Results and Discussion 11
1.4.1 Discharge Coefficient 12
1.4.1.1 Variation of C_d with Pipe Diameter 12

1.4.1.2 Variation of C_d with Material Properties 12
1.4.1.3 Variation of C_d with β 13
1.4.2 Pressure Loss Coefficient 13
1.4.2.1 Variation of K_{or} with Pipe Diameter 13
1.4.2.2 Variation of K_{or} with Material Properties 16
1.4.2.3 Variation of K_{or} with β 16
1.5 Conclusion 16
Keywords 16
References 17

1.1 INTRODUCTION

Adequate knowledge and information of flow rates of various process streams plays a very important role in any chemical process industry, specifically when flow rate directly affects the purity of product and plant efficiency. Orifice meters are the most common type of meters used for flow measurement by various industries. Even though these meters have large pressure drop/losses and hence correspondingly large pumping cost, they are very simple in construction with no moving parts and are easy to install and replace. They are less costly compare to other devices and can be used for wide range of fluid flow rates. They can be used for flow measurement of gases, liquids, and slurries and can be operated at extreme operating conditions.

Orifice meter works on the Bernoulli's principle, which states the relationship between the pressure of the fluid and the fluid velocity. It can be used for measurement of either volumetric or mass flow rate of fluids. A large amount of literature and experimental work has been carried out for characterization of orifice meter over a wide range of beta ratio and Reynolds number [1–5]. A standard orifice plate is one of a variety of obstruction-type flow meters that is used extensively to measure the flow rate of fluid in a pipe; it consist of a thin plate placed inside the pipe. Plate has a hole in it mostly at the center. Orifices are also used for many engineering applications as restriction plates to reduce pressure or restrict flow, in air conditioning and water pipe system, hydraulic systems, etc. Even though studies in orifice plates have been done, gaps in the data still exist.

Except for rare cases, most of the literature and research work data has been focused on the discharge and pressure loss coefficient analysis for Newtonian fluid. However, studies in the field of non-Newtonian fluids have not been extensive, despite their importance in the field of polymer processing, flow of petroleum products, biomedical engineering, biochemical engineering, food processing, and mineral processing plants, where the liquid involved shows non-Newtonian character. In such applications, the flow remains laminar even at large flow rates [2–4].

1.2 THEORY

Orifice plate is a thin differential pressure-producing device, which is usually placed inside a pipe for flow rate measurement. As fluid passes through the orifice, there is a slight pressure build up on the upstream side but as fluid passes through the hole, fluid pressure decreases and velocity increases. Following parameters decides the characteristics of the flow flowing through orifice [3]:

1. Reynolds number;
2. Edge geometry of the orifice;
3. Ratio of orifice bore diameter to pipe diameter (β ratio);
4. Ratio between orifice plate thickness to bore diameter.

1.2.1 DISCHARGE COEFFICIENT

Variation in Reynolds number affects the discharge coefficient for all flow measuring devices (which works on the principle of differential pressure generation), because of which they can be used only over a certain range of flow rates. A particular orifice geometry can be used only over a particular Reynolds number constancy limit beyond which a large and abrupt change in discharge coefficient value takes place. A little downstream from the orifice plate, flow reaches at its maximum convergence, which is called *vena contracta*, where velocity reaches the maximum and pressure at its minimum. After this, flow expands and velocity decreases and pressure increases. Ratio between orifice diameter and pipe diameter is generally kept as 0.5, though it can be varied between 0.2 and 0.8. When fluid passes through the pipe irreversible energy loss takes place because of which actual discharge is always lesser than the theoretical discharge.

Discharge coefficient provides the correlation between actual and theoretical discharge [6].

$$C_d = \frac{Q_{actual}}{Q_{theoretical}}$$

Discharge coefficient (C_d) value also depends upon geometry of the orifice.

As shown in Figure 1.1 an obstruction to the fluid, passing through a pipe having diameter *D*, is provided by orifice plate with a central hole having bore diameter *d*. When fluid passes through orifice, on upstream side of constriction fluid velocity increases with decrease in pressure. The ratio between orifice diameter *d* and inner pipe diameter *D* is represented by beta ratio *β*.

$$\beta = \frac{d}{D} \tag{1}$$

For most of the commercial used orifice meters *β* ratio lies between 0.25 and 0.75. As per Bernoulli's equation, for flow through a pipe across two cross sectional position 1 and 2,

$$\frac{p_1}{\rho g} + \frac{v_1^2}{2g} + z_1 = \frac{p_2}{\rho g} + \frac{v_2^2}{2g} + z_2 \tag{2}$$

If the pipe is in horizontal position, $z_1 = z_2$

$$\frac{p_1}{\rho g} + \frac{v_1^2}{2g} = \frac{p_2}{\rho g} + \frac{v_2^2}{2g} \tag{3}$$

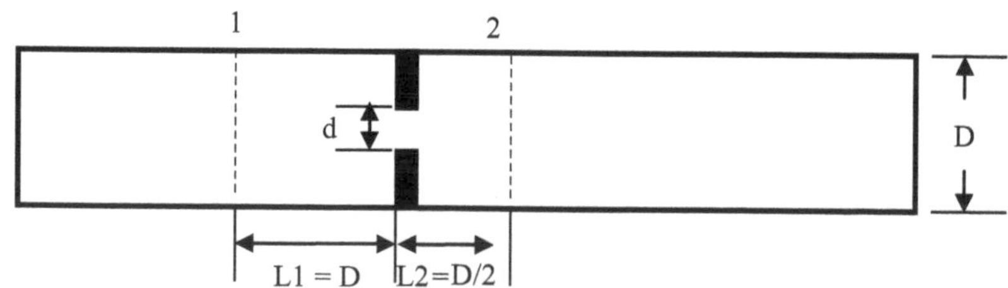

FIGURE 1.1 Schematic of orifice meter showing pressure tapings.

For incompressible flow across cross-sections 1 and 2 (Figure 1.1), by applying conservation of mass,

$$v_1A_1 = v_2A_2 \tag{4}$$

where A_1 and A_2 are the cross-sectional areas at location 1 and 2, and v_1 and v_2 are the average velocities at respective locations. At location 1 and 2 cross sectional areas are given as $A_1 = \pi D^2/4$ and $A_2 = \pi d^2/4$. After substituting the values of A_1, A_2 and β in Eq. (4),

$$v_1 = \beta^2 v_2 \tag{5}$$

On solving the Bernoulli equation for v_2,

$$v_2 = \sqrt{\frac{2(p_1 - p_2)}{\rho(1-\beta^4)}} \tag{6}$$

From the value of average velocity v_2 at location 2, volumetric flow rate $Q_{Theoretical}$ can be calculated as,

$$Q_{Theoretical} = A_2 v_2 = A_2\sqrt{\frac{2(p_1 - p_2)}{\rho(1-\beta^4)}} \tag{7}$$

Since there will always be irreversible losses when fluid flows through the pipe Q_{Actual} (actual flow rate) will always be less than $Q_{Theoretical}$. Discharge coefficient C_d gives correlation between actual and theoretical discharge.

$$Q_{actual} = C_d Q_{theoretical} \tag{8}$$

Hence, the actual volumetric flow rate will be,

$$Q_{Actual} = A_2 C_d\sqrt{\frac{2(p_1 - p_2)}{\rho(1-\beta^4)}} \tag{9}$$

1.2.2 PRESSURE LOSS COEFFICIENT

Miller (1990) defines the pressure loss coefficient as a ratio of total pressure drop in a straight pipe with no fittings and the real fitting installed.

The orifice pressure loss coefficient can be obtained in terms of pressure drop by:

$$k_{or} = \frac{\Delta P_{or}}{\frac{1}{2}\rho v_1^2}$$

1.3 NUMERICAL METHODOLOGY

Flow through an orifice can be described by two governing differential equations:

Continuity equation:

$$\nabla . U = 0$$

Navier–Stokes equation:

$$\rho \frac{DU}{Dt} = -\nabla p + \mu \nabla^2 U + \rho g$$

1.3.1 GEOMETRY AND GRID DETAILS

Consider the two-dimensional axis-symmetric flow of an incompressible non-Newtonian fluid with a uniform velocity U_∞ (Figure 1.2).

Orifice having plate thickness of 2 mm for 2 inch and 4 inch diameter pipe have been modeled. Pipe length of 20D has been taken on upstream side of the orifice and 30D on downstream side to take the consideration of the flow characteristics such as fully developed flow profile (upstream side), recirculation zone (downstream side), low pressure regions, etc., which get developed in the domain due to orifice.

Six geometries were used for analysis, which include parameters like pipe diameter and β ratio (orifice diameter to the pipe diameter ratio). For defining the geometry pipe diameters of 2 inch and 4 inch were taken into consideration. Laminar flow regime was taken for analysis. Reynolds number was varied from 20 to 100. Three sets of β ratios (0.3, 0.4, and 0.6) were used. This provides us six sets of geometries (Figure 1.3).

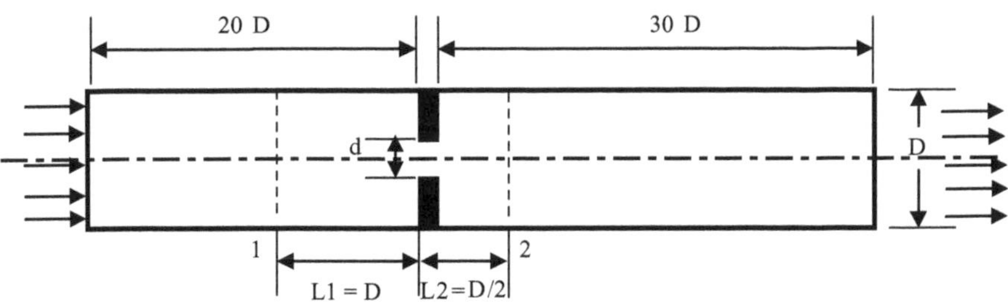

FIGURE 1.2 Schematic of orifice meter showing pressure tapings with flow direction.

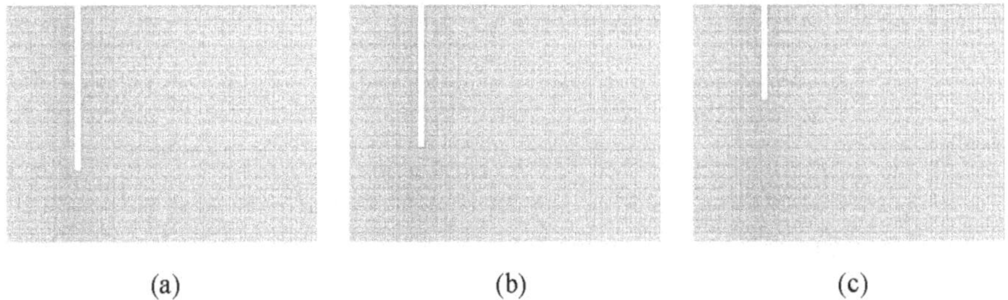

FIGURE 1.3 Representation of grid near the orifice for beta values 0.3, 0.4, and 0.6, respectively.

The mesh size near the orifice was maintained in the present work at 0.33 mm. Mesh refinement has been provided for the regions where large mesh resolution is required.

1.3.2 MATERIAL PROPERTIES

Dilute solutions of CMC (carboxy methyl cellulose) and Kaolin were used as non-Newtonian fluids [5]. The air and liquid temperatures used were closed to atmospheric temperature, 31±2°C. Two aqueous solutions of CMC of approximate concentrations 4% and 8% w/w and Kaolin of 39% w/w were used as the non-Newtonian liquid. Material properties were defined in the material panel, which allows inputting values for the properties, which were relevant to the problem. These properties include density, consistency index, flow behavior index, and effective viscosity.

The flow behavior is defined using Re_{MR}, for example, Metzner-Reed Reynolds number [1] given by:

$$Re_{MR} = \frac{8\rho V^2}{K(\frac{8V}{D})^{n'}}$$

The combination of material properties, pipe diameter, β ratios, and Reynolds number studied are shown in Table 1.1.

1.3.3 BOUNDARY CONDITIONS

The physically realistic boundary conditions for this flow are written as follows:

- At the inlet boundary: The conditions of the uniform flow is imposed, for example,

$$U_x = U_\infty; U_y = 0$$

TABLE 1.1 Parameter Ranges Used

Materials	D [in]	β	*Re*
CMC (4%) ρ = 1023 kg/m³, K' = 0.44, n' = 0.75	2, 4	0.3, 0.4, 0.6	20, 50, 100
CMC (8%) ρ = 1043 kg/m³, K' = 8.3, n' = 0.6	2, 4	0.3, 0.4, 0.6	20, 50, 100
Kaolin (39%) ρ = 1324 kg/m³, K' = 16, n = 0.15	2, 4	0.3, 0.4, 0.6	20, 50, 100

- On the orifice and pipe wall: The standard no-slip condition is used, for example,

$$U_x = 0; U_y = 0$$

- On axis: The axis boundary condition is specified on the axis boundary
- At exit boundary: In FLUENT zero diffusion flux for all flow variables, represented by the default outflow boundary condition option is used. In physical terms, this condition implies that the conditions of the outflow plane are extrapolated from within the domain without exerting any influence on the upstream flow conditions.

1.3.4 SOLVER DETAILS

FLUENT (version 12.1) has been used for solving governing equations. Gambit has been used for generating structured 'quadrilateral' cells of non-uniform grid spacing. For solving incompressible flow on the above arrangement, two-dimensional, laminar segregated solver has been used. For discrediting the convective terms in the momentum equations second order upwind scheme has been used. For solving the pressure–velocity decoupling SIMPLE algorithm (semi-implicit method for the pressure linked equations) has been used.

Relative convergence criteria of 1e-06 for the continuity and *x*- and *y*-components of velocity were prescribed in this work. Also, pressure upstream and downstream of the orifice at the tapings was monitored for convergence.

1.4 RESULTS AND DISCUSSION

The main focus of this work is determination of discharge coefficient and loss coefficient for steady state flow of incompressible non-Newtonian fluids through orifice meter for different values of β ratios and Reynolds number. The Reynolds number range used in the present study is fixed at $20 \leq Re \leq 100$, which leads a predominant laminar flow regime.

1.4.1 *DISCHARGE COEFFICIENT*

Variation of discharge coefficient for varying values of β and pipe diameters as a function of Reynolds number has been plotted as shown in Figure 1.4. In the present study, maximum value of discharge coefficient occurs for CMC (4%) at β = 0.6 and Re = 20.

1.4.1.1 Variation of C_d with Pipe Diameter

The discharge coefficient for a 4″ diameter pipe is always less than that for a 2″ diameter pipe for all ranges of parameters studied. This may be attributed to decrease in viscous forces for the 4″ diameter pipe as compared to 2″ diameter pipe.

1.4.1.2 Variation of C_d with Material Properties

For all β values the discharge coefficient decreases with decreasing value of power law index and increasing consistency index. The more shear-thinning fluid has a lower value of discharge coefficient as the apparent viscosity and hence the viscous forces are lower.

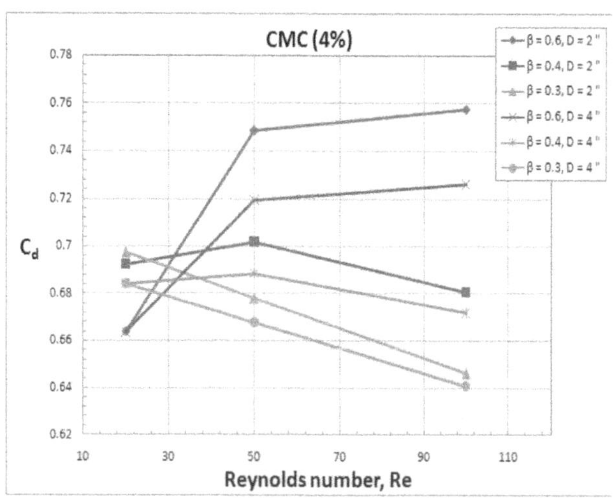

FIGURE 1.4 Variation of C_d for varying values of β and pipe diameter for CMC (4%).

1.4.1.3 Variation of C_d with β

The discharge coefficient decreases with decreasing value of β for Re = 50 and 100. For CMC (4%) and CMC (8%), discharge coefficient at Re = 20 and β = 0.6 is lower as compared to other β values (Figures 1.4–1.6).

1.4.2 PRESSURE LOSS COEFFICIENT

The variation of loss coefficient for varying values of β and pipe diameters as a function of Reynolds number has been plotted as shown in Figures 1.7–1.9. For the range of pipe Reynolds number studied the loss coefficient does not vary much with Re for a fixed value of β and material properties.

1.4.2.1 Variation of k_{or} With Pipe Diameter

The effect of pipe diameter on loss coefficient is very less as compared to other factors like β ratios.

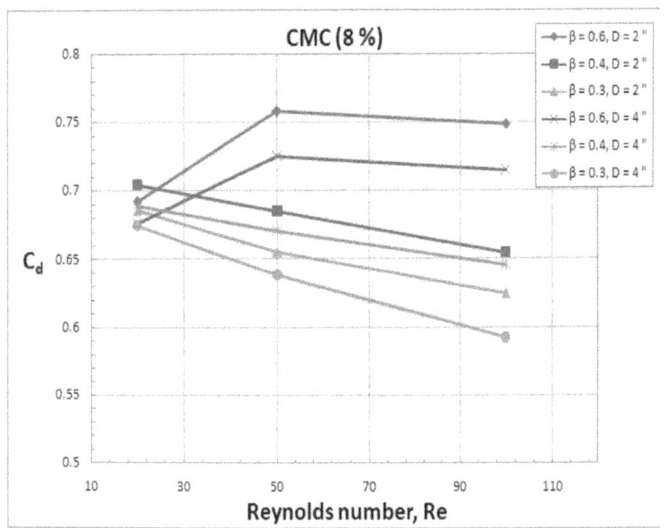

FIGURE 1.5 Variation of C_d for varying values of β and pipe diameter for CMC (8%).

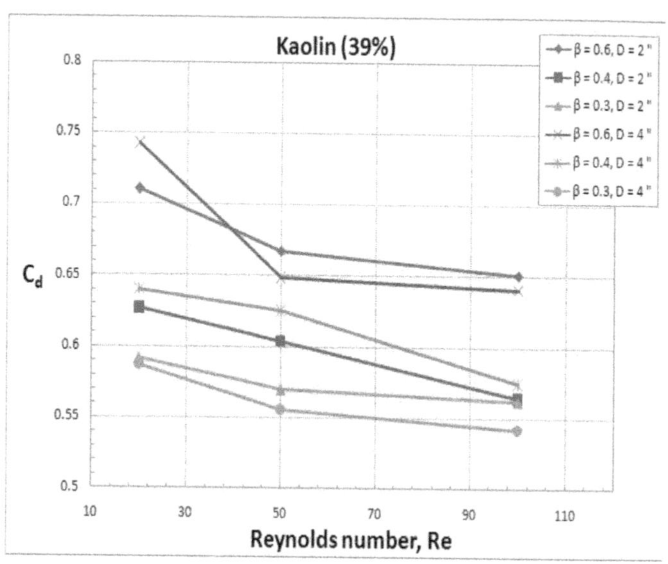

FIGURE 1.6 Variation of C_d for varying values of β and pipe diameter for Kaolin (39%).

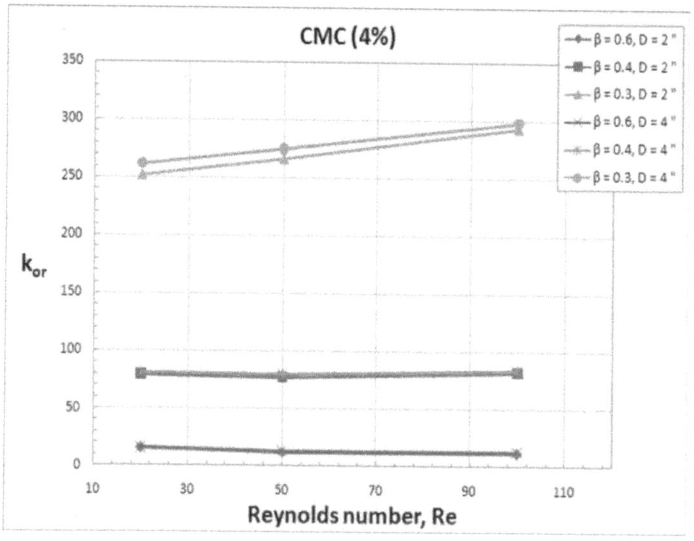

FIGURE 1.7 Variation of K_{or} for varying values of β and pipe diameter for CMC (4%).

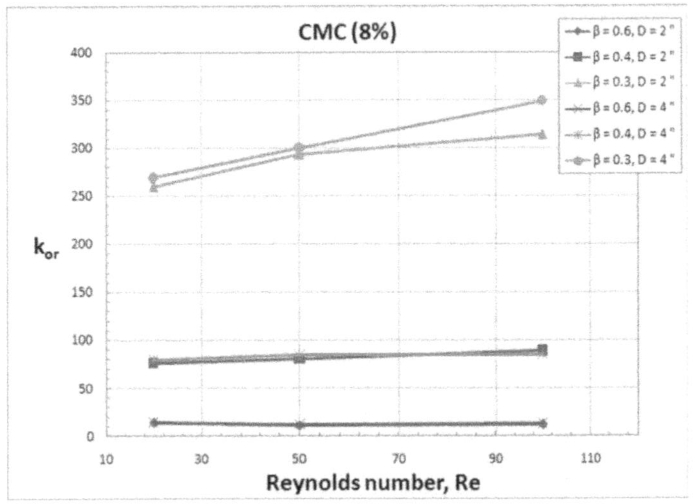

FIGURE 1.8 Variation of K_{or} for varying values of β and pipe diameter for CMC (8%).

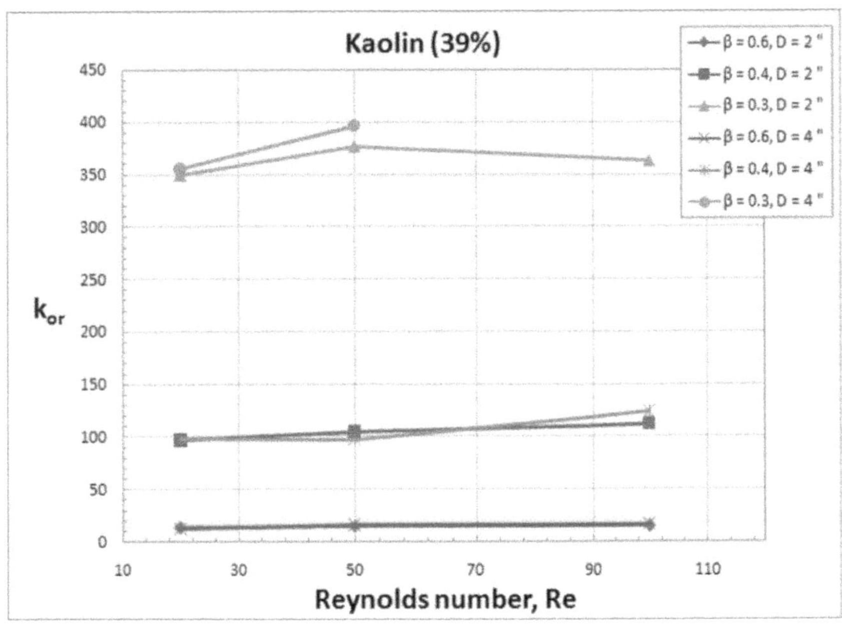

FIGURE 1.9 Variation of K_{or} for varying values of β and pipe diameter for Kaolin (39%).

1.4.2.2 Variation of k_{or} With Material Properties

For a fixed β value, variation in material properties does not significantly change the loss coefficient values. For CMC (4%) and CMC (8%) with β = 0.4 and 0.6, loss coefficient has almost constant values around 80 and 12, respectively. For Kaolin (39%) the loss coefficient values are higher as compared to CMC (4%) and CMC (8%).

1.4.2.3 Variation of k_{or} With β

For all the material properties and range of Reynolds number studied the pressure loss coefficient values increases with decreasing β (Figures 1.7–1.9).

1.5 CONCLUSION

The effect of pipe diameter, β ratio and material properties on discharge coefficient and loss coefficient was studied for six configurations of orifice meter for a Reynolds number range $20 \leq Re \leq 100$. From the results obtained, it is evident that for constant β ratio and with increase in pipe diameter discharge coefficient decreases. It is also observed that for all the fluids for same pipe diameter the discharge coefficient increases with the increase in β ratio. With regards to loss coefficient, it is seen that only changes in β ratio have a significant impact on the values. For all the material properties and range of Reynolds number studied the loss coefficient values increased with decreasing value of β.

KEYWORDS

- **Discharge coefficient**
- **K (consistency index)**
- ***n* (flow behavior index)**
- **Non-Newtonian fluid**
- **Power-law fluid**

REFERENCES

1. Chhabra, R. P., & Richardson, J. F. In *Non-Newtonian Flow and Applied Rheology. 2nd edn.* Oxford: Butterworth-Heinemann, 2008.
2. Butteur M. N. Non-Newtonian pressure loss and discharge coefficients for short square-edged orifice plates. *Diss. Cape Peninsula University of Technology,* 2011.
3. Fester, V. G., Chowdhury, M. R., & Iudicello, F., Pressure loss and discharge coefficients for non-Newtonian fluids in long orifices. In *British Hydromechanics Research Group 18th International Conference on Slurry Handling and Pipeline Transport HYDROTRANSPORT*, 2010, *18*, 22–24.
4. Bohra, L. K., Flow and Pressure Drop of Highly Viscous Fluids in Small Aperture Orifices. 2004.
5. Shah, M. S., Joshi, J. B., Avtar, S. K., Prasad, C. S. R., & Shukla, D. S., Analysis of flow through an orifice meter: CFD simulation. *Chem. Eng. Sci.* 2012, *71*, 300–309.
6. Naveenji, A., Malavarayan, S., & Kaushik, M., CFD analysis on discharge coefficient during non-Newtonian flows through orifice meter. *Int. J. Eng. Sci. Technol.* 2010, *2*, 3151–3164.

CHAPTER 2

NUMERICAL SIMULATION OF FIXED BED LIQUID CHROMATOGRAPHY IN MULTIPLE COLUMNS

A. NAG and B. C. BAG

Defence Research and Development Establishment, Ministry of Defence, Government of India Mahanagar Palika Marg, Nagpur–440001, India

CONTENTS

2.1 Introduction 19
2.2 Model Equations 21
2.3 Process Operation Scheme 23
2.4 Simulation Strategy 24
2.5 Numerical Test 25
2.6 Results and Discussion 26
  2.6.1 Simulation for Binary Mixture Separation 26
  2.6.2 Simulation for Multiple Stage Columns 30
2.7 Conclusion 32
Keywords 33
References 33

2.1 INTRODUCTION

Chromatographic methods are being used routinely for chemical analysis since 1950s, and for automated analysis of process streams in process control (process chromatography). The uses of chromatographic methods for analysis are well known and are being used for a variety of cases. As for example, a high-performance liquid-chromatographic procedure for 5-hydroxy-3-indoleacetic acid is described and compared with a colorimetric method in which 1-nitroso-2-naphthol is used [1]. A simple and cost-effective assay for urinary 5-hydroxyindole acetic acid (5-HIAA) is reported [2]. Chromatographic separation was used for separation of Uroporphyrins I and III on a microporasil column using n-heptane, glacial acetic acid, acetone and water [3]. A number of examples are available where chromatographic methods were used for the fractionation and separation for compounds of similar or closely related physical properties [4, 5]. Off late chromatographic separation has become one of the main tools for separation of racemic mixtures to obtain chiral drugs containing only one enantiomer [6, 7]. The effects of adsorption characteristics of a displacer on the displacement efficiency of desorption chromatography were also reported [8].

A chromatographic technique for automated analysis of process streams in process control is known as process chromatography. Though innumerable examples exist for the analysis but use of chromatography as a commercial separation process is limited. Commercial use of chromatography is called production or large-scale chromatography to distinguish it from its smaller, laboratory scale relative, preparative chromatography. Because of the rapid development of preparative and large-scale chromatography for bio-separations, there has been a demand for adequate mathematical modeling of various chromatographic processes. Unlike analytical liquid chromatography, preparative and large-scale chromatography often involves various mass transfer resistances [9]. Large number of mathematical models as well as experimental verification studies was carried out on column chromatography to visualize the breakthrough curve [10–12]. Most of the earlier works mainly involve experimental research to visualize the specified system for effective separation of the target compound. But the results are inadequate for general visualization of optimum process scheme. Present simulation strategy throws a light on column sequencing and range selection for process parameters to scale up of a fixed bed chromatography column.

The present work aims to a purely application oriented analysis of column breakthrough curves for actual production process. The ultimate goal is increased production rate with high purity by manipulating process condition and sequence of the column. This column requires regeneration time and hence the production is discontinuous. The time lag may be reduced by using additional column in the production and make it continuous. In addition to the use of two columns at a time the first one is equivalent to a striping section and the latter as an enriching section. Single column for higher purity product always reduce the production rate. As our conventional distillation column principle the concept of two sections has been used in the present study. Simulation carried out for two columns separately and prediction of the final scenario of a process condition has not been found in the literature. Current work also aims to visualize nature of breakthrough curves generated by finite element method used to solve the nonlinear-coupled partial differential equations of flow field and diffusion kinetics. During two dimensional convective transport process the relative role of concentration dilution rate and operating time with various Peclet and Biot number range on production and quality of the product has been studied in detail by analyzing the area under the breakthrough curves. Numerical results are presented in the terms of dimensionless concentration and time for a range of Peclet and Biot number within fixed bed height and diameter of the column. The average outlet concentrations are also tabulated for the various inlet concentrations to visualize best dilution rate. Finally, optimum time sequencing and inlet concentration based on specified Peclet and Biot number has been assessed via larger production and higher purity level.

2.2 MODEL EQUATIONS

The present study has been carried out for a two-dimensional domain filled with fluid saturated porous media. The fluid inside porous bed is considered as incompressible, Newtonian, and isothermal. The flow is assumed to be laminar with negligible size exclusion effect. All the physical properties are assumed to be constant and mass transfer effect is independent of mixing effect. It is further assumed that axial dispersion coefficient is single lumped parameter for all axial mixing mechanism. Fluid moves through the void space inside porous bed, hence the velocity is defined by intrinsic velocity. The Peclet number (Pe) and Biot number (Bi) is the key dimensionless number governing the pattern of breakthrough curve. The detail of the multi-component rate model equation is found in various literatures [13–15]. The various model equation used are given below.

Continuity equation in dimensionless form for bulk fluid phase:

$$\frac{1}{Pe_{Li}}\frac{\partial^2 C_{bi}}{\partial z}+\frac{\partial C_{bi}}{\partial z}+\frac{\partial C_{bi}}{\partial \tau}+\varsigma_i(C_{bi}-C_{pi,R=1})=0 \tag{1}$$

Continuity equation in dimensionless form inside the macro-pores considering spherical particle:

$$\frac{\partial}{\partial \tau}\left[(1-\varepsilon_p)C_{pi}+\varepsilon_p C_{pi}\right]-\eta_i\left[\frac{1}{r^2}\frac{\partial}{\partial r}\left(r^2\frac{\partial C_{pi}}{\partial r}\right)\right]=0 \tag{2}$$

Interstitial velocity is determined by

$$u=\frac{4Q}{\pi\varepsilon_b D^2} \tag{3}$$

Dimensionless constants are defined as

$$C_{pi}=\frac{c_{pi}}{c_{oi}};C_{bi}=\frac{c_{bi}}{c_{oi}};C^*_{pi}=\frac{c^*_{pi}}{c_{oi}};r=\frac{R}{R_p};\tau=\frac{ut}{L}$$

$$Z=\frac{z}{L};Pe_{Li}=\frac{uL}{D_{bi}};B_i=\frac{K_i R_p}{\varepsilon_p D_{pi}};\eta_i=\frac{\varepsilon_p D_{pi}}{2uR_p}L;\varsigma_i=\frac{3B_i\eta_i(1-\varepsilon_b)}{\varepsilon_b} \tag{4}$$

Initial condition ($t=0$), $\tau=0$

$$C_{bi}=C_{bi}(0,z)=0;C_{pi}=C_{pi}(0,r,z)=0 \tag{5}$$

Boundary conditions:

$$Z=0,\ \frac{\partial C_{bi}}{\partial z}=Pe_{Li}\left[C_{bi}-\frac{C_{fi}(\tau)}{C_{oi}}\right];\ \text{At } z=1,\ \frac{\partial C_{bi}}{\partial z}=0 \tag{6}$$

(Where $\frac{C_{fi}(\tau)}{C_{oi}}=1$ for frontal adsorption;)

$$\text{At r} = 0,\ \frac{\partial C_{pi}}{\partial r} = 0;\ \text{At r} = 1,\ \frac{\partial C_{pi}}{\partial r} = Bi_i(C_{bi} - C_{pi,r=1}) \tag{7}$$

where C^*_{pi} is dimensionless concentration of component *i* in the solid phase for multicomponent Langmuir isotherm.

$$C^*_{pi} = \frac{a_i C_{pi}}{1 + \sum_{j=1}^{N_s} b_j C_{pj}};\ C^*_{pi} = \frac{a_i C_{pi}}{1 + \sum_{j=1}^{N_s} (b_j C_{oj}) C_{pj}} \tag{8}$$

The area under the curve is the major analysis tool for operating parameter selection and denoted by Σ*A* and Σ*B*.

$$\text{where} = \sum A = \int_0^{\tau} \frac{c}{c_0} d\tau \tag{9}$$

$$C_{av} = \int_0^t \frac{c}{t} dt = \sum A\left(\frac{c_0 L}{ut}\right) = \sum A\left(\frac{c_0}{\tau}\right) \tag{10}$$

Production of the target component is correlated by:

$$\text{Production} = \left(\sum AC_{a0}\right)\frac{QL}{u} = \left(\sum AC_{a0}\right)\frac{\pi\varepsilon_b LD^2}{4} \tag{11}$$

2.3 PROCESS OPERATION SCHEME

Figure 2.1 describes the arrangement for separation of impure reaction mixture by fixed bed liquid chromatography columns. Two columns are operated continuously for efficient separation where the first column (A) is equivalent to a stripping section and the second one (C) is enriching section. One additional column (B) is kept for alternate operation when column A is eluted the column B is kept for regeneration. The main objective is to get higher flow rate in the first column (A) where the purity of the target compound is compromised for higher production. The material is collected in a collection vessel and finally it is charged to second column (C) for highly pure product. While second column is operated the first one is washed with solvent. Similarly many sets of columns may be used to achieve a desired production rate.

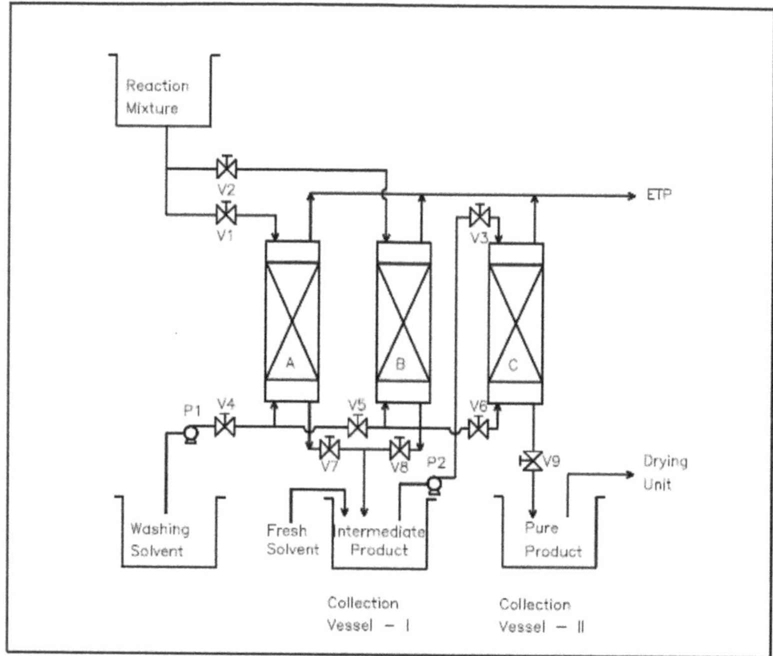

FIGURE 2.1 Model figure; column sequencing for high production rate.

2.4 SIMULATION STRATEGY

A binary impure reaction mixture with average molecular weight of 200, if diluted in 10 liter of solvent in 60 gm and 40 gm weight the average concentration becomes $(Ca/Cb)_{av} = 0.03/0.02$ in kmol/m^3. The estimation of Pe and Bi number range is checked by simulating the stripping column both in low concentration and high concentration $(Ca/Cb)_{av} = 0.03/0.02$ and 0.3/0.2 in kmol/m^3. The optimum amount of dilution required is first estimated by varying average concentration ratio in the inlet for fixed Pe and Bi range. Now changing τ with fixed average concentration is carried out to find the outlet average concentration. The average concentration is calculated as per Eq. (10). Finally, check the outlet concentration at the enriching column. FE technique has been used to solve the coupled nonlinear PDEs. The computation is carried out in MATLAB [16] software. The Langmuir constants for component A and B as described in validation figure (Figure 2.2) is used for whole simulation process.

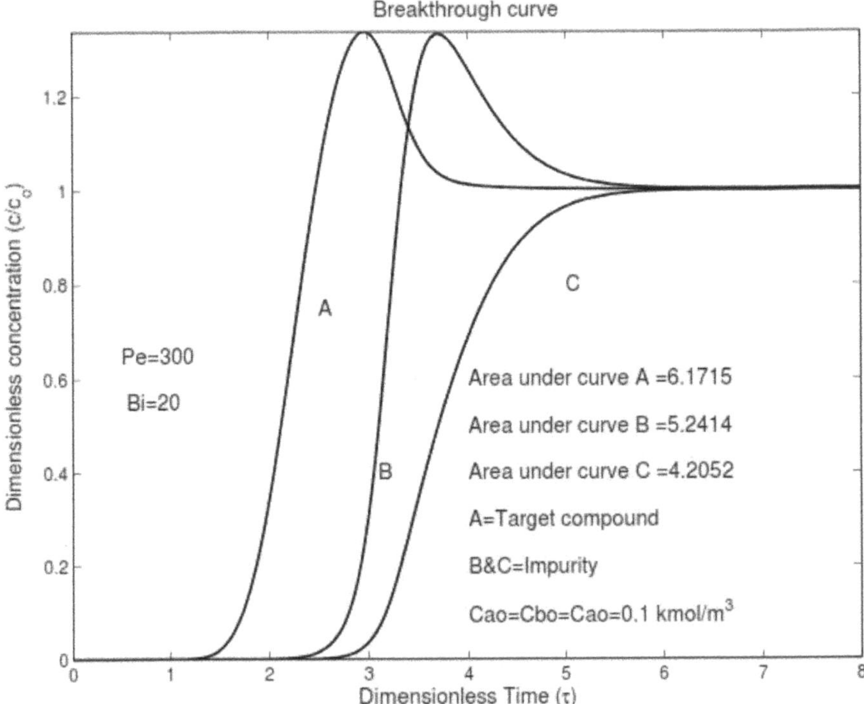

FIGURE 2.2 Validation [14] for ternary mixture breakthrough curve involving competitive Langmuir isotherm (at Pe = 300, η = 1, Bi = 20, a_i = 1, 10, 20, b_i = 2, 20, 40, C_{i0} = 0.1).

2.5 NUMERICAL TEST

The code has been validated (Figure 2.2) first with reported values [14]. Figure 2.2 shows the concentration profiles corresponding to a ternary component system for frontal adsorption involving competitive Langmuir isotherm (at Pe = 300, η = 1, Bi = 20, a_i = 1, 10, 20, b_i = 2, 20, 40, Ci_0 = 0.1 kmol/m^3). The effluent history of a frontal analysis where feed is continuously charged gives the breakthrough curve of each component assuming negligible size exclusion effect. The algorithm and computational technique for solving the PDE's used in the present work has been discussed in details in the literature [13, 14]. The computational domain for bulk phase and particle phase has been discretized by finite element and orthogonal collocation method, respectively. Finally generated ODE's are solved by ODE 15 MATLAB solver. The disadvantage of handling bigger matrices is simplified by using sparse matrices concept. The area under the breakthrough curve was estimated using MATLAB module TRAPZ, which uses trapezoidal numerical integration [18]. The calculations are in excellent agreement with the literature result.

2.6 RESULTS AND DISCUSSION

2.6.1 *SIMULATION FOR BINARY MIXTURE SEPARATION*

Figures 2.3(a–d) show the breakthrough curves of binary mixture in dimensionless concentration vs dimensionless time at specified Pe and Bi both in low and high dilution. Figure 2.3(a) shows effluent history for single column for Pe = 50 and Bi = 5 for binary system with Ca_0 = 0.3 and

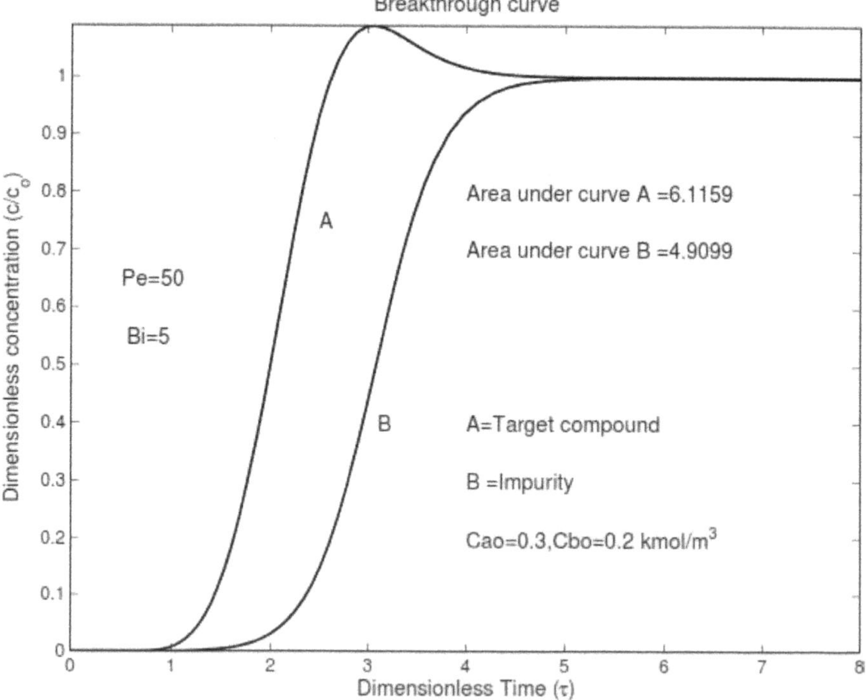

FIGURE 2.3(A) Effluent history for single column for Pe = 50 and Bi = 5 for binary system with Ca_0 = 0.3 and Cb_0 = 0.2 kmol/m^3.

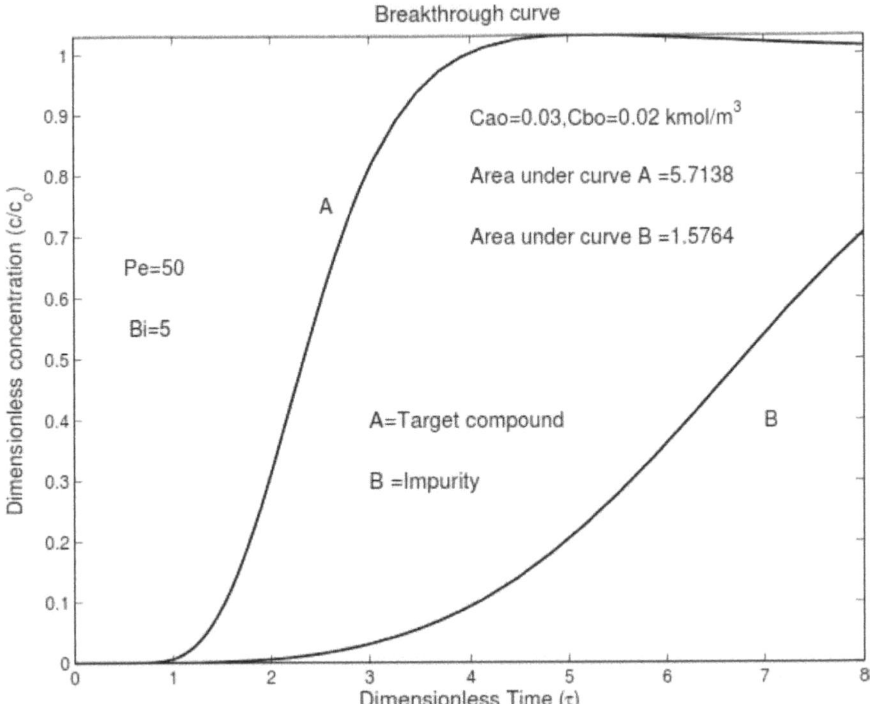

FIGURE 2.3(B) Effluent history for single column for Pe = 50 and Bi = 5 for binary system with $Ca_0 = 0.03$ and $Cb_0 = 0.02$ kmol/m³.

$Cb_0 = 0.2$ kmol/m³, where A is the target compound and B is the impurity. It is observed that the leveling off time is $\tau = 5$ in the break through curve. It indicates at time $\tau = 5$, the outlet concentration of both the components are same and hence no further purification/separation is possible. Area under the curve A and B are 6.1159 and 4.9099, respectively, upto $\tau = 8$ and visible amount of pure A is obtainable within range $\tau = 1.5$.

Effluent history for single column for Pe = 50 and Bi = 5 for binary system with $Ca_0 = 0.03$ and $Cb_0 = 0.02$ kmol/m³ is presented in Figure 2.3(b). Keeping same Pe and Bi range, the concentration is diluted and it observed that a considerable amount of separation is possible and also the leveling off time is increased ($\tau = 14$) compared to previous case ($\tau = 5$) (see Figures 2.3a and 2.3b). It means that column needs to be operated for longer duration for more product recovery from the mixture. It is interesting to observe that time for obtaining pure compound $\tau = 2$ (approx.) is

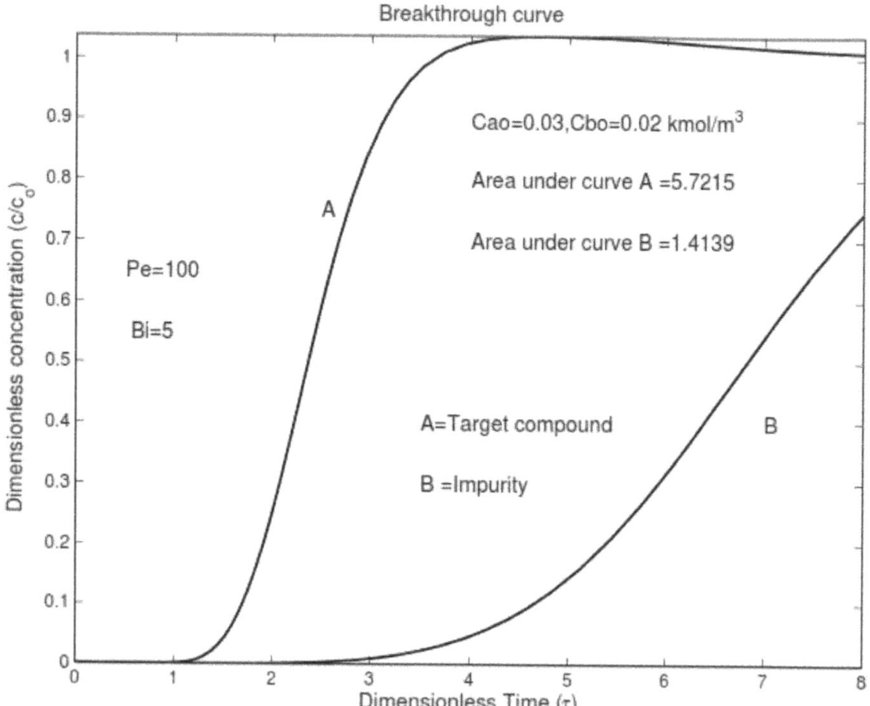

FIGURE 2.3(C) Effluent history for single column for Pe = 100 and Bi = 10 for binary system with average inlet concentration $Ca_0 = 0.03$ and $Cb_0 = 0.02$ kmol/m^3.

increased than the previous case (see Figures 2.3b and 2.3a) hence production of pure component will be increased as production rate is directly proportional to area under the curve. Note that area under the curve A and B are 5.7138 and 1.5764, respectively, upto $\tau = 8$. As a whole the separation band is also increased for lower dilution hence it is advantageous for higher purity product recovery but production rate will decrease.

Figure 2.3(c) presents the effluent history for single column at Pe = 100 and Bi = 10 for binary system with $Ca_0 = 0.03$ and $Cb_0 = 0.02$ kmol/m^3. The value of ΣA and ΣB are 5.7215 and 1.4139, respectively, upto $\tau = 8$ in this case whereas ΣA and ΣB are 5.7138 and 1.5764 respectively upto $\tau = 8$ for Pe = 50 and Bi = 5 (see Figures 2.3c and 2.3b). This curve (Figure 2.3a) is qualitatively similar to Figure 2.3(b) and similar explanation follows. As the operating cost for higher Pe number is higher hence it is not advisable to operate the stripping column in this range because for

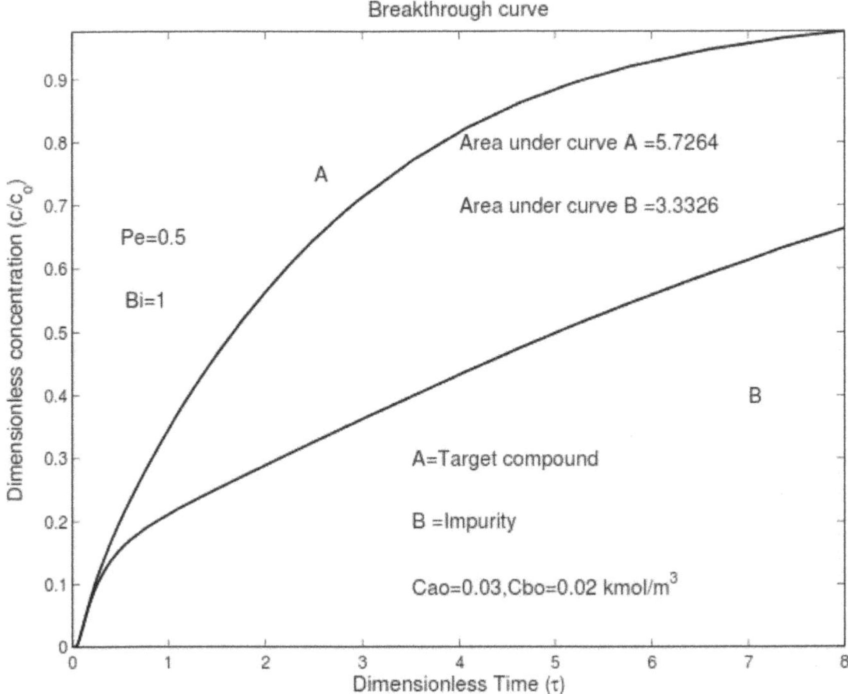

FIGURE 2.3(D) Effluent history for single column with range of Pe = 0.5 and Bi = 1 for binary system and average inlet concentration $Ca_0 = 0.03$ and $Cb_0 = 0.02$ kmol/m^3.

higher product recovery we need to run the column for maximum τ. It is interesting to observe that leveling off time ($\tau = 12$) of the breakthrough curve in this case will be moderately high on the other hand the time for pure product recovery is also higher as $\tau = 2.5$. Hence, pure component isolation with high production may be obtained if this Pe and Bi number range is used in the enriching column.

Figure 2.3(d) shows Effluent history for single column with range of Pe = 0.5 and Bi = 1 for binary system with $Ca_0 = 0.03$ and $Cb_0 = 0.02$ kmol/m^3. In most of the industrial application this range of Pe and Bi is observed during low flow rate and smaller column length. Leveling off time ($\tau = 20$) of the breakthrough curve will be very high in this case resulting higher product recovery but impure product will be obtained as minimum τ for pure product recovery is very low <0.5. Note that area under the curve A and B are 5.7264 and 3.3326, respectively, upto $\tau = 8$ signify higher product recovery but very insignificant separation. Hence, this range may be used for stripping column but absolutely discarded for enriching column.

2.6.2 SIMULATION FOR MULTIPLE STAGE COLUMNS

Table 2.1 shows calculation of discrete values of inlet and outlet average concentration where simulation has been carried out for optimum selection of dilution rate for fixed Biot number (Bi = 5) and Peclet number (Pe = 50) in the stripping column upto $\tau = 8$ in all cases. Reaction mixture coming from plant may be richer in target compound or at least equimolar. Hence, two-concentration ratio has been chosen (ratio I multiple of 0.03/0.02, rich in A and ratio II multiple of 0.1/0.1, equimolar mixture in kmol/m^3). The values of $\sum A/\tau$, $\sum B/\tau$ and fractional $\sum A$ have been estimated as a pointer

TABLE 2.1 Simulation for Optimum Selection of Dilution for Bi = 5, Pe = 50 in the Stripping Column upto $\tau = 8$ in All Cases (Ratios I and II Multiple of 0.03/0.02 and 0.1/0.1, Respectively)

Ratio of average concentration	(Ca/Cb) av inlet	∑A	∑B	∑A/(∑A+∑B)	(Ca/Cb)av outlet
Ratio I Multiple of 0.03/0.02	0.015/0.01	5.6141	1.2233	0.82	0.01/0.001
	0.03/0.02	5.7138	1.5764	0.78	0.02/0.004
	0.06/0.04	5.8530	2.3976	0.71	0.044/0.012
	0.09/0.06	5.933	3.102	0.656	0.066/0.023
	0.12/0.08	5.9852	3.6083	0.623	0.089/0.03
	0.15/0.1	6.0219	3.9769	0.60	0.11/.04
	0.18/0.12	6.0498	0.7562	0.58	0.13/0.06
	0.21/0.14	6.0719	0.7589	0.57	0.159/0.07
Ratio II Multiple of 0.1/0.1	0.01/0.01	5.6082	1.2023	0.82	0.007/0.001
	0.03/0.03	5.7837	1.9182	0.75	0.021/0.007
	0.05/0.05	5.8893	2.6942	0.68	0.03/0.016
	0.07/0.07	5.9536	3.2972	0.64	0.05/0.028
	0.09/0.09	5.9976	3.7325	0.61	0.06/0.042
	0.1/0.1	6.0152	3.9056	0.60	0.075/0.048
	0.2/0.2	6.1111	4.8606	0.55	0.152/0.121
	0.3/0.3	6.1513	5.2626	0.53	0.23/0.197

for separation efficiency. Finally $(Ca/Cb)_{av}$ outlet dilution is estimated from $(Ca/Cb)_{av}$ inlet as per equation [10]. It is observed that the fraction $\sum A$ is monotonically decreasing with lower dilution irrespective of ratio I and ratio II. If the stripping column is operated for more than 0.1/0.1 (ratio II) or 0.12/0.08 (ratio I) the output dilution will not be advantageous to be used in the second column for pure compound isolation. If lower dilution is required to be used in enriching column (irrespective of both the cases) a provision for additional solvent addition (Figure 2.1) is required to further dilute the composition further for pure product recovery. In that case solvent recovery cost has to be sacrificed.

Table 2.2 gives simulation results for estimating outlet concentration in stripping column; with inlet concentration $Ca_0 = 0.03$ kmol/m^3, $Cb_0 = 0.02$ kmol/m^3 and Pe = 50, Bi = 5 with various τ. The value of $\sum A/\tau$, $\sum B/\tau$, fractional $\sum A$ and $(Ca/Cb)_{av}$ outlet has been estimated and tabulated for detail breakthrough curve analysis. It is important to have the information about the discrete operating time for higher product recovery and absolutely pure component separation margin. It is observed that at lower τ

TABLE 2.2 Simulations for Estimating Outlet Concentration in Stripping Column; with Inlet Concentration $Ca_0 = 0.03$ kmol/m^3, $Cb_0 = 0.02$ kmol/m^3, Bi = 5, Pe = 50 with Various τ

T	∑A	∑B	∑A/(∑A+∑B)	∑A/τ	∑B/τ	(Ca/Cb)av outlet
1	6.6e-4	2.6e-4	0.72	0.006	0.0026	Very Dilute
1.5	0.01	3.8e-4	0.991	0.057	0.0038	
2	0.114	0.002	0.9827	0.057	0.001	
2.5	0.3389	0.0068	0.9803	0.13	0.002	0.003/0.0004
3	0.6924	0.0179	0.9747	0.23	0.005	0.006/0.0001
3.5	1.1344	0.0394	0.9664	0.324	0.011	0.009/0.0002
4	1.6225	0.0764	0.955	0.4	0.019	0.012/0.0003
4.5	2.1301	0.1350	0.9403	0.47	0.03	0.014/0.0006
5	2.6444	0.2213	0.9227	0.52	0.044	0.015/0.0008
5.5	3.1595	0.3415	0.9024	0.57	0.062	0.017/0.001
6	3.6739	0.5010	0.8799	0.611	0.083	0.018/0.0016
6.5	4.1867	0.7036	0.8561	0.644	0.108	0.019/0.002
7	4.6976	0.9511	0.8316	0.67	0.135	0.02/0.003
7.5	5.2066	1.2429	0.8072	0.69	0.165	0.02/0.003
8	5.7138	1.5764	0.7837	0.7125	0.197	0.02/0.003

(τ<3) absolutely pure compound is separated at very lower dilution on the other hand at $\tau = 7$ and above the dilution is almost constant simultaneously the purity goes below 83%. That means further separation is not occurring at this Peclet and Biot number range, hence $\tau = 6$ may be optimum time factor. The stripping column may be operated up to $\tau = 6$ for moderately pure (87.99%) compound with higher product recovery rate and the outlet average concentration (0.018/0.0016) will be sufficiently dilute to be used in the enriching column for 99.9% pure compound.

2.7 CONCLUSION

Detail analysis of the curve area gives the pathway for proper operating condition. It is important to select higher value of ΣA for scale up, for fixed L and D. Flow rate and bed height need to be adjusted to get high Peclet (Pe) and Biot (Bi) number range. High Pe number is costly operation, on the other hand low Pe will not give desired purity; specifically for enriching section. The higher the τ higher the value of ΣA and that results in higher production but actual selection of τ is based on the limit upto which the column is operative with maximum efficiency and no clogging is occurring. While column A is operated for elution, column B can be kept under regeneration mode. The production rate in the enriching column is very low for $\tau = 1.5$ and $\Sigma A = 0.01$ compared to $\Sigma A = 5.7$ upto $\tau = 8$ in stripping section with 78.37% purity. Parallel column addition in enriching section will give high production with high purity. Dilution of the impure mixture is also dependent on the product and solvent recovery cost. In complex cases column optimization can be well handled by design of experiments (DOE) tools. As Peclet and Biot number is interrelated, the range chosen here is practicably achievable. It is difficult to get pure product at low Peclet and Biot number. The compounds which are highly separable (high ratio of Langmuir constant) by nature gives higher production rate automatically and for complex cases flow rates and dilution might be kept low. Scale out is the best option for complex cases. Scope is there to research further with specified system and by changing of adsorbent and solvent estimating Langmuir constants experimentally. Finally establish the most optimum time, dilution and operable Peclet and Biot number range for process scale up by using the detail analysis of breakthrough curve as discussed in the present work.

KEYWORDS

- **Breakthrough curve analysis**
- **Fixed bed chromatography**
- **Frontal adsorption**
- **Multi-component rate model**
- **Multistage column system**
- **Process scale up**

REFERENCES

1. Draganac, P. S., Steindel, S. J., & Trawick, W. G., *Clinic. Chem.* 1980, *26*, 910.
2. Perry, H., & Keevil, B., *Ann. Clin. Biochem.* 2008, 45, 149.
3. Bommer, J. C., Burnham, B. F., Carlson, R. E., & Dolphin, D., *Ana. Biochem.* 1979, *95*–444.
4. Heikkilä, H., Lewandowski, J., & Kuisma, J, *Chromatographic Separation Method, US Patent.* 2007, 7229558.
5. Tanimura, M., & Ikemoto, M., *Chromatographic Separation Process, US Patent.* 2002, 6482323.
6. Negawa, M., & Shoji, F., *J. Chromatography A.* 1992, *590*, 113.
7. Haag, J., Wouwer, A. V., Lehoucq, S., & Saucez, P., *Control Eng. Practice.* 2001, *9*, 921.
8. Gu, T., Tsai, G. J., & Tsao, G. T., *Biotech. Bioeng.* 1991, *37*, 65.
9. Gu, T., Tsai, G. J., & Tsao G. T., *AIChE J.* 1991, *37*, 1333.
10. Khosravanipour, M. A., Sarshar, M., Javadian, S., Zarefard, M. R., & Amirifard, Z., *Sep. Purif. Technol.* 2011, *79*, 72.
11. Silva Marta, S. P., Mota José, P. B., & Rodrigues Alírio, E., *Sep. Purif. Technol.* 2012, *90*, 246.
12. Ahmad, R. T., Nguyen, T. V., Shim, W. G., Saravanamuthu, V., Moon, H., & Kandasamy, J., *Sep. Purif. Technol.* 2012, *98*, 46.
13. Özdural, A. R., *Modeling Chromatographic Separation, Comprehensive Biotechnology* (Second Edition). 2011, *2*, 681.
14. Gu, T., *Mathematical Modeling and Scale-up of Liquid Chromatography.* Springer, Berlin, 1995.
15. Bird, R. B., Steward, W. E., & Lightfoot, E. N., *Transport Phenomena.* John Wiley. New York, 1960.
16. Mathworks, *Using MATLAB Version 7.8.* The Mathworks, Massachusetts, 2009.
17. Chapra, S. C., & Canale, R. P., *Numerical Methods for Engineers.* WCB/McGraw-Hill, Boston, 1998.

CHAPTER 3

STATISTICAL MODELING FOR ADSORPTION OF CONGO RED ONTO MODIFIED BENTONITE

T. MOHAN RAO[1] and V. V. BASAVA RAO[2]

[1]*Department of Chemical Engineering, Bapatla Engineering College, Bapatla, India*

[2]*Faculty of Technology, University College of Technology, Osmania University, Hyderabad, India*

CONTENTS

3.1 Introduction 36
3.2 Experimental Methods and Materials 36
 3.2.1 Adsorbent 36
 3.2.2 Adsorbate 37
3.3 Results and Discussion 38
 3.3.1 Effect of Operating Parameters 38
  3.3.1.1 Effect of Amount of Adsorbent 38
  3.3.1.2 Effect of Contact Time 38
  3.3.1.3 Effect of pH 39
  3.3.1.4 Effect of Temperature 40
 3.3.2 Statistical Modeling and Analysis 40
3.4 Conclusion 50
Keywords 51
References 51

3.1 INTRODUCTION

Synthetic dye stuff from various industries like textile, leather, plastic and paper pollute water bodies and act as an eco-toxic hazard contributing to serious environmental problem. Azo dyes like Congo red (CR) has been known to cause allergic reactions, various toxic effects on human and aquatics [1–5] and must be removed before they discharge into water streams.

Dyes have complex structure and resistant to various degradation methods posing technological challenge for many decades. Compared to several conventional methods of separation like oxidation, membrane separation, precipitation, coagulation and electro dialysis for color de contamination, Adsorption is proven to be most popular process due to its simplicity, high efficiency, easy recovery and reusability of the adsorbent [6, 7].

Numerous adsorbents have been developed so far, and activated carbon is most popular among all, due to its high adsorption capacity. But it is expensive and non-economical to regenerate [8, 9]. This led to search for alternative materials, which are cost effective and can replace activated carbon. In recent years extensive research is carried out for low cost adsorbents. Various naturally occurring materials and industrial waste materials were tested. Among all, Bentonite is proven to be a promising material for adsorption process in its pure and modified form [9–11].

In this work raw Bentonite was modified by interacting with Acid solution to synthesize a new absorbent (Modified Bentonite) to remove Congo red dye from aqueous solution. The objective of the present work is to develop a successful model for CR removal onto modified Bentonite with the help of statistical tools.

3.2 EXPERIMENTAL METHODS AND MATERIALS

3.2.1 ADSORBENT

The Bentonite was activated by adding concentrated H_2SO_4 (1:1 w/v) with constant stirring. The material was kept in a hot air oven at 110°C for 12 h. This material was washed with distilled water and was soaked in 2%

$NaHCO_3$ solution overnight to remove the residual acid. Then the material was washed with distilled-water, until the pH of the adsorbent reached slightly above 7. Finally, it was dried in a hot air oven at 110°C for 4 h. The particle size was determined by sieving the dried material and it is 125 μm. The sieved adsorbent was stored in an airtight container for further experiment.

3.2.2 ADSORBATE

Congo red is the first synthetic azodye produced that is capable of dyeing cotton directly. A stock solution of the dye with a concentration of 1000 mg L^{-1} was prepared with Millipore water and it is diluted to get the working solutions. Adsorbent dosage was measured accurately with an analytical balance (SHIMADZU – AX200). PH of the solution is measured with a digital pH meter (ELICO-L1–612) and varied by using 0.1 N HC1 and 0.1 N NaOH Solutions. Solution with added adsorbent is agitated with Remimake Temperature Controlled Orbital Shaker (REMI – CIS 24 BL). At the end the samples were collected and centrifuged to remove the suspended solid particles using REMI C 24 centrifuge. The clear liquid was collected and analyzed with UV-Visible Spectrophotometer (SYSTRONICS-117) at a wavelength of 498 nm.

All determinations were performed in triplicate per experiment. The amount of CR adsorbed onto modified Bentonite was measured in terms of uptake, qt (mg of CR adsorbed per one gram of adsorbent) and percent removal %R which are calculated using the following equations:

$$qt = \frac{(C_0 - C_t)}{m}V \tag{1}$$

$$\%R = \frac{(C_0 - C_t)}{C_0}100 \tag{2}$$

where C_0 is the initial CR concentration (mg/L), C_t is the concentration of CR at a time t, (mg/L), V is the volume of dye solution (L) and m is the mass (g) of the adsorbent.

3.3 RESULTS AND DISCUSSION

3.3.1 EFFECT OF OPERATING PARAMETERS

3.3.1.1 Effect of Amount of Adsorbent

The amount of dye uptake prominently depends on the surface area available. Experiments were conducted to examine the extent of its dependence and fix the optimum dosage. Different amounts of adsorbent were added to 50 mL of 100 mg/L solution at 30°C and the results were displayed in Figure 3.1. Dye removal percent increased with increasing amount of adsorbent from 0.02 to 0.1 g and slowed down up to 0.2 g. The further increase in the amount of adsorbent did not affect the removal significantly. Hence, 0.1 g is taken to be the optimum dosage and used in further studies.

3.3.1.2 Effect of Contact Time

Experiments were conducted by adding 0.1 g adsorbent to 50 mL of 100 mg/L solution and agitated at fixed temperature. Samples withdrawn at different time intervals and were analyzed. This was repeated at temperatures (20, 40, and 50°C) and the results were shown in Figure 3.2.

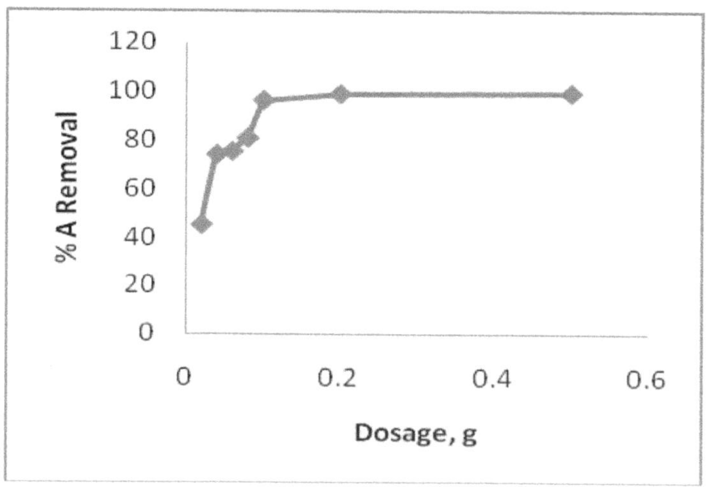

FIGURE 3.1 The effect of adsorbent dosage on the color removal.

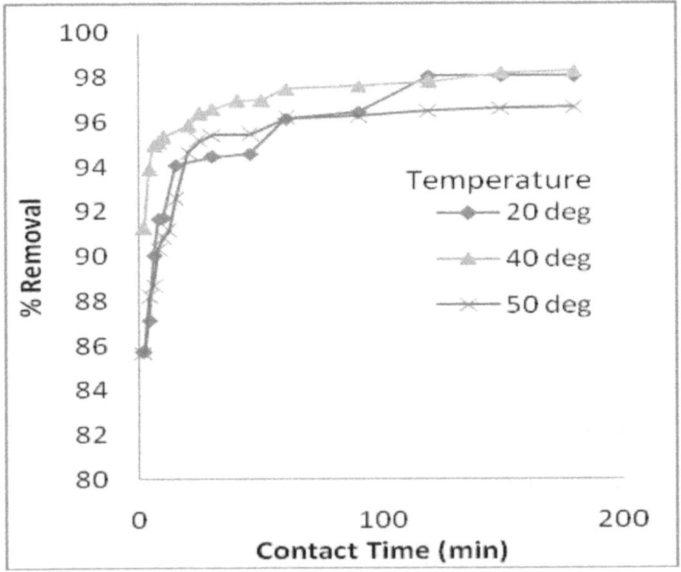

FIGURE 3.2 Effect of agitation time and temperature on the adsorption.

Adsorption was rapid in the beginning up to first 50 min and thereafter it is stabilized. Finally, it attained equilibrium at 180 min. There is no considerable change in % adsorption at different temperatures, but it is slightly less at 50°C (96%) compared to other temperatures (98%).

3.3.1.3 Effect of pH

Effect of pH in the range of 2–12 was examined with 100 mg/L solution. The equilibrium dye uptake at these conditions are determined and tabulated (Table 3.1). There is a slight decrease in uptake with pH. The insignificant change may be due to retention of H^+ or OH^- of clay fraction made the solution tend to neutral. Similar results were observed in case of Congo red removal with modified-Bentonite [7] and adsorption of Ni, Zn and Pb onto Bentonite [12].

TABLE 3.1 Effect of pH on the Adsorption Capacity of Modified Bentonite

pH	2	4	6	8	10	12
Uptake	49.99	49.74	46.32	47.03	48.41	47.61

3.3.1.4 Effect of Temperature

The effect of temperature on adsorption of CR onto MB was presented in Figure 3.3. % removal increased with increase in solution temperature from 20°C to 50°C. This shows that the process is endothermic in nature.

The increase in percentage removal at higher temperature may be due to a greater kinetic energy acquired, resulting in an easier diffusion from the bulk solution onto the surface of MB.

3.3.2 STATISTICAL MODELING AND ANALYSIS

The factors that influence the sorption process are effluent concentration (Co), dosage (Do), temperature (T), pH, time and speed of agitation. The agitation speed is fixed at 150 rpm. Equilibrium time can be estimated from equilibrium studies and is taken a 4 hrs. The negligible influence of pH is observed in parameter studies. Based on these facts first three parameters (Co, Do & T) are considered for process modeling. Univariate analysis is very tedious with innumerable experiments. It also suffers with the disadvantage of overlooking the influence of interactions. Statistical

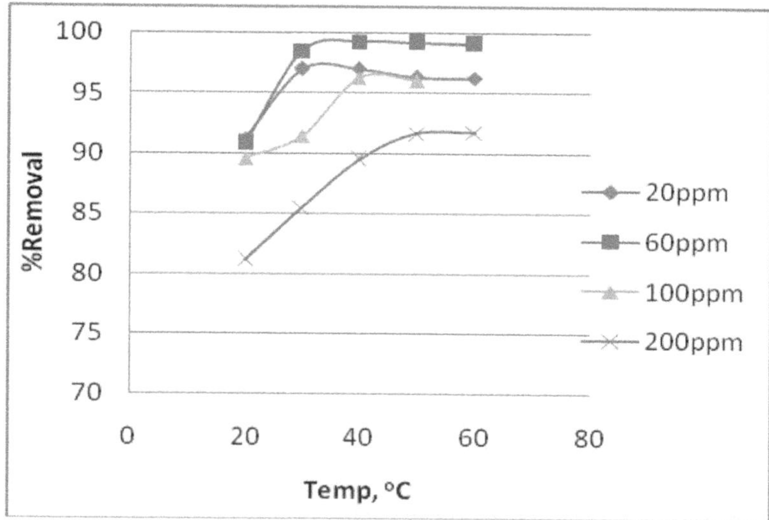

FIGURE 3.3 The effect of solution temperature on adsorption.

design of experiments provides better prospects of analyzing individual parameters as well as their interactions. In general 2 factorial methods are used to eliminate parameters that have negligible influence [13]. In this case was replaced by parameter study presented in Section 3.1. A standard RSM design called central composite design requires a minimum number of experiments and is widely used in experimental studies [14].

Generally, the CCD consists of 2p axial runs and pc center runs along with 2P factorial runs [15]. The axial points facilitate rotatability, ensuring constant variance of the model prediction at all points equidistant from the design center [16]. Replicates of the test at the center provide an independent estimate of the experimental error and hence are very important. The recommended number of tests at the center is six to three variables [17]. Total number of runs (N_R) required in totality is estimated as:

$$N_R = 2^p + 2p + pc = 2^3 + 2 \times 3 + 6 = 20 \quad (3)$$

Response surface methodology was applied to the experimental data using statistical software, Design-expert 8.0.7.1 (trial version). Statistical terms and their definitions used in the Design-expert software are well defined elsewhere [18]. The details of experimental domain and the levels of each factor are given in the Table 3.2. Experiments were conducted according to the design matrix generated and the measured responses were presented in Table 3.3.

Various models were tested with the experimental data to obtain the regression equations. And the model adequacy was tested with sequential F-test, lack-of-fit test and other adequacy measures were used for selecting the best model [19, 20]. Analyzing the measured responses, the fit summary output indicated that the quadratic polynomial model was significant for the present system. General form of the quadratic model is

TABLE 3.2 Experimental Range and Levels of Independent Variables

Variables	Code	Symbol	−α	−1	0	1	+α
Initial Concentration, mg/L	A	Co	20	218.64	510	801.36	1000
Dosage, g	B	Do	0.01	0.109	0.255	0.401	0.5
Temperature, °C	C	T	20	26.08	35	43.92	50

TABLE 3.3 Experimental Design Matrix

Run	Concentration, mg/L	Dosage, g	Temperature, °C	Uptake, mg/g
1	510.00	0.500	35.00	49.5185
2	510.00	0.010	35.00	424.691
3	801.36	0.109	43.92	281.609
4	510.00	0.255	35.00	85.3098
5	510.00	0.255	35.00	84.7756
6	20.00	0.255	35.00	3.36538
7	218.64	0.109	43.92	74.6764
8	510.00	0.255	50.00	84.8943
9	218.64	0.401	26.08	26.3385
10	510.00	0.255	35.00	85.292
11	510.00	0.255	20.00	77.5938
12	1000.00	0.255	35.00	159.182
13	801.36	0.109	26.08	161.098
14	801.36	0.401	26.08	88.2989
15	801.36	0.401	43.92	91.7675
16	510.00	0.255	35.00	85.7597
17	218.64	0.401	43.92	26.6857
18	218.64	0.109	26.08	63.7337
19	510.00	0.255	35.00	85.2802
20	510.00	0.255	35.00	84.8528

$$q = b_0 + \sum_{i=1}^{n} b_i x_i + \sum_{i=1}^{n} b_{ii} x_i^2 + \left(\sum_{i=1}^{n-1} \sum_{j=1+i}^{n} b_{ij} x_i x_j \right) + e_i \tag{4}$$

where q is CR uptake, b_0 = constant coefficient, b_i, b_{ii}, b_{ij}, are the interaction coefficients of linear, quadratic and second order terms respectively. n is the number of factor and e_i is the error [21, 22].

The uptake of CR varied from 3.36 (minimum) to 424.69 mg/g (maximum) in the present study. The ratio of minimum to maximum uptake of Congo red removal was 126.39, which was greater than 10 suggesting that transformation was required in the present system. Since CR removal in the present investigation represented right skewness with non-zero positive values, a natural log transformation was applied to the experimental data.

Sequential model Sum of Squares (Table 3.4), Lack of Fit Tests (Table 3.5) and Model Summary Statistics (Table 3.6) were carried out to check the adequacy of the model for CR removal by the MB. p-values for the regressions were lower than 0.01 (see Table 3.4) suggesting a model of quadratic. The model summary statistics (Table 3.6) indicates higher regression coefficient ($R^2 = 0.9284$) for the quadratic model with the minimum standard deviation (0.36).

TABLE 3.4 Sequential Model Sum of Squares

Source	Sum of Squares	df	Mean Square	F Value	p-value prob>F	Remarks
Mean vs Total	373.65	1	373.65			
Linear vs Mean	13.07	3	4.36	13.90	0.0001	Suggested
2FI vs Linear	0.082	3	0.027	0.072	0.9737	
Quadratic vs 2FI	3.64	3	1.21	9.36	0.0030	Suggested
Cubic vs Quadratic	1.18	4	0.30	15.68	0.0025	Aliased
Residual	0.11	6	0.019			
Total	391.74	20	19.59			

TABLE 3.5 Lack of Fit Tests

Source	Sum of Squares	df	Mean Square	F Value	p-value prob>F	Remarks
Linear	5.01	11	0.46	25870.12	<0.0001	Suggested
2FI	4.93	8	0.62	34986.21	<0.0001	
Quadratic	1.30	5	0.26	14701.97	<0.0001	Suggested
Cubic	0.11	1	0.11	6412.41	<0.0001	Aliased
Pure Error	8.809E-005	5	1.762E-005			

TABLE 3.6 Model Summary Statistics

Source	Std. Dev.	Adjusted R-Squared	Predicted R-Squared	R-Squared	Press	Remark
Linear	0.56	0.7228	0.6708	0.4992	9.06	Suggested
2FI	0.62	0.7273	0.6015	0.3394	11.95	
Quadratic	0.36	0.9284	0.8639	0.4547	9.86	Suggested
Cubic	0.14	0.9937	0.9802	−0.3771	24.90	Aliased

The significance of the regression models and individual model coefficients and the lack of fit are tested using the same statistical package. The resulting ANOVA for the reduced quadratic models summarize the analysis of variance of each response and shows the significant model terms. Table 3.7 shows the ANOVA result for CR removal onto the MB, with a model F- value of 14.40 implying that the model is significant (at $p < 0.05$). In this case, A(Co), B(Do) are highly significant model terms, A^2 and B^2 are significant model terms, while model values greater than 0.10 indicated that the model terms were not significant.

The "Lack of Fit F-value" of 14701.97 implies it is significant. There is only a 0.01% chance that a "Lack of Fit F-value" this large could occur due to noise. The "Pred R-Squared" of 0.8639 is close to the "Adj R-Squared" of 0.9284 as one might normally expect.

The desirable signal to noise ratio in "Adeq Precision" measurement is greater than 4. For CR removal by MB a ratio of 16.132 indicates an adequate signal. This model can be used to navigate the design space.

TABLE 3.7 Analysis of Variance Table

Source	Sum of Squares	Df	Mean Square	F-Value	p-value Prob>F	Remarks
Model	16.79	9	1.87	14.40	0.0001	Significant
A-Co	9.16	1	9.16	70.73	<0.0001	
B-Do	3.85	1	3.85	29.71	0.0003	
C-T	0.062	1	0.062	0.48	0.5049	
AB	4.521×10^{-3}	1	4.521×10^{-3}	0.035	0.8555	
AC	0.023	1	0.023	0.17	0.6848	
BC	0.055	1	0.055	0.43	0.5281	
A^2	2.56	1	2.56	19.73	0.0013	
B^2	0.75	1	0.75	5.77	0.0372	
C^2	7.281×10^{-3}	1	7.28×10^{-3}	0.056	0.8174	
Residual	1.30	10	0.13			
Lack of Fit	1.30	5	0.26	14701.97	<0.0001	Significant
Pure Error	8.809×10^{-5}	5	1.762×10^{-5}			
Cor Total	18.09	19				

The final mathematical model in terms of the actual factors as determined by Design-expert software is shown below:

In terms of coded factors:

$$\ln q = 4.44 + 0.82A - 0.53B + 0.067C + 0.024A.B + 0.053A.C$$
$$-0.83B.C - 0.42A^2 + 0.23B^2 + 0.022C^2 \quad (5)$$

In terms of coded factors:

$$\ln q = 3.29066 + 7.01189\times10^{-3}C_0 - 7.16079D_0 - 6.34375\times10^{-3}T$$
$$+5.60067\times10^{-4}C_0.D_0 + 2.04670\times10^{-5}C_0T - 0.064008D_0.T \quad (6)$$
$$-4.96075\times10^{-6}C_0^2 + 10.72858\times10^{-4}D_0^2 + 2.82558\times10^{-4}T^2$$

The normality of the data can be checked by plotting a normal probability plot of the residuals. If the data points on the plot fall fairly close to the straight line, then the data are normally distributed [21]. The normal probability plot of the residuals for CR was shown in Figure 3.4. It can be seen that the data points were fairly close to the straight line and it indicates that the experiments come from a normally distributed population.

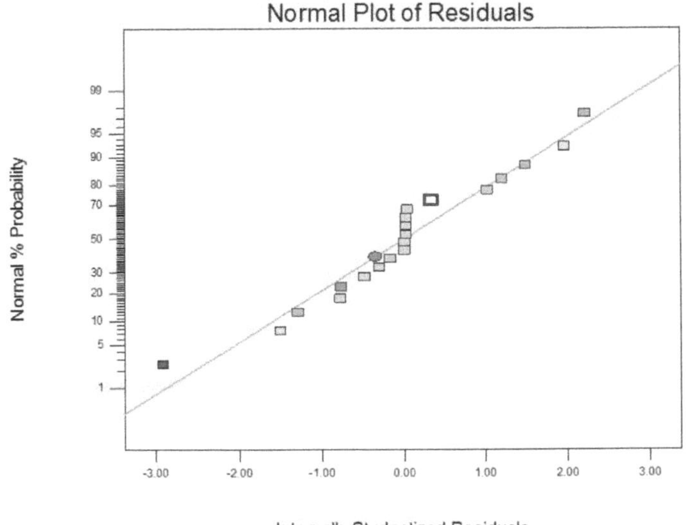

FIGURE 3.4 The studentized residuals and normal percentage probability plot of adsorption of CR.

Figure 3.5 shows the studentized residuals versus predicted conversion percent. The general impression is that the plot should be a random scatter, suggesting the variance of original observations is constant for all values of the response. If the variance of the response depends on the mean level of Y, then this plot often exhibits a funnel-shaped pattern [23, 24]. This is also an indication that there was no need for the transformation of the response variable.

Figure 3.6 shows the relationship between the actual and predicted values of ln q for CR removal onto MB. It can be seen in Figure 3.6 that the developed models were adequate because the residuals for the prediction of each response are minimum, since the residuals tend to be close to the diagonal line.

Uptake response surface graphs and contour plots for the adsorption of Congo red are shown in Figures 3.7–3.9. An increase in uptake is observed at higher concentrations and lower dosages (Figure 3.7). Figure 3.8 shows the interactive effect of temperature and initial concentration of CR uptake onto MB. It is clear that the removal of CR increased with temperature confirmed that the sorption process for CR uptake with MB was endothermic in nature. In Figure 3.9 the combined effect of temperature and dosage at constant initial feed was studied. Increase of uptake observed at higher temperatures and lower dosage.

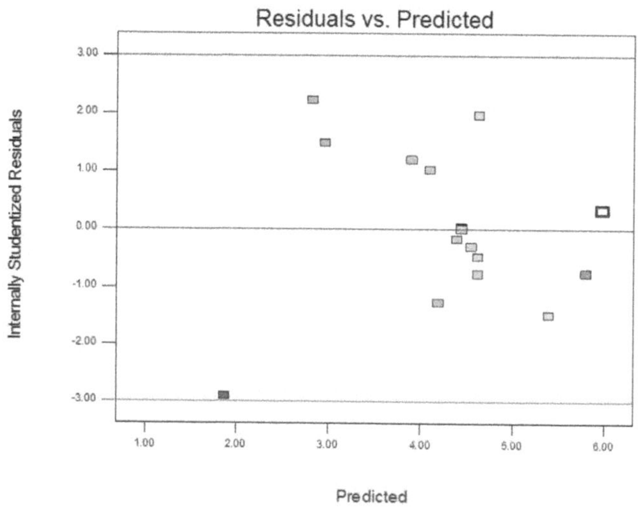

FIGURE 3.5 The studentized residuals versus predicted conversion percent.

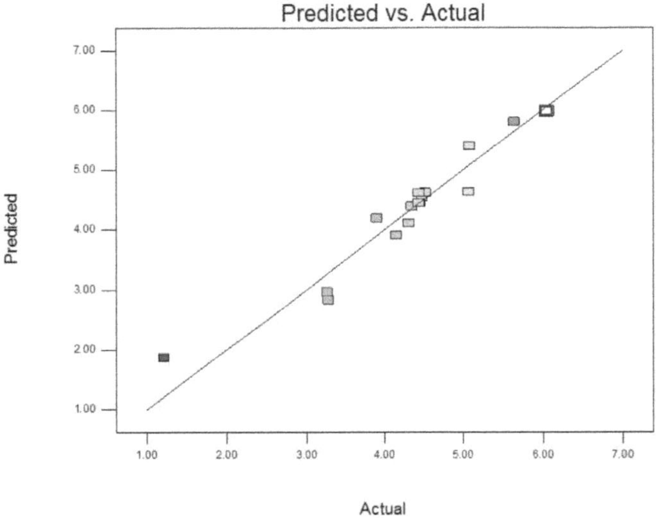

FIGURE 3.6 The actual and the predicted adsorption of Congo red uptake.

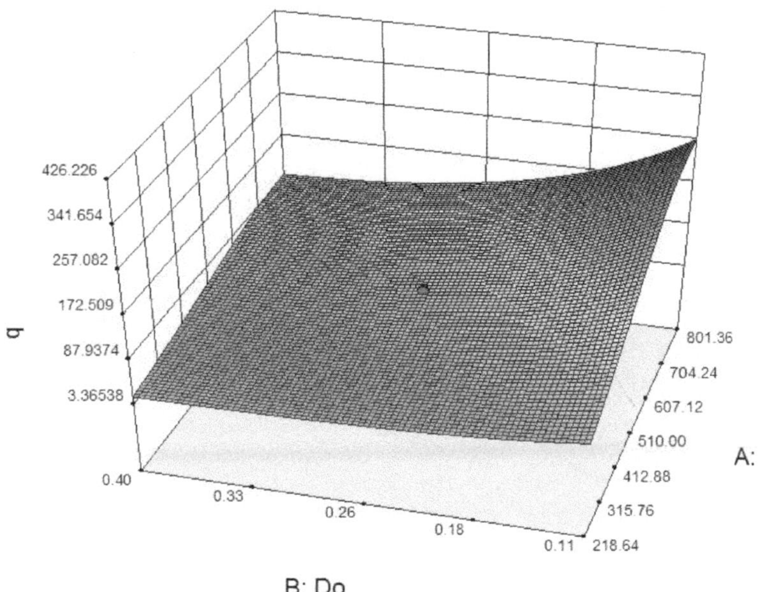

FIGURE 3.7 Continued

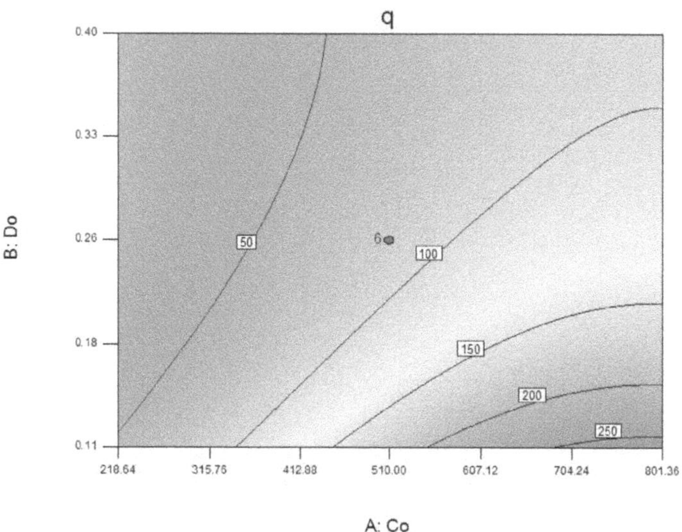

FIGURE 3.7 The combined effect of initial concentration and adsorbent dose on adsorption of CR. (a) 3D Response surface Graph, (b) Contour Plot.

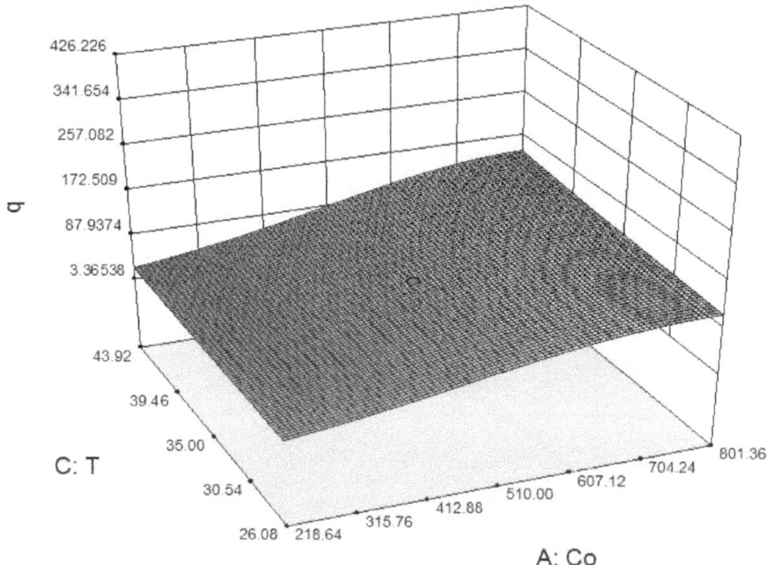

FIGURE 3.8 Continued

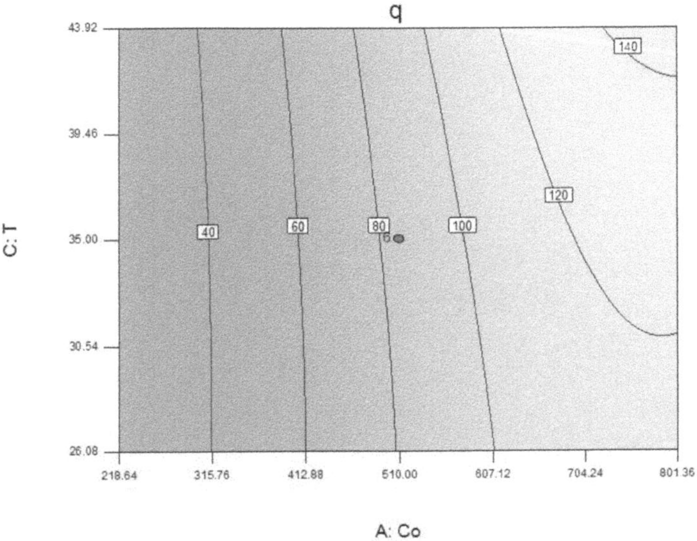

FIGURE 3.8 The combined effect of initial concentration and temperature on adsorption of CR. (a) 3D Response surface Graph, (b) Contour Plot.

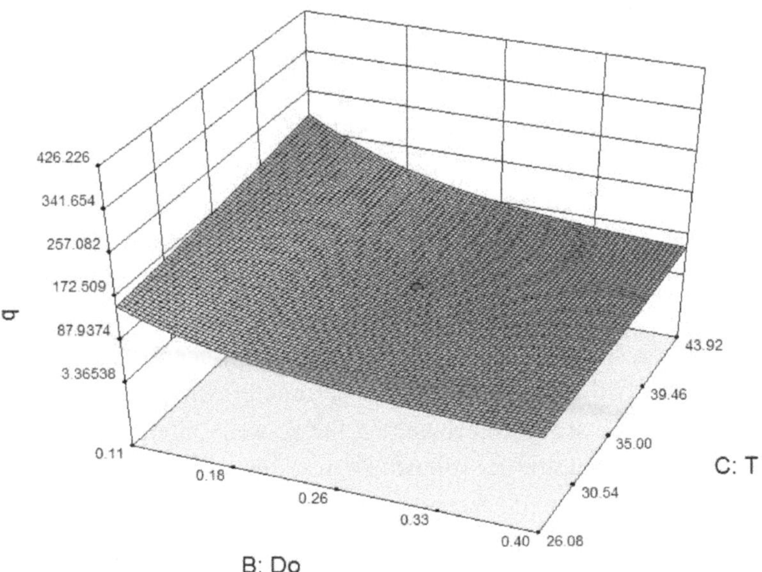

FIGURE 3.9 Continued

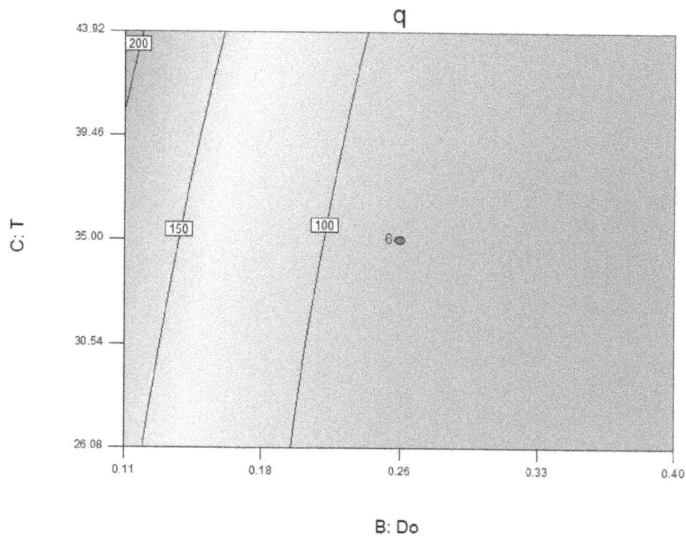

FIGURE 3.9 The combined effect of temperature and adsorbent dose on adsorption of CR. (a) 3D Response surface Graph, (b) Contour Plot.

3.4 CONCLUSION

Bentonite was modified and used to remove Congo red from dye effluent water. It is proven to be an effective adsorbent. The equilibrium time required is determined as 4 hrs. Univariative parametric studies were conducted and ineffective variable (pH) was eliminated. Dye uptake increased with temperature indicating endothermic nature of the process. The effects of various parameters and their interactions were studied using response surface methodology. Effluent concentration and dosage were identified as most influential parameters and the second order effects of the same are also influencing the process. Statistical quadratic model was generated and its robustness was tested as per the standards given in the literature. 3D Response surface graphs and contour plots were generated and analyzed.

KEYWORDS

- **Congo Red**
- **Modified Bentonite**
- **RSM**

REFERENCES

1. Wang, L., Li, J., Wang, Y., Zhao, L., & Jiang, Q., Adsorption capability for Congo red on nanocrystal HneMFe204 (M=Mn, Fe, Co, Ni) spinel ferrites. *Chemical Engineering Journal* 2010, doi: 10.1016/j.cej.-2011.10.088.
2. Somasekhara Reddy, M. C., Sivaramakrishna, L., & Varada Reddy, A., The Use of an Agricultural Waste Material, Jujuba Seeds for the Removal of Anionic Dye (Congo Red) from Aqueous Medium. *J Hazard Mater* 2012, *203*, 118–127.
3. Sabnis, R. W., Handbook of Biological Dyes and Stains. Synthesis and Industrial Applications (John Wiley & Sons, New Jersey, Canada, Springer, 2010, pp. 106–107).
4. Han, R., Ding, D., Xu, Y., Zou, W., Wang, Y., Li, Y., & Zou, L., Use of rice husk for the adsorption of Congo red from aqueous solution in column mode. *Bioresource Technol.* 2008, *99*, 2938–2946.
5. Raymundo, A. S., Zanarotto, R., Belisario, M., Pereira, M. G., Ribeiro, J. N., & Ribeiro, A. V. F. N., Evaluation of sugar-cane bagasse as bioadsorbent in the textile wastewater treatment contaminated with carcinogenic Congo red dye. *Brazilian Arch. Biology Technol.* 2010, *53*, 931–938.
6. Hu, Q. H., Xu, Z. P., & Qiao, S. Z., A novel color removal adsorbent from hetero coagulation of cationic and anionic clays. *J. Colloid Interface Sci.* 2007, *308*, 191–199.
7. Lian, L., Guo, L., & Wang, A., Use of $CaC1_2$ modified bentonite for removal of Congo red dye from aqueous solutions. *Desalination* 2009, *249*, 797–801.
8. Sanghi, R., & Bhattacharya, B., Review on decolonization of aqueous dye solutions by low cost adsorbents. *Color. Technol.* 2002, *118*, 256–269.
9. Kumar, A., Kumar, S., Kumar, S., & Gupta, D. V., Adsorption of phenol and 4-nitrophenol on granular activated carbon in basal salt medium: Equilibrium and kinetics. *J. Hazard. Mater.* 2007, *147*, 155–166.
10. Xin, X., Si, W., Yao, Z., Feng, R., Du, B., Yan, L., & Wei, Q., Adsorption of benzoic acid from aqueous solution by three kinds of modified Bentonites. *J. Colloid Interface Sci.* 2011, *359*, 499–504.
11. Erdem, B., & Ozcan, A. S., Adsorption and solid phase extraction of 8-hydroxyquinoline from aqueous solutions by using natural bentonite, *Applied Surface Science* 2010, *256*, 5422–5427.
12. Ayari, F., Srasra, E., & Trabelsi-Ayadi, M., Characterization of bentonitic clays and their use as adsorbent. *Desalination* 2005, *185*, 391–397.

13. Gottipati, R., & Mishra, S., Process optimization of adsorption of Cr(VI) on activated carbons prepared fromplant precursors by a two-level full factorial design. *Chemical Engineering Journal* 2010, *160,* 99–107.
14. Cronje, K. J., Chetty, K., Carsky, M., Sahu, J. N., & Meikap, B. C., Optimization of chromium (VI) sorption potential using developed activated carbon from sugarcane bagasse with chemical activation by zinc chloride. *Desalination* 2011, *275*, 276–284.
15. Myers, R. H., Response Surface Methodology, Allyn and Bacon, New York, 1971.
16. Box, G. E. P., & Hunter, J. S., Multi factor experimental designs for exploring response surfaces. *Ann Math Statist* 1957, *28*, 195–241.
17. Box, G. E. P., & Hunter, J. S., The 2k p fractional factorial designs, parts I and II, *JTechnometrics* 1961, *3*, 311–458.
18. Muthukumar, M., Mohan, D., & Rajendran, M., Optimization of mix proportions of mineral aggregates using Box-Behnken design of experiments. *Cem. Concr. Compos.* 2003, *25*, 751–758.
19. Benyounis, K. Y., Olabi, A. G., & Hashmi, M. S. J., Effect of laser welding parameters on the heat input and weld-bead profile. *J. Mater. Process. Technol.* 2005, *164*, 978–985.
20. Tarangini, K., Kumar, A., Satpathy, G. R., & Sangal, V. K., Statistical Optimization of Process Parameters for Cr (VI) Biosorption onto Mixed Cultures of Pseudomonas aeruginosa and Bacillus subtilis. *Clean* 2009, *37*, 319–327.
21. Box, G. E. P., Hunter, J. S., Multi factor experimental designs for exploring response surfaces. *Ann. Math. Statist.* 1957, *28*, 195–241.
22. Ozer, A., Gurbuz, G., Calimli, A., & Korbahti, B. K. Investigation of nickel (II) biosorption on Enteromorphaprolifera: optimization using response surface analysis. *J. Hazard. Mater.* 2008, *152*, 778–788.
23. Mahalik, K., Sahu, J. N., Patwardhan, A. V., & Meikap, B. C., Statistical modeling and optimization of hydrolysis of urea to generate ammonia for flue gas conditioning. *J. Hazard. Mater.* 2010, *182*, 603–610.
24. Myers, R. H., & Montgomery, D. C., *Response Surface Methodology: Process and Product Optimization using Designed Experiments. 2nd Ed.*, USA: John Wiley & Sons, 2002.

PART II

ENVIRONMENTAL APPLICATIONS

CHAPTER 4

SURFACE ALTERED ALUMINO-SILICATE RESIN (ZEOLITE-Y) FOR REMEDIATION OF OIL SPILLAGE

S. U. MESHRAM,[1] C. M. SHAH,[2] and H. J. BALANI[2]

[1]*Assistant Professor, Department of Applied Chemistry, Laxminarayan Institute of Technology, RTM Nagpur University, Amravati Road, Nagpur–33, India*

[2]*Research Assistant, Department of Oil, Fats and Surfactants Technology, Laxminarayan Institute of Technology, RTM Nagpur University, Amravati Road, Nagpur–33, India*

CONTENTS

4.1 Introduction 56
4.2 Causes of Oil Spillage 57
4.3 Adverse Effects of Oil Spill 58
4.4 Remediation Techniques for Prevention of Oil Spill 58
4.4.1 Cellulose Aerogel Method 59
4.4.2 Underwater Dispersants 59
4.4.3 Controlled Burns 59
4.4.4 Booms and Skimmers 60
4.4.5 Gelling Agents 60
4.4.6 Oil Spill Hair Mats 60
4.4.7 Adsorbent Materials 60

4.5 Existing Techniques to Detect the Oil Spill 61
4.6 Materials and Methodology 62
4.6.1 Materials 62
4.6.2 Method For Synthesis of Surface Modified Zeolite-Y 63
4.7 Instrumental Investigations of SMZ-Y 64
4.8 Batch Adsorption Studies 64
4.9 Result and Discussion 66
Keywords 68
References 68

4.1 INTRODUCTION

An oil spill is the release of a liquid petroleum hydrocarbon into the environment, especially marine areas, due to human activity, and is a form of pollution. The term is usually applied to marine oil spills, where oil is released into the ocean or coastal waters, but spills may also occur on land. Oil spills may be due to releases of crude oil from tankers, offshore platforms, drilling rigs and wells, spills of refined petroleum products (such as gasoline, diesel) and their by-products. It may also cause due to heavier fuels used by large ships such as bunker fuel, or the spill of any oily refuses or waste oil. Spilt oil penetrates into the structure of the plumage of birds and the fur of mammals, reducing its insulating ability, and making them more vulnerable to temperature fluctuations and much less buoyant in the water [1]. The major categories responsible for oil spill mainly includes:

(i) **Natural seeps**: This type is generally due to fissures in the ocean bed and eroding sedimentary rock. Here the hydrocarbons, which ooze out of the seabed, contribute upto 46% approximately. One of the best-known areas where this happens is Coal oil Point along the California Coasts near Santa Barbara. An estimated 2,000–3,000 gallons of crude oil is released naturally from the ocean bottom everyday just a few miles offshore from the beach.

(ii) **Operational discharge**: This type of oil spill is reported from consumption of oil mainly during various ongoing activities in edible and non-edible oil industries such as transportation of oil, leakage of oil from containers and other land based sources. Also spill occurs due to loss of oil through engine due to wear and tear from transportation vehicles. It has been reported that about 37% of oil spill generated due to this.

(iii) **Accidental spills**: It is the most common oil spill reported due to accidental spill from ships. It often causes due to unsuitable weather conditions and due to human error. It accounts approximately 12% [2]. Cleanup and recovery from an oil spill is difficult and depends upon many factors, including the type of oil spilled, the temperature of the water (affecting evaporation and biodegradation), and the types of shorelines and beaches involved. Spills may take weeks, months or even years to clean up [3].

4.2 CAUSES OF OIL SPILLAGE

When an oil spill occurs, responders consider factors such as oil toxicity, rate of formation of oil slick, and time period for breakdown of oil. And even weather conditions and location of the spill are taken into consideration [4]. The majority of oil spillage occurred due to Petroleum oil, which mainly includes various categories of oil. ***Class A*** oil is the most toxic but least persistent of all oils. This oil spreads quickly when spilled and has a strong odor. Class A oils include high-quality light crude oils as well as refined products such as gasoline and jet fuel. Toxic components of gasoline include benzene, a known carcinogen, and hexane, which can damage nervous systems in humans and animals. However, ***Class B*** oils are known as "non-sticky" oils. They are less toxic than class A oils but more likely to adhere to surfaces. Refined products such as kerosene and other heating oils are considered into this class. Class B oils are highly flammable and will burn longer than class A oils. Another category is called as ***Class C*** oil, which do not spread quickly or penetrate sand and soil as easily as lighter oils. It produces a sticky film; can severely contaminate intertidal zones, leading to expensive, long-term cleanups. Moreover, ***Class D*** crude oil is solid and has the least toxicity. The biggest environmental concern posed by class D oil occurs if the oil is heated and hardens on a surface, making cleanup nearly impossible. The Synthetic oils and oils derived from plant or animal fats which are commonly known as Non-Petroleum oils regulated by the EPA because they cause contamination if released into the environment. Non-petroleum oils are slow to break down and easily penetrate soil, causing long-lasting damage to an affected area. Largest Oil spill till Date reported is in Kuwait at Kuwaiti Oil Fires on November 1991 the Oil spill was of 136,000 (thousands) Tone of Crude Oil. From a report in 2002 by National Research Council (NRC) the total inputs of oil is 470,000–8.4 million tons per year out of which nearly 12% contributes to accidental oil spills. Also a another report by Group of Experts on Scientific Aspects of Marine Environmental Protection (GESAMP) the total oil input is nearly about 2.3 million tons per year off which 24% contributes to accidental spills [5].

4.3 ADVERSE EFFECTS OF OIL SPILL

Oil spills have a number of effects on the environment and economy. On a basic level, oil will damage waterways, marine life and plants and animals on the land. Oil penetrates into the structure of the plumage of birds and the fur of mammals, reducing its insulating ability, and making them more vulnerable to temperature fluctuations and much less buoyant in the water. Oil can impair a bird's ability to fly, preventing it from foraging or escaping from predators. As they preen, birds may ingest the oil coating their feathers, irritating the digestive tract, altering liver function, and causing kidney damage. Together with their diminished foraging capacity, this can rapidly result in dehydration and metabolic imbalance. The majority of birds affected by oil spills die from complications without human intervention [6, 7]. Heavily furred marine mammals exposed to oil spills are affected in similar ways. Oil coats the fur of sea otters and seals, reducing its insulating effect, and leading to fluctuations in body temperature and hypothermia. Oil can also blind an animal, leaving it defenseless. Animals can be poisoned, and may die from oil entering the lungs or liver. Exonn oil spills which occurred in 1989 an approximately 40,000 tons of oil was spilled, which oiled 30,000 birds. The negative effects of on oil spill may eventually fade away, but in many cases it will be matter of several years, even decades, before an area or ecosystem has fully recovered from a spill that caused extensive damages [8].

4.4 REMEDIATION TECHNIQUES FOR PREVENTION OF OIL SPILL

The current methods for prevention of oil spill mainly include:

4.4.1 CELLULOSE AEROGEL METHOD

This method is commonly applied for treatment of oil spill samples using the waste generated from paper industries. However, polypropylene is commonly used for crude oil spill cleaning, but it has been noticed that it has low absorption capacity with lack of biodegradability. In this technique ultra-light and highly porous material was successfully prepared from paper waste cellulose fibers. The material was functionalized with methyltrimethoxysilane (MTMS) to enhance its hydrophobicity and oleophilicity. It was observed that the viscosity of the crude oil is the main factor affecting their absorption onto the aerogel. The strong affinity of the MTMS coated recycled cellulose aerogel the oil makes the aerogel good absorbent for crude oil spill cleaning [9].

4.4.2 UNDERWATER DISPERSANTS

The use of underwater dispersants was purely experimental when the BP oil spill first occurred. Dispersants enhance the emulsification and dissolution of oil into the water, which augments degradation and prevents slick formation, reducing the amount of oil that reaches shorelines [10, 11]. 1.84 M gallons of chemical dispersant was applied to oil released in the subsurface and to oil slicks at the surface [12]. Functionalized carbon black (CB) nanoparticles were found to be nontoxic as well as hydrophobic, also CB was shown to absorb benzene [13]. Limited toxicology data are available regarding oil dispersant exposure to coral species as it can cause bleaching of corals. Corexit EC9500A is a commonly applied dispersant well known for its use after the Deepwater Horizon spill [14].

4.4.3 CONTROLLED BURNS

Controlled burns have been conducted throughout the Gulf oil spill. Fireproof boom corrals leaked oil into smaller, denser pockets that can be ignited remotely from the air and burned off. The process of burning removes large portions of oil from the water's surface, keeping it away from the shoreline. During the 2010 British Petroleum/Deepwater Horizon Gulf oil spill, an estimated one of every 20 barrels of spilled oil was deliberately burned off to reduce the size of surface oil slicks and minimize impacts of oil on sensitive shoreline ecosystems and marine life.

The black smoke that rose from the water's surface than 1 million pounds of black carbon (soot) pollution into the atmosphere [15].

4.4.4 BOOMS AND SKIMMERS

Another technique is using booms and skimmers to remove oil from the water's surface. Booms are used to collect oil in concentrated areas, while skimmers separate the crude from the water. Skimming the oil from the surface requires a fleet of skimming vessels and expensive equipment also the efficiency of skimmers is highly dependent on sea conditions and the presence of debris, which can both pose serious roadblocks to these techniques [16].

4.4.5 GELLING AGENTS

A gelling agent is a chemical used to solidify spilled oil, making it easier to collect. Using the motion of the sea, the gelling agent turns the oil into a rubbery substance that can be easily removed from water with nets, suction devices or skimmers. It is a mixture of nonionic surfactants (48%) and anionic surfactants (35%) in a solvent medium containing light hydrocarbon distillates and (1-, 2-butoxy-1-methylethoxy) propanol [17].

4.4.6 OIL SPILL HAIR MATS

Early on in the BP crisis, human hair was discussed as a possible material for sopping up the oil. Matter of Trust, a nonprofit environmental organization, has partnered with thousands of salons all over the world to use their clippings, which would normally get swept up off the floor and thrown in the trash. Instead, Matter of Trust collects it, stuffs it into mesh or nylon casings and creates improvised containment booms to control oil spills. Hair is very efficient at gathering oil, so it acts as a good adsorbent [18].

4.4.7 ADSORBENT MATERIALS

Sorbents are chemicals or materials that can capture liquids or gases. Absorbents incorporate a substance throughout the body of the absorbing material, for example, polypropylene, which absorbs oil. Adsorbents adhere substances over the surface of the adsorbing material, for example, activated carbon.

Sodium silicate ($Na_2O/SiO_2 = 1:3.3$) as a precursor prepa silica aerogel. This prepared silica aerogel is modified by mercaptopropyl trimethoxysilane [19]. Fly ash, a coal combustion by product with a predominantly aluminosilicate composition, is modified to develop an inexpensive sorbents for oil spill remediation. X-type zeolite sorbent with different surface functionalization propyl-, octyl-, octadecyl-trimethoxy siliane [20]. Adsorption by Crude oil, Mechanism behind the crude oil sorption is low micronaire (immature) cotton. Low micronaire cotton is employed, because of its finer structure and wax content which can absorb higher amounts of oil than regular-grade cottons [21, 22]. As this remediation for cleaning of oil spill are not economic and moreover the oil cannot be recovered which causes loss of oil as well as marine destruction so the technique which we are presenting in this paper is low costs surface altered adsorbents for remediation of oil spillage.

4.5 EXISTING TECHNIQUES TO DETECT THE OIL SPILL

The concentration of oil spill in water can be monitored by aerial surveillance, which includes monitoring of both the position and character of slick. Positions can be calculated by computer based oil spill trajectory models (OSTM). Amount of oil spill can also be approximately determined as shown in (Figure 4.1). Certain oil spill response and cleanup steps are tabulated as shown in (Figure 4.2) [23]. The first method consists of sorption and desorption using solvents such as petroleum ether and n-hexane. However, it was deduced that petroleum ether is quite flammable hence further studies has been made using n-hexane [24]. The extraction is conducted using Soxhlet method according to method prescribed by IS 3025 (Part 39) [25].

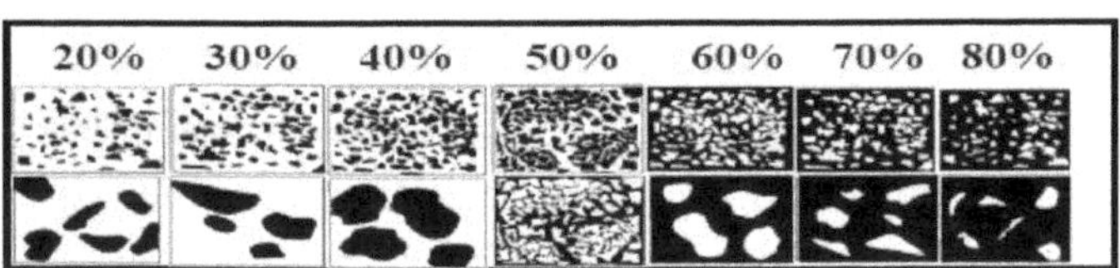

FIGURE 4.1 Percent cover of oil spill.

Response Stage	Description of Monitoring
Stage 1 Pre Spill	This includes true baseline monitoring and may be long term and large scale. "Control" sites can be well established. Study design can be modified and refined over time. Generally, such monitoring is undertaken in areas of high risk or on resources that are sensitive to spills or are of protection or conservation priority.
Stage 2 Post Spill – Pre Impact	Monitoring done at this stage is reactive and must often be designed and implemented at short notice to collect a "snapshot" of pre-impact conditions. Establishment of reliable "control" sites is difficult.
Stage 3 Post Impact – Pre Cleanup	Monitoring of oil-impacted shorelines, waters or resources. Examples include monitoring of oil behaviour and persistence in uncleaned shorelines or monitoring of immediate damage due to oil (not cleanup).
Stage 4 Cleanup	Monitoring that occurs through a cleanup activity. For example, monitoring the success or the effect of cleanup on shorelines, water quality or biological resources.
Stage 5 Post Cleanup - Pre Response Termination	Monitoring of resources, water or shorelines after cleanup activities have ceased but before the response has been terminated. These are usually short-term programmes. This would include final assessments of cleaned shorelines, perhaps as an agreed precondition to terminating a response.
Stage 6 Post Response	This includes all monitoring that occurs after the formal end of a response. Such studies may be short, medium or long-term.

Stage	Stage 1	Stage 2	Stage 3	Stage 4	Stage 5	Stage 6
Response	Pre Spill	Response (Post Spill)				Post Response
Impact	Pre Impact		Post Impact			
Cleanup	Pre Spill	Pre Cleanup		Cleanup	Post Cleanup	
	Spill	Impact	Start of Cleanup	End of Cleanup	End of Response	

FIGURE 4.2 Description of monitoring according to the stage of Incident (*Courtesy*: Oil Spill Monitoring Handbook, 2003. ISBN 0-642-70992-0).

4.6 MATERIALS AND METHODOLOGY

4.6.1 MATERIALS

The source material, for example, Pyrophilite clay was obtained from mines near Jhansi, Madhya Pradesh, India. The other chemicals such as sodium hydroxide flakes, HDTMABr salt and n-hexane has been procured from Merck India. The Petroleum products responsible for oil spillage such as diesel and kerosene have been purchased form commercial sources. Similarly a marine water sample was collected from Arabian Sea Mumbai, to make artificial oil spill source. The jar test apparatus (make: Secor) was used for gelation. A specially designed autoclave made up of stainless steel having a capacity of 350 mL was used in hydrothermal crystallization at autogenous pressure. A rotary orbital shaker (make: Murhopye scientific company) and magnetic Stirrer with controller (make: Elico) were used to modify the surface of Zeolite-Y in aqueous system.

4.6.2 *METHOD FOR SYNTHESIS OF SURFACE MODIFIED ZEOLITE-Y*

Method for producing SMZ-Y is mainly categorized into two stages. In the first stage Zeolite-Y resin is prepared using Pyrophilite clay as a source. The synthesis route is based upon fusion technique of clay with alkali at 500–600°C for a period of 1.5 to 2 h to extract the silica and alumina in the form of sodium silicate and aluminate. The fused mass was further cooled, milled with de-ionized water and subjected for agitation. The aged mass was later hydrothermally crystallized using Stainless steel autoclave at 95–100°C and separated using vacuum pump. The residue was washed repeatedly using de-ionized water till it reaches to desired pH and dried overnight. The second stage comprises of Surface alteration of Zeolite-Y using a quaternary ammonium salt (HDTMABr) as a surface modifier. The overall steps included in synthesis of SMZ-Y is depicted in (Figure 4.3).

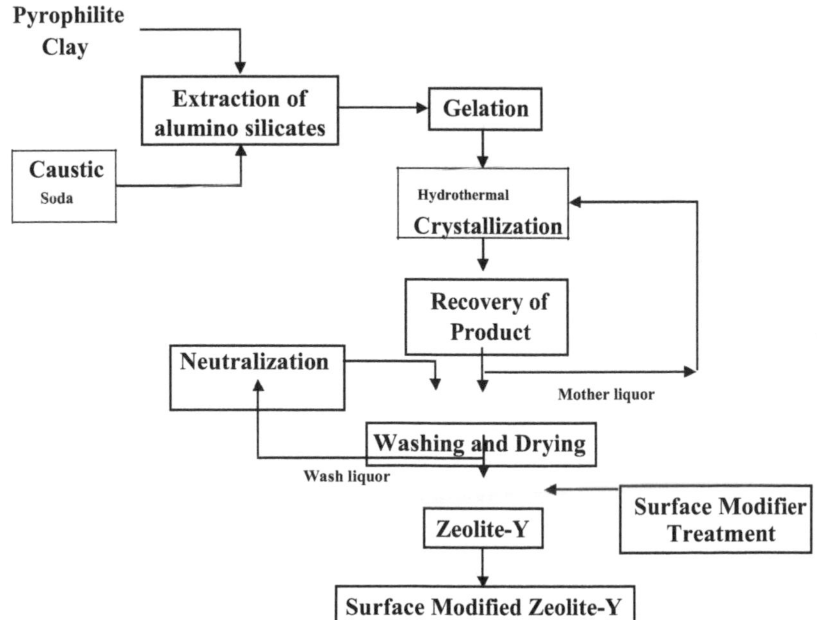

FIGURE 4.3 Schematic route for synthesis of Surface Modified Zeolite-Y.

4.7 INSTRUMENTAL INVESTIGATIONS OF SMZ-Y

Powder XRD analysis was employed to monitor the crystalline phase of Zeolite-Y, using CuKα as source of X-rays Model Panalytical. The sum x'perta total of relative pro intensities of *d*-spacing values in angstroms Å of Commercial Zeolite as compared to clay based Zeolite-Y has been used as basis for estimation of percent crystallinity. The *d*-spacing values used are as follows: 14.15, 8.73, 7.46, 5.69, 4.78, 4.39, 3.79, 2.87, and 2.65.

The surface morphology of the zeolite was examined by Jeol-840-A scanning electron microscope (SEM) Morphological characterization was carried out using SEM Model Jeol-840.

Surface Morphology of SMZ-Y reveals that the crystals of Zeolite-Y have been successfully altered as shown in (Figure 4.5) also the tetrahedral geometry was retained with little distortion at the edges which does not affect the characteristics of the material.

4.8 BATCH ADSORPTION STUDIES

A stimulated oil spill sample was prepared by spiking of 5 mL of petroleum product (diesel and kerosene) in seawater, which was procured from suitable source. To this solution, variable doses of SMZ-Y from 2.0 g/L to 10 g/L was employed. Subsequently the doses of adsorbent were added and

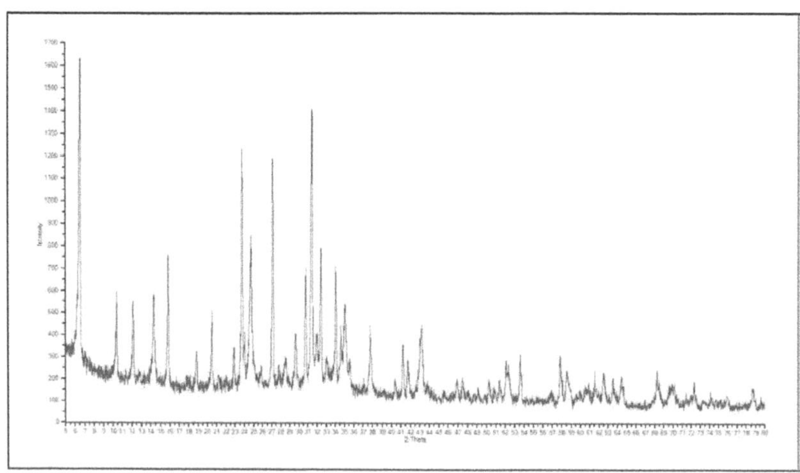

FIGURE 4.4 XRD of pyrophilite based Zeolite-Y.

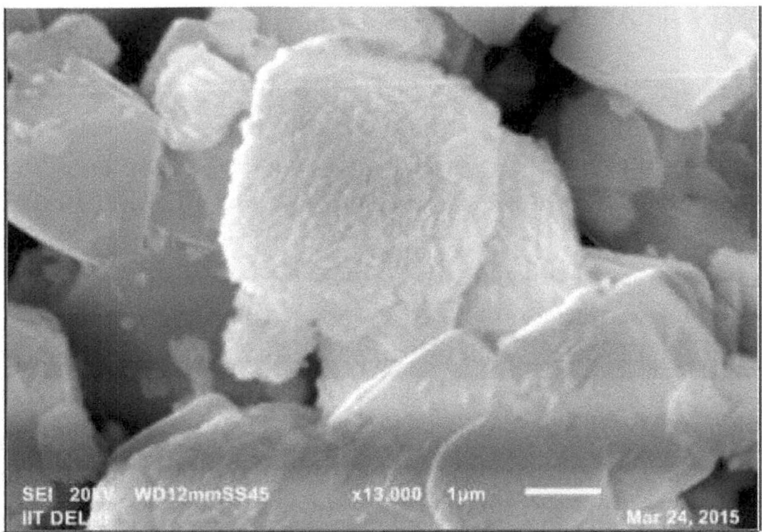

FIGURE 4.5 SEM of SMZ-Y.

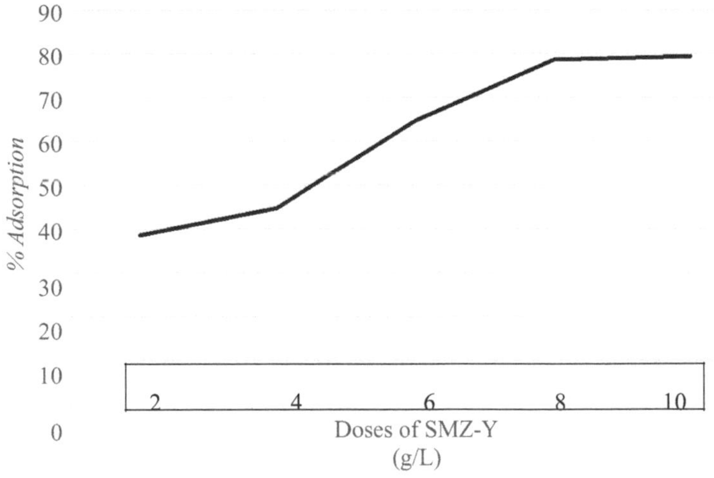

FIGURE 4.6 Adsorption with varying dose of SMZ-Y.

agitated with the help of gyratory shaker for a contact period of 60 min. Later, the sample was centrifuged and the top fraction was decanted and oil adsorbed SMZ-Y sample (bottom product) was further separated with the aid of solvent by suitable means and the quantification of oil has been calculated by subjecting it to Soxhlet Extraction.

4.9 RESULT AND DISCUSSION

To investigate the adsorption efficiency of SMZ-Y, batch adsorption studies were carried out using variable doses of SMZ-Y against fixed volume of oil spill sample for 1h. The maximum oil adsorption efficiency (75–80%) was observed with an optimum dose of 8 g/L. The adsorption of spilled oil can be determined qualitatively by IR Spectrometry and quantitatively by Soxhlet extraction.

Solvent from the sample was evaporated and the residue left was desorbed oil. The detection of oil sample in solvent has been analyzed by subjecting the solvent towards Infrared Spectroscopy (IR) in the range of 400–4000 cm^{-1}. The peak obtained at 2930 cm^{-1} proves the presence of CH2 bond, which is a common characteristic of oil [26].

Figure 4.5 (a) shows the IR of hexane and (b) shows oil in hexane, which was recovered after the Soxhlet extraction process where the basic characteristic peak of hexane is found to be intact with certain changes in some peaks, which shows the presence of oil in solvent.

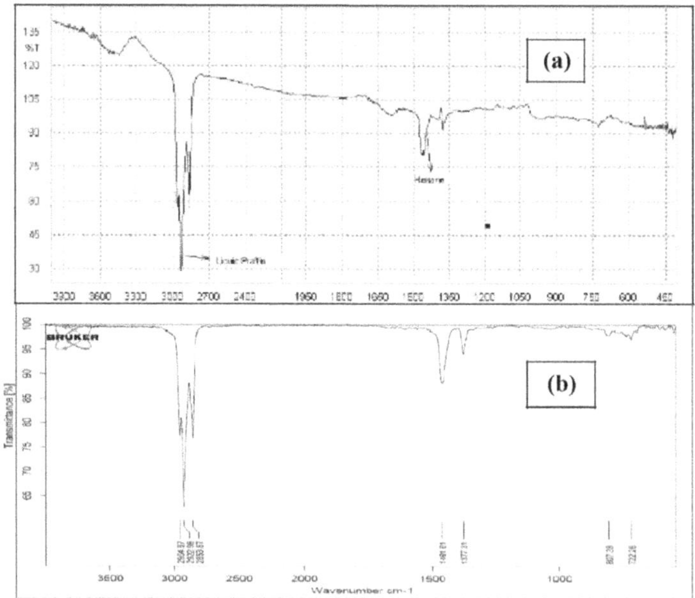

FIGURE 4.7 Comparative investigation of IR patterns for adsorption studies (a) *n*-hexane (b) oil in hexane.

Chanin et al. [27] proposed using Freon 113 as a solvent. They showed that its use gives results that are virtually the same as *n*-hexane, and overcomes the flammability problems. Taras and Blum [28] confirmed these results. More importantly, Taras and Blum showed extraction efficiency can be greatly improved if sodium chloride is added to the oil sample at a concentration of 5 gL^{-1}. The use of salt overcomes the problems of low extraction efficiency with liquid/liquid extraction. The high salt concentration apparently coagulates the emulsified oil by 15 double-layer compression. The next important development was reported by Gruenfeld [29] who showed that the extracted oil and grease in the solvent can be measured by spectrophotometry. The use of spectrophotometry overcomes two important problems with the oil and grease analysis. First, it extends the nominal limits of detection of oils and grease to levels below 0.5 ugL^{-1}. Second, the evaporation of solvent is not required in the IR spectrophotometry technique, which reduces the loss of low molecular weight compounds, an unfortunate shortcoming of all the gravimetric techniques [30]. Furthermore, the oil adsorption capacity per gram of adsorbent can be calculated as stated in Eq. (1).

$$\text{Oil Adsorption Capacity (OAC)} = (\text{Aas} - \text{Ai})/\text{Ai} \qquad (1)$$

where Aas is the weight of sorbent with oil at the end of sorption test, and Ai is the initial dry sorbent weight, and the quantity (Aas – Ai) is the net oil adsorbed (all weights are measured in grams).

The aforesaid method reveals an adsorption efficiency of SMZ-Y upto 75–80%, which is quite encouraging. Moreover to improve its efficacy, certain parameters have to optimized such as amount of surfactant loading on Zeolite-Y, Contact period, Type of agitation and pH, etc. Further the adsorbent synthesized is in the form of powder, which has to be converted into certain operational form such as pellets and fibrils by addition of suitable binder. Likewise, recovery of adsorbent as well as spilled oil using desorption studies are also in conduit to make the overall process economically viable and commercially feasible.

KEYWORDS

- **Adsorption**
- **Oil spillage**
- **Remediation**
- **Surface modified Zeolite-Y**

REFERENCES

1. Lingering lessons of the Exxon Valdez oil spill. *http://commondreams.org*. 2004–03–22, 2014.
2. *U.S. Geological Survey (USGS),* U.S. National Academy of Science by National Research Council (NRC).
3. *Hindsight and Foresight, 20 Years After the Exxon Valdez Oil Spill.* NOAA 2010–03–16, 2014.
4. https://www.epa.gov/emergency-response/types-crude-oil. *Emergency Response, United States Environmental Protection Agency.*
5. Hogan, C. M., Stromberg, N., *Magellanic Penguin, http://GlobalTwitcher.com,* 2008.
6. Dunnet, C. D., Conan, G., & Bourne, W., Oil pollution, *Philosophical Transactions of the Royal Society of London,* 1982.
7. Untold sea bird mortality due to marine oil pollution, elements online environmental magazine.
8. http://oils.gpa.unep.org/facts/whildlife.html.
9. Nguyen, S. T., Feng, J., Le, N. T., Le, A. T., Nguyen, H., Tan, V., & Duong, H. M., *Ind. Eng. Chem. Res.* 2013, *52*, 18386–18391.
10. Chapman, H., Purnell, K., Law, R. J., & Kirby, M. F., The use of Chemical dispersants to combat oil spills at sea: A review of practice And research needs in Europe. *Mar. Pollut. Bull.* 2007, *54*, 827−838.
11. Wise, J., & Wise, J. P., A review of the toxicity of chemical Dispersants. *Rev. Environ. Health* 2011, *26*, 281−300.
12. White, H. K., Lyons, S. L., Harrison, S. J., Findley, D. M., Liu, Y., & Kujawinski, E. B., *Environ. Sci. Technol. Lett.,* 2014, *1*, 295–299.
13. Rodd, A. L., Creighton, M. A., Vaslet, C. A., Rangel-Mendez, J. R., Hurt, R. H., & Kane, A. B., *Environ. Sci. Technol;* 2014, *48*, 6419–6427.
14. Studivan, M. S., Hatch, W. I., & Mitchelmore, C. L., *SpringerPlus,* 2015, *4*(80).
15. National Oceanic and Atmospheric Administration (NOAA), 2011, *http://noaanews.noaa.gov/stories2011/20110920_gulfplume.html.*
16. Sakthivel, T., Reid, D. L., Goldstein, I., Hench, L., & Seal, S., *Environmental Sci. Journal*; 2013.

17. Venkataraman, P., Tang, J., Frenkel, E., McPherson, G. L., He, J., Raghavan, S. R., Kolesnichenko, V., Bose, A., & John, V. T., *ACS Appl. Mater. Interfaces* 2013, *5*, 3572–3580.
18. Matter of Trust – Eco-Enthusiasts for Renewable Resources, matteroftrust.org.
19. Sakthivel, T., Reid, D. L., Goldstein, I., Hench, L., & Seal, S., *Environmental Sci. Journal*, 2013.
20. Cui, X. L., Price, J. B., Calamari, T. A., Hemstreet, J. M., & Meredith, W. Cotton wax and its relationship with fiber and yarn properties – Part I: Wax content and fiber properties. *Text. Res. J.* 2002, *72*, 399–404.
21. Singh, V., Kendall, R. J., Hake, K., & Ramkumar, S., *Ind. And Engg. Chem. Research*; 2013, *52*, 6277–6281.
22. *Oil Spill Monitoring Handbook*, ISBN 0642709920, AMSA, MSA of New Zealand.
23. Chanin, G., Chow, E. H., Alexander, R. B., & Powres, *Journal of the Water Pollution Control Federation,* 1967, *39*, 1892–1895.
24. *Indian Standard Methods for Methods of Sampling and Test (Physical and Chemical) for Water and Wastewater Part 39 Oil and Grease*; IS 3025; 1991.
25. Taras, M. J., & Blum, K. A., *Journal of the Water Pollution Control Federation*; 1968, *40*, 404–411.
26. Gruenfeld, M., *Environmental Science and Technology*; 1973, *7*, 636–39.
27. Stenstrom, M. K., Fam, S., & Silverman, G. S., *USA Environmental Technology Letters*; 1986, *7*, 625–636.

CHAPTER 5

REMOVAL OF CR(VI) BY USING SWEETLIME PEEL POWDER IN A FIXED BED COLUMN

N. M. RANE,[1] S. P. SHEWALE,[1] A. V. KULKARNI,[1] and R. S. SAPKAL[2]

[1]*Department of Chemical Engineering, MIT Academy of Engineering, Alandi, Pune, India*

[2]*UDCT, Sant Gadge Baba Amravati University, Amravati, India. E-mail: nmrane@chem.maepune.ac.in*

CONTENTS

5.1 Introduction 72
5.2 Experimental Part 74
5.2.1 Material Preparation and Characterization 74
5.2.1.1 Preparation of Adsorbent 74
5.2.1.2 Chemical Preparation 74
5.2.1.3 Apparatus 75
5.2.1.4 Adsorbent Characterization 75
5.2.1.4.1 SEM (Scanning Electron Microscope) Images 75
5.2.1.4.2 FTIR Analysis of Sweetlime Peel Powder Before and After Adsorption 76
5.2.1.4.3 BET (Brunauer–Emmett–Teller) Method 76

5.3 Adsorption Study 77
5.4 Breakthrough Curve Modeling 77
5.4.1 The Bohart-Adams Model 78
5.4.2 The Bed Depth Service Time (BDST) Model 78
5.4.3 The Yoon–Nelson Model 79
5.5 Results and Discussion 79
5.5.1 Column Studies 79
5.5.1.1 Effect of Flow Rate 79
5.5.1.2 Effect of Bed Height 80
5.5.1.3 Effect of Initial Cr(VI) Concentration 81
5.5.1.4 Application of Different Breakthrough Models 82
5.5.1.4.1 Bohart–Adams Model 82
5.5.1.4.2 Bed Depth Service Time (BDST) Model 82
5.5.1.4.3 Application of Yoon–Nelson Model 83
5.6 Conclusion 84
Keywords 85
References 85

5.1 INTRODUCTION

Water, a very important element on the earth is vital for the survival of plants, animals, human beings, other living organism, etc. About 14,000 billion cubic meters (BCM) of water, which is estimated by UN, as the total amount of water on earth is enough to cover the earth with a layer of 3000 meters depth. But the availability of fresh water is of very small proportion of 2.7%. Out of this, 75.2% is held frozen in polar regions and another 22.6% is present as ground water. Due to rapid industrialization and population explosion the available fresh water quality and quantity has declined to a greater extent [1]. Also many heavy metals are released to these water bodies, which are well above the tolerance limit permitted in aquatic environment leading to further deterioration.

Many processes involved in industries like electroplating, dye, leather tanning, photography and cement produces large quantities of effluents, which contains the toxic metal like chromium. Hexavalent chromium is one of the major toxic heavy metal pollutants in wastewater. The toxicity of Cr(VI) is rendered due to its high mobility. It is considered acutely toxic, carcinogenic and mutagenic to living organisms and hence more hazardous than other heavy metals. So the treatment of Cr(VI) from wastewater streams has become one of the most important ongoing environmental issues faced by all countries in today's world [2].

The heavy metal ions from aqueous solution can be removed by employing a number of different methods such as electrochemical reduction [3] chemical precipitation [4], ion exchange [5], reverse osmosis [6] and bio-reduction [7]. After studying all these methods, adsorption technique can be noted as economically favorable and technically feasible to separate Cr(VI) from aqueous solutions as, the requirement of operative controls are minimal [8]. Generally, heavy metal adsorption using biomaterials can reduce capital cost by 20%, operational cost by 36%, and total treatment cost by 28%, as compared to other treatment processes [9]. A number of low cost non-conventional adsorbents have been used to evaluate the extent of treatment of Cr(VI) contaminated wastewater [10].

This study involves the use of peel of sweetlime in powder form as adsorbent. Our previous studies were focused on the Cr(VI) adsorption capacity in batch experiments [11]. From those studies, our selected adsorbent for the treatment of Cr(VI) contaminated aqueous solutions was found to be highly effective by showing high adsorption capacity due to ligand exchange mechanism. The adsorbent was found to be suitable for the treatment of both mine water and industrial effluents which contains higher concentrations of Cr(VI). Fixed bed column is preferable for industrial wastewater treatment [12]. Therefore, attempt is made to obtain the experimental data from laboratory scale fixed-bed column, which can be used in designing an adsorption column for industrial application. In continuation of our previous work [13], the present study was focused on evaluating the performance of the sweetlime peel powder for the effective removal of Cr(VI) from aqueous solutions in a fixed-bed column. The Cr(VI) uptake capacity of the sweetlime peel was studied as a function of various operating conditions such as initial Cr(VI) concentration, flow rate and bed height. The best fit of the experimental data was calculated in three different kinetic models [14, 15].

5.2 EXPERIMENTAL PART

5.2.1 MATERIAL PREPARATION AND CHARACTERIZATION

5.2.1.1 Preparation of Adsorbent

Sweetlime peel, which contains citric acid was collected from fruit center. The collected peels were kept for drying for a 3-week period at room temperature. After drying the next step was grinding and screening. Using household simple mixer grinder, the dried sweetlime peel was grinded and converted to a fine powder (Figure 5.1). After grinding, the sweetlime powder was screened through the sieve shaker and selected 180 micron pore size powder material as a adsorbent for the removal of Cr(VI) from wastewater [15].

5.2.1.2 Chemical Preparation

A stock synthetic standard solution of potassium dichromate was used to prepare the adsorbate solutions of required strength. Diphenyl carbazide, H_2SO_4, acetone, etc. analytical grade chemicals were used.

FIGURE 5.1 Sweetlime peel powder 180-micron pore size.

5.2.1.3 Apparatus

Micropipette, packed bed glass column, conical flask, volumetric flask, Whatman filter paper, funnels etc.

5.2.1.4 Adsorbent Characterization

5.2.1.4.1 SEM (Scanning Electron Microscope) Images

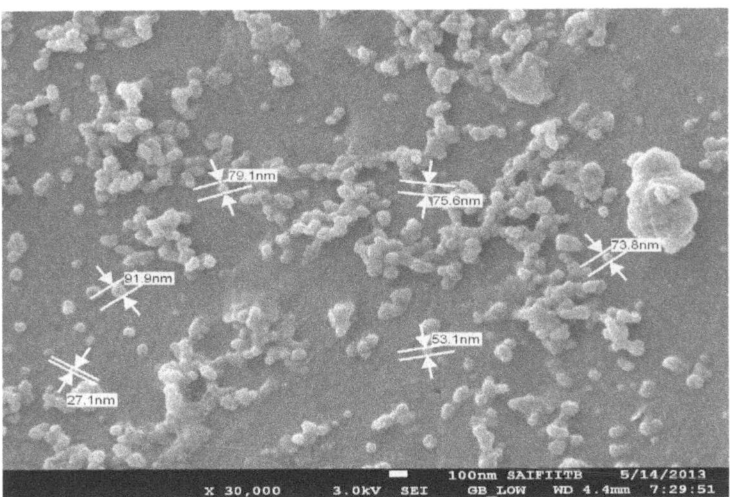

FIGURE 5.2 SEM image of sweetlime peel powder before adsorption.

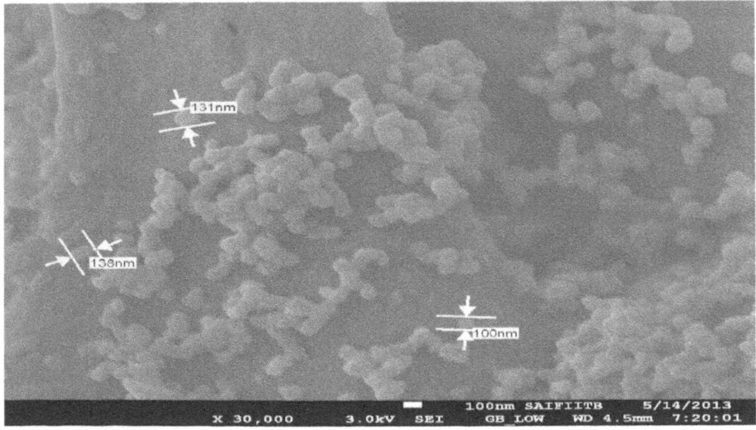

FIGURE 5.3 SEM image of sweetlime peel powder after adsorption.

5.2.1.4.2 FTIR Analysis of Sweetlime Peel Powder Before and After Adsorption

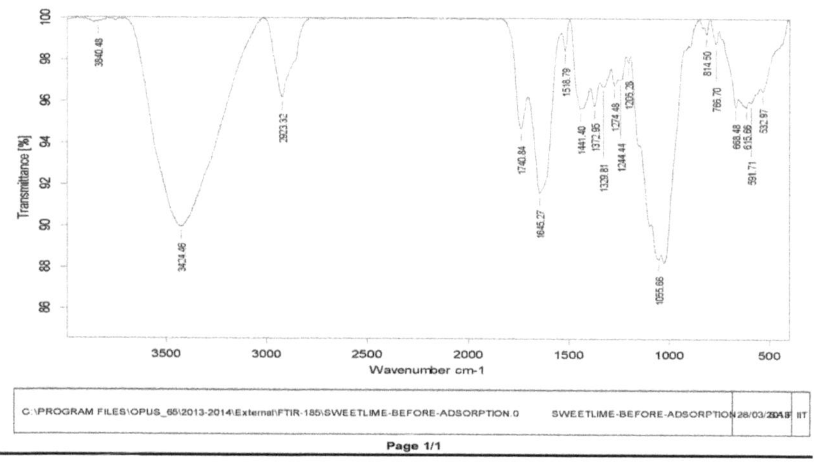

FIGURE 5.4 FTIR analysis for sweetlime peel powder before adsorption.

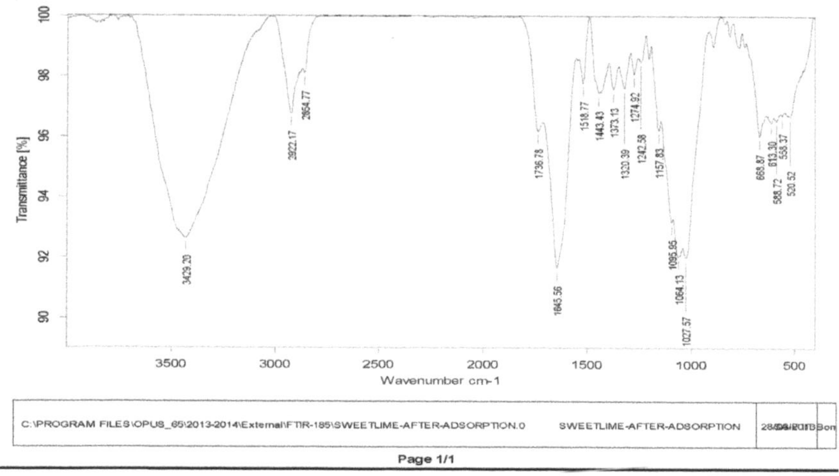

FIGURE 5.5 FTIR analysis for sweetlime peel powder after adsorption.

5.2.1.4.3 BET (Brunauer–Emmett–Teller) Method

The BET method is widely used in surface science for the calculation of surface areas of solids by physical adsorption of gas molecules (Table 5.1).

TABLE 5.1: Surface Area of Sweetlime Peel Powder

Sample	Test	Observation
Sweetlime peel powder	Surface Area	0.68 m^2/gm

5.3 ADSORPTION STUDY

The Cr(VI) solution of known concentration was passed with the help of peristaltic pump at the bottom of the column. The treated Cr(VI) solution was collected at regular intervals from the top of the column. The schematic diagram of experimental setup used for column studies is as shown in Figure 5.6. The flow rate, bed height and initial Cr(VI) concentration were checked regularly. The analysis of the effluent was measured spectrophotometrically by using (Thermofischer 840-210800) UV-VIS spectrophotometer [16]. The experiment was conducted till there were no further change in Cr(VI) concentration at the inlet and outlet of the column. Repetition of the experiment was done for the change in various parameters.

5.4 BREAKTHROUGH CURVE MODELING

The breakthrough curve is tested for the various readings obtained from the packed bed column. The important characteristics such as time for

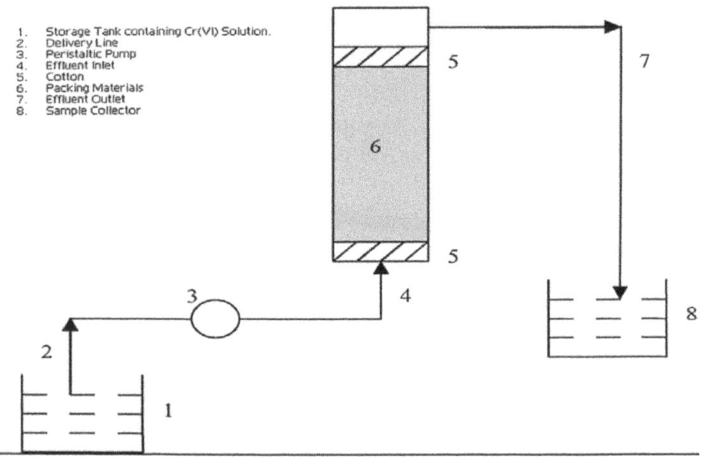

FIGURE 5.6 Experimental setup for column studies.

breakthrough appearance and shape of breakthrough curve were evaluated for dynamic response of an adsorption column. From the breakthrough curve, the concentration-time profiles were studied. The various kinetic models such as Bohart–Adams, BDST, and Yoon–Nelson were verified. So in this study the attempt has been made to find out the best-fit model, which describes the adsorption kinetics in the column.

5.4.1 THE BOHART-ADAMS MODEL

A basic equation to explain the relationship between C_t/C_o and t in a continuous flowing system for the adsorption of Cr(VI) on sweetlime peel powder was given by Bohart and Adams. The basis for the model is surface reaction theory and it assumed that equilibrium is not instantaneous. So, the rate of adsorption was proportional to both the residual capacity of the activated carbon and the concentration of the sorbing species. The initial part of breakthrough curve was described by this model. The mathematical equation of the model can be given as:

$$\ln\left(\frac{C_t}{C_0}\right) = k_{AB}C_0t - k_{AB}N_0\left(\frac{z}{U_0}\right)$$

where C_0 is inlet adsorbate concentration, C_t is outlet adsorbate concentrations, *z*(cm) is the bed height, U_0 (cm/min) is the superficial velocity, N_0 (mg/L) is the saturation concentration and K_{AB} (L/mg.min) is the mass transfer coefficient. The time period required for obtaining the breakthrough curve in present study was considered from the beginning to the end [14].

5.4.2 THE BED DEPTH SERVICE TIME (BDST) MODEL

The modified form of Bohart–Adams model [15] is known as BDST model were used to check the uptake capacity of the adsorption column. The assumptions were made that forces like intra-particular diffusion and external mass transfer resistance are negligible and adsorption kinetics is controlled by surface chemical reaction between the solute in the solution and unused adsorbent. The performance of the column was verified for different process variables. The BDST model is given as,

$$C_0 t = \left(\frac{N_0 h}{t}\right) - \left(\frac{1}{k}\right) \ln\left(\frac{C_0}{C_t} - 1\right)$$

where C_0 is inlet concentration and C_t is effluent concentration, K (mg/min) is the adsorption rate constant, N_0 (mg/g) is the adsorption capacity, h (cm) is the bed height, t (min) is the service time to breakthrough and u (cm/min) is the specific velocity [14, 15].

5.4.3 THE YOON–NELSON MODEL

Yoon and Nelson model is a relatively simple and is based on the adsorption of gases with respect to activated charcoal. In this model the important assumption is that the rate of decrease in the probability of adsorption for each adsorbate molecule is proportional to the probability of adsorbate adsorption and the probability of adsorbate breakthrough on the adsorbent [16]. The linearized model for a single component system is given as:

$$\ln\left[\frac{C_t}{C_0 - C_t}\right] = k_{YN} t - \tau k_{YN}$$

where k_{YN} (min^{-1}) is the rate constant and τ is the time required for 50% adsorbate breakthrough [14, 16].

5.5 RESULTS AND DISCUSSION

5.5.1 COLUMN STUDIES

5.5.1.1 Effect of Flow Rate

From the previous batch studies, the conclusion was drawn that the pH of the Cr(VI) solution was to be maintained at 1.25. The adsorption experiments were carried out by varying the flow rate as 3 mL/min,

5 mL/min, and 7 mL/min to find out its effect on breakthrough curve. The other parameter such as bed height, Cr(VI) concentration, pH of Cr(VI) solution were kept constant.

The effect of variation in flow rate on exit concentration of chromium solution from column is as shown in Figure 5.7. It can be observed from the figure that with the decrease in flow rate, the adsorption efficiency is increased. The reason behind it is, with the lower flow rate the residence time of the adsorbate is more and hence resulting in more time for adsorbate to bond with the metal efficiently. Sufficient residence time of the solute in the column has to be ensured for the adsorption equilibrium to be reached at the given flow rate, otherwise Cr(VI) solution leaves the column before equilibrium occurs.

5.5.1.2 Effect of Bed Height

The Cr(VI) solution with concentration 10 mL/L, pH 1.25 and flow rate 3 mL/min was passed through the adsorption column by varying the bed height as 2 cm, 3 cm and 4 cm. Figure 5.8, represents the performance of breakthrough curve at various bed heights. As we increase the bed height, the throughput column has increased. This has happened due to higher contact time. As we increase the bed height, the throughput column has increased.

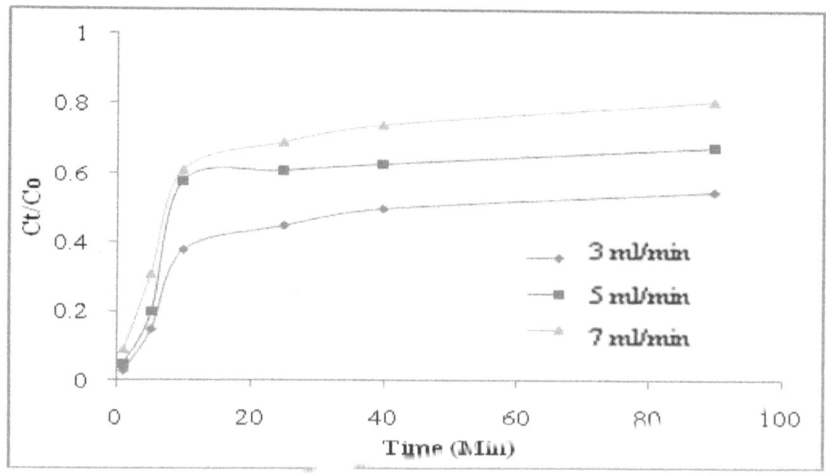

FIGURE 5.7 Effect of flow rate on Cr(VI) adsorption.

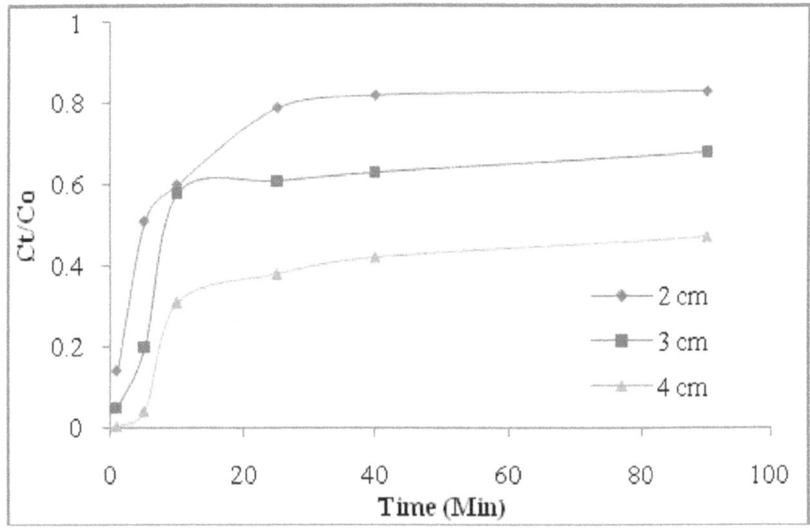

FIGURE 5.8 Effect of bed height on Cr(VI) adsorption.

This has happened due to higher contact time. With the increase in bed height in the fixed bed column it was observed that the metal uptake capacities were increased. This is due to increased surface area of the adsorbent, which provided more binding sites for the adsorption. It is also observed that with the increase in bed height, the breakthrough time is also increased.

5.5.1.3 Effect of Initial Cr(VI) Concentration

The Cr(VI) solution of various concentration such as 6 mL/L, 8 mL/L and 10 mL/L were prepared and passed through the column to determine the effect of adsorbate concentration on the performance of breakthrough curve. This is shown in Figure 5.9. During these experiments other parameters like pH (1.25), bed height (3 cm) and flow rate (3 mL/min) were kept constant. It is observed from the experimental study that with increasing inlet Cr(VI) concentration, the uptake and total Cr(VI) adsorbed were increased. The increase in uptake capacity of the adsorbent may be due to high inlet Cr(VI) concentration providing higher driving force for the transfer process to overcome the mass transfer resistance.

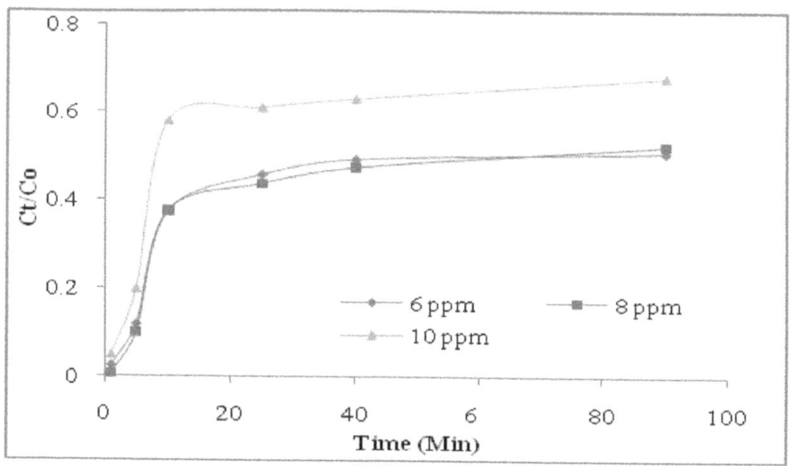

FIGURE 5.9 Effect of initial Cr(VI) concentration.

5.5.1.4 Application of Different Breakthrough Models

5.5.1.4.1 *Bohart–Adams Model*

The experimental data were fitted in Bohart–Adams model. The purpose of this is to estimate the characteristic parameters such as N_o (maximum adsorption capacity) and K_{AB} (mass transfer coefficient) (Table 5.2). In the current study, period of time was considered from the beginning to the end of breakthrough curve. At different bed heights, $\ln(C_t/C_o)$ versus time(t) was plotted. The slope and intercept of the curve gives the mass transfer coefficient and saturation concentration, respectively. It is observed from the curve that with the increase in bed height mass transfer coefficient increases.

5.5.1.4.2 *Bed Depth Service Time (BDST) Model*

BDST curve was plotted by using the experimental data obtained from the column studies. "K" and "N_o" values were calculated from the slope and intercept of the plot between $\ln[(C_o/C_t)-1]$ versus time(t) at different bed height. The estimated values of characteristic parameters like "K," "N_o" and the other statistical parameters are presented in Table 5.3. The comparison was made between the breakthrough curves predicted from BDST model with the experimental breakthrough curve.

TABLE 5.2 Bohart–Adams Model Calculations for Sweetlime Peel Powder

Parameter	Value	$K_{AB\ (Lit/mg.min)}$	$N_{0\ (mg/lit)}$
Bed Height	2 cm	0.0207	49.54
	3 cm	0.1141	28.43
	4 cm	0.1148	23.82
Flow Rate	3 mL/min	0.049	18.17
	5 mL/min	0.112	28.43
	7 mL/min	0.114	44.22
Initial Conc.	6 ppm	0.186	28.43
	8 ppm	0.114	24.62
	10 ppm	0.096	17.94

TABLE 5.3 BDST Model Calculations for Sweetlime Peel Powder

Parameter	Value	K (mg/min)	N_0 (mg/gm)
Bed Height	2 cm	0.043	3.35
	3 cm	0.121	2.74
	4 cm	0.132	0.81
Flow Rate	3 mL/min	0.122	2.06
	5 mL/min	0.132	2.33
	7 mL/min	0.072	3.35
Initial Concentration	6 ppm	0.212	3.43
	8 ppm	0.132	3.35
	10 ppm	0.115	2.42

5.5.1.4.3 Application of Yoon–Nelson Model

A simple theoretical model developed by Yoon–Nelson was applied to investigate the breakthrough behavior of chromium on adsorbent. The values of "K_{YN}" and "τ" were estimated from the plot between $\ln[(C_t/C_o)–C_t]$ versus time (t) at different bed heights (Table 5.4).

The values of "K_{YN}" were found to decrease with increase in the bed height whereas the corresponding values of "τ" decreases. The predicted uptake values from the model, experimental values along with the values of "K_{YN}" and "τ" and statistical parameters are listed. From the values it

TABLE 5.4 Yoon–Nelson Model Calculations for Sweetlime Peel Powder

Parameter	Value	K_{YN} (min^{-1})	τ (min)
Bed Height	2 cm	1.065	5.179
	3 cm	0.696	4.045
	4 cm	0.638	2.703
Flow Rate	3 mL/min	0.682	4.892
	5 mL/min	0.696	4.045
	7 mL/min	0.706	3.351
Initial Concentration	6 ppm	0.707	5.000
	8 ppm	0.696	4.444
	10 ppm	0.680	4.040

can be seen that simulation of the entire breakthrough curve was effective by the Yoon Nelson Model at higher bed heights.

5.6 CONCLUSION

From the studies we can conclude that the selected sweetlime peel powder acts as very effective adsorbent for the efficient removal of Cr(VI) from aqueous solutions. It was found out that the maximum removal of Cr(VI) in a fixed bed adsorption column was found to be at pH 1.25, initial concentration 10 mg/L, flow rate of 3 mL/min and bed height 2 cm. The effect of various adsorption parameters like flow rate, bed height and adsorbent concentration for the removal of Cr(VI) were studied. It was found from the experimental study that with the increase in adsorbant dose (bed height), the removal of Cr(VI) was found to be increased. The reverse trend was observed when adsorbate concentration and flow rate were increased. It was also found that the uptake capacity of the adsorbent was decreased with increase in adsorbent dose.

The fit of various models such as Bohart–Adams, BDST and Yoon–Nelson were tested by applying experimental data obtained from dynamic studies performed in fixed bed column. From the statistical analysis it was found that Bohart–Adams model well fits through the experimental data.

It was also observed from the studies that the adsorption capacity of sweetlime peel powder is well above as compared to other similar literature reported adsorbents.

KEYWORDS

- **Sweetlime Peel Powder**
- **Fixed Bed Column**
- **Cr(VI)**
- **Yoon – Nelson Model**
- **Bohart Adams model**
- **BDST Model**

REFERENCES

1. Feasibility Report on Abatement of pollution due to chromite mining and processing industries in Orissa, Phase-I, Regional Research Laboratory, Bhubaneswar, 1996, 121.
2. Rawat, M., Moturi, M. C. Z., & Subramanian, V., Inventory compilation and distribution of heavy metals in wastewater from small-scale industrial areas of Delhi, India. *J. Environ. Monit.* 2003, *5*, 906–912.
3. Mukhopadhyay, B., Sundquist, J., & Schmitz, R. J., Removal of Cr(VI) from Cr-contaminated groundwater through electrochemical addition of Fe(II). *J. Environ. Manage.* 2007, *82*, 66–76.
4. Paterson, J. W., Wastewater Treatment Technology, Ann Arbour Science, Michigan, 1975, 43–58.
5. Petruzzelli, D., Passino, R., & Tiravanti, G., Ion exchange process for chromium removal and recovery from tannery wastes. *Ind. Eng. Chem. Res.* 1995, *34*, 2612–2617.
6. Ozaki, H., Sharma, K., & Saktaywin, W., Performance of an ultra-low pressure reverse osmosis membrane (ULPROM) for separating heavy metal: effects of interference parameters. *Desalination* 2002, *144*, 287–294.
7. Sultan, S., & Hasnain, S., Reduction of toxic hexavalent chromium by Ochrobactrum intermedium strain SDCr-5 stimulated by heavy metals. *Bioresour. Technol.* 2007, *98*, 340–344.
8. Garg, V. K., Gupta, R., & Kumar, R., Adsorption of Cr from aqueous solution on treated saw dust. *Bioresour. Technol.* 2004, *92*, 78–81.
9. Loukidou, M. X., Zouboulis, A. I., Karapantsios, T. D., & Matis, K. A., Equilibrium and kinetic modeling of Chromium(VI) biosorption by Aeromonas caviae. *Colloid. Surf. A Physicochem. Eng. Aspects* 2004, *242*, 93–104.

10. Baral, S. S., Das, S. N., Rath, P., & Chaudhury, G. R., Cr(VI) removal by calcined bauxite. *Biochem. Eng. J.* 2007, *34*, 69–75.
11. Baral, S. S., Das, S. N., & Rath, P. Hexavalent chromium removal from aqueous solution by adsorption on treated sawdust. *Biochem. Eng. J.* 2006, *31*, 216–222.
12. Baral, S. S., Das, S. N., Rath, P., Chaudhury, G. R., & Swamy, Y. V., Removal of Cr(VI) from aqueous solution using waste weed, Salvinia cucullata. *Chem. Ecol.* 2007, *23*, 105–117.
13. Baral, S. S., Das, S. N., Chaudhury, G. R., Swamy, Y. V., & Rath, P., Adsorption of Cr(VI) using thermally activated weed Salvinia Cucullata. *Chem. Eng. J.* 2008, *139*, 245–255.
14. Baral, S. S., Das, N., Ramulu, T. S., Sahoo, S. K., Das, S. N., & Chaudhuary, G. R., Removal of Cr(VI) by thermally activated weed Salvinia Cucullata in a fixed-bed column. *J. Hazard Mater* 2009, *161*, 1427–1435.
15. Kundu, S., & Gupta, A. K., Analysis and modeling of fixed bed column operations on As(V) removal by adsorption onto iron oxide-coated cement (IOCC). *J. Colloid Interface Sci.* 2005, *290*, 52–60.
16. Chun, L., Hongzhang, C., & Zuohu, L. M., Adsorptive removal of Cr (VI) by Fe-modified steam exploded wheat straw. *Process Biochem.* 2004, *39*, 541–545.

CHAPTER 6

REMOVAL OF CR(VI) FROM WASTEWATER USING RED GRAM HUSK AS ADSORBENT

V. S. WADGAONKAR[1] and R. P. UGWEKAR[2]

[1]*Petrochemical Engineering Department, Maharashtra Institute of Technology, Pune, Maharashtra, India*

[2]*Chemical Engineering Department, Laxminarayan Institute of Technology, Nagpur, Maharashtra, India*

CONTENTS

6.1 Introduction 88
6.2 Materials and Methods 89
6.3 Adsorption Kinetics 90
6.4 Result and Discussion 91
6.4.1 Effect of Contact Time 91
6.4.2 Effect of Chromium Concentration on Adsorption Process 92
6.4.3 Effect of pH 92
6.4.4 Effect of Temperature 93
6.4.5 Effect of Agitation Speed 93
6.5 Conclusion 94
Keywords 94
References 94

6.1 INTRODUCTION

Pure water is very important for human life. All agriculture, industrial and commercial sector need water of required quality. Pollution of water is due to all these sectors. Waste from all sectors is entering freshwater everyday. The demand for water is increasing while water availability is less because of improper waste disposal. The removing pollutant from all water resources is costly so there is need for new technologies, which are less expensive, require less maintenance and are energy effective. Chromium is widely used in electroplating, alloying, leather tanning, corrosion protection, etc. [1]. Chromite mining and processing units produce waste and effluent containing chromium in two oxidation states, for example, Cr(VI) and Cr(III) in aqueous solution. Cr(VI) oxidizing properties is dangerous due to carcinogenicity [2]. The potable water allows only 0.05 mg/lit Cr(VI) because of toxicity [3]. Higher concentrations are in the industrial and mining effluents, which crosses permissible limit.

The toxic metal pollution is major problem due to increase in population, industrial operations and household materials. It can be hazardous at low concentration. Health of plants, animals and humans are affected when water contains heavy metals. Wastewater contains toxic metals and separating efficiently presents great challenge. In recent years removal heavy metals from wastewater by using low cost adsorbents is important for environmental concern. Various adsorbents are used in the market for this purpose. Discharging effluent into environment is harmful so there is need to treat effluent because pollutants are present in the water. Attention is increased due to the potential health hazard created by heavy metals in the environment.

The health risks of heavy metal are different. Chromium causes irritation, nausea and vomiting at low level exposure, kidney, liver, circulatory and nerve tissue damage at long-term exposure [8, 13]. Pollution of Cr(VI) mainly consists two processes: converting Cr(VI) to Cr(III) to make it harmless and removal of Cr(VI) as it is. The reduction and precipitation technique is adopted for treatment of wastewater containing Cr(VI). Since process having several drawbacks such as solid-liquid separation and disposal of sludge.

Chemical precipitation of metals, reverse osmosis and other methods are incapable for removing heavy metals at trace concentration in large volume of solution. The new separation technologies have importance for cost effective method for the removal of heavy metals. Adsorption is best substitute for all methods for such conditions. Adsorption is one of the few alternatives available for such situation. The membrane system has encountered problems like scaling, fouling and blocking. The ion exchange system is not suitable due to the cost of commercial ion exchange resins. The adsorption technique is cost effective and technically easy to handle as the requirement of the control system is minimum. In last decades, a number of low price non-conventional adsorbents have been used to treat Cr(VI) contaminated wastewater [3–7].

6.2 MATERIALS AND METHODS

Red Gram husk was obtained from crop. To remove dirt and other matter it was washed in running tap water. The husk was heated in oven at 80°C for 24 h. Then it is grinded in a mixer to convert into powder. To obtain specific size it was ground and screened (using screen with mesh size 100). Plastic stopper bottles (containers) were used for preservation of adsorbent. Humidity problem is handled by keeping bottles in desiccators before time of use.

Analytical grade chemicals such as potassium dichromate, sulfuric acid, sodium hydroxide, 1,5-diphenyl carbazide, acetone were procured from Sourav Scientific and used as received without further purification. Distilled water was used for all the experimental runs. Adsorption experiments were carried out using four different amount of adsorbent to calculate the efficiency of red gram husk in single solution. Adsorption experiments were performed by changing contact time, adsorbent dose, agitation speed and adsorbate concentration. The adsorptions were conducted by using 250 mL conical flask and the total volume of reaction mixture was 125 mL. After specific time, filtration was done using Whitman No. 40 filter paper to avoid probable problem of turbidity. UV-visible spectrophotometer was used to determine concentration of chromium.

As adsorption dose increase, percentage adsorption increased. More surface area is available for increase in adsorbent dose hence more the adsorption capacity of adsorbent. Agitation speed is having little effect on adsorption.

Different isotherm models, such as Freundlich and Langmuir Isotherm were used for fitting experimental data. Langmuir adsorption isotherm was used to calculate the optimum adsorption capacity of the adsorbent. The contact time between adsorbate and adsorbent is very important in wastewater treatment. It was found that adsorbent prepared from Red Gram can be effectively used for removal of Cr (VI) from wastewater.

6.3 ADSORPTION KINETICS

Removal of Cr(VI) is elaborated in the literature using pseudo first-order kinetics. To study the adsorption kinetics of heavy metal ions, the kinetic parameters for the adsorption process were carried out for contact time ranging from 1 to 180 min observing the removal of the Cr(VI) (Figures 6.1 and 6.2).

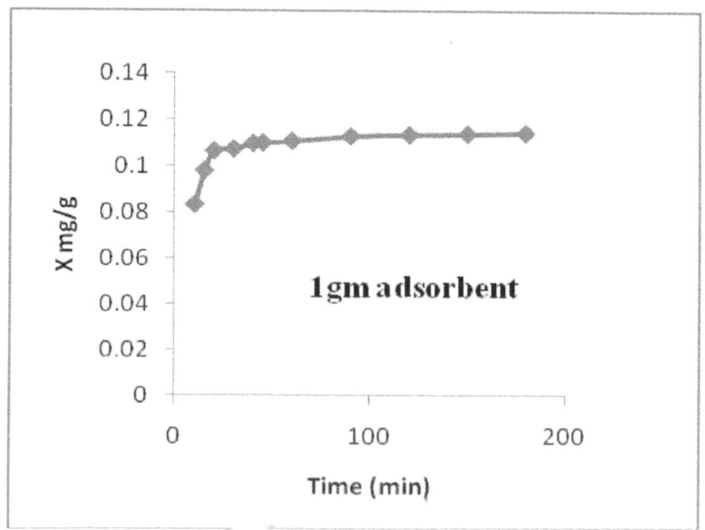

FIGURE 6.1 Adsorption kinetics for 1 gm red gram powder.

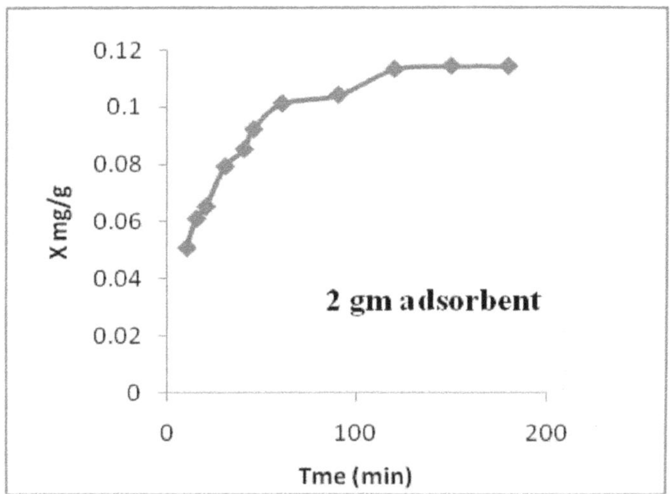

FIGURE 6.2 Adsorption kinetics for 2 gm red gram powder.

6.4 RESULT AND DISCUSSION

Contact time has significant effect on percentage adsorption. Analysis of percentage adsorption versus initial concentration was done by varying contact time. It was clear that percentage adsorption reduced with increase in initial concentration of the adsorbate. But holding capacity raised with increase in initial concentration, which may be due to the high number of Cr(VI) ions in solution for adsorption. To overcome all mass transfer resistances of the metal ions higher initial adsorbate concentration is required because of higher driving force. Due to this higher chances of collisions between Cr(VI) and active sites. This leads to increase in uptake of Cr(VI) for required amount of treated adsorbent [12].

6.4.1 EFFECT OF CONTACT TIME

Percentage adsorption versus adsorption time is shown in the plot. Figure 6.3 indicates adsorption rises linearly with time. After 1 h, percentage Cr(VI) removal is 54.45, then there is reduction in the chromium removal because of continuous adsorption and desorption process. Formation of monolayer on the surface of adsorbent indicates that all the

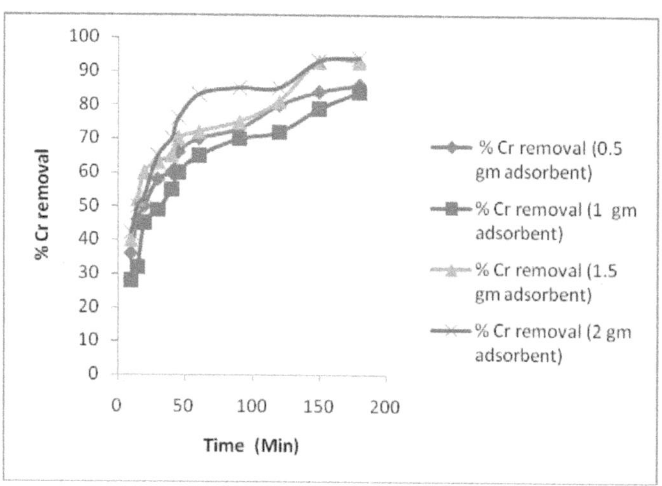

FIGURE 6.3 Adsorption kinetics for red gram husk powder.

curves were smooth. In the early stage, slope of the plot was one and it reduced with time. It indicates that the rate of adsorption was high in the initial stages but slowly reduced and became constant when equilibrium was reached.

6.4.2 EFFECT OF CHROMIUM CONCENTRATION ON ADSORPTION PROCESS

A physio-chemical aspect of sorption process provides necessary information for evaluation of adsorption process as major operation. The adsorption isotherm is carried out by changing the initial concentration of Cr(VI). Figure 6.4 shows the Langmuir isotherm for adsorption for red gram husk powder.

6.4.3 EFFECT OF pH

There was little increase in the percentage of adsorption as the pH increased and it was highest at pH 3. When pH is again increased, there was decrease in percentage of adsorption was observed. The reason for this may be weakening of electrostatic force of attraction between oppositely charged adsorbate and adsorbent and finally causes the decrease in

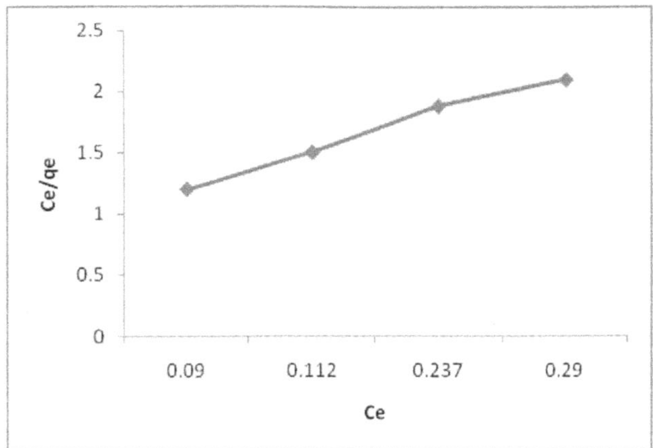

FIGURE 6.4 Langmuir Isotherm for adsorption for red gram husk powder.

adsorption capacity. When the pH was increased beyond 6.0, reduction in the percentage adsorption was observed. This is due to competition between (OH–) and chromate ions (CrO_4^{2-}) [12]. Weakening of electrostatic force between adsorbate and adsorbent with increasing pH resulted the net surface potential of the adsorbent decreased. This finally causes lowering of adsorption capacity.

6.4.4 EFFECT OF TEMPERATURE

Adsorption process is affected by temperature parameter. The study was carried out as a function of temperature for Cr(VI) adsorption. As temperature rises there is decrease in percentage adsorption. This is due to increase in the available thermal energy which causes desorption. Desorption was caused by higher temperature which provides higher mobility of the adsorbate. At low pH acid chromate ions ($HCrO_4^-$) are the dominant species shown by stability diagram of Cr(VI)–H_2O system.

6.4.5 EFFECT OF AGITATION SPEED

Adsorption experiments were conducted by initial Cr(VI) concentration of 100 mg/L. Agitation speeds were changed from 40 r.p.m. to 100 r.p.m. It is seen that agitation speed has little effect on adsorption process.

6.5 CONCLUSION

Effective removal Cr(VI) from aqueous solutions, Red Gram husk naturally available adsorbent was shown in this paper. For initial concentration of 100 mg/lit, the equilibrium time is 2 h for the adsorbent. At pH 2.5–3 maximum adsorption was taking place. Percentage of adsorption was decreased with increase in temperature, which shows that process is exothermic in nature. Low temperatures favor the adsorption process. The red gram husk powder is an ideal adsorbent for removal of Cr(VI) from aqueous solutions because it is easy available and having high efficiency for removal of Cr(VI).

KEYWORDS

- **Adsorption isotherm**
- **Chromium (VI)**
- **Isotherm model**

REFERENCES

1. Erdam, M., Altundagan, H. S., & Tumen. F., Removal of hexavalent chromium by using heat activated bauxite. *Min. Eng.* 2004, *17*, 1045–1052.
2. Sarma, Y. C., Cr(VI) from industrial effluents by adsorption on an indigenous low cost material. *Colloid Surf. A. Phyiochem. Eng. Aspects* 2003, *215*, 155–162.
3. Selvaraj, K., Manonmani, S., & Pattabhi, S., *Bioresour. Technol.* 2003, *89*, 207–211.
4. Pradhan, J., Das, S. N., & Thakur, R. S., Adsorption of hexavalent chromium from aqueous solution by using activated red mud. *J. Colloid Interf. Sci.* 1999, *217*, 137–141.
5. Sarma, Y. C. Cr(VI) from industrial effluents by adsorption on an indigenous low cost material. *Colloid Surf. A. Phyiochem. Eng. Aspects* 2003, *215*, 155–162.
6. Tewaria, N., Vasudevana, P., & Guhab, B. K., Study on biosorption of Cr (VI) by *Mucor hiemalis*. *Biochem. Eng. J.* 2005, *23*, 185–192.
7. Raji, C., & Anirudhan, T. S. *Water Res.* 1998, *32*, 3772.
8. Das, D. D., Mohapatra, R., Pradhan, J., Das, S. N., & Thakur, R. S., Adsorption of Cr(VI) from aqueous solution using activated cow dung carbon. *J. Colloid Interf. Sci.* 2000, *232*, 235–240

9. Sarin, V., & Pant, K. K., Removal of chromium from industrial waste by using eucalyptus bark. *Bioresour. Technol.* 2005, *97*, 15–20.
10. Singh, K. K., Rastogi, R., & Hasan, S. H., Removal of Cr(VI) from wastewater using rice bran. *J. Colloid Interf. Sci.* 2005, *290*, 61–68.
11. Kobya, M., Demirbas, E., Senturk, E., & Ince, M., Adsorption of heavy metal ions from aqueous solutions by activated carbon prepared from apricot stone. *Bioresour. Technol.* 2000, *96*, 1518–1521.
12. Baral, S. S., Das, S. N., & Rath, P., Hexavalent chromium removal from aqueous solution by adsorption on treated sawdust. *Biochemical Engineering Journal* 2006, *31*, 216–222.
13. Hamadi, N. K., Chen, X. D., Farid, M. M., & Lu, M. G. Q., Adsorption kinetics for the removal of chromium (VI) from aqueous solution by adsorbents derived from used tires and sawdust. *Chem. Eng. J.* 2001, *84*, 95–105.

CHAPTER 7

LOW COST ADSORBENTS IN THE REMOVAL OF Cr(VI), Cd AND Pb(II) FROM AQUEOUS SOLUTION

P. SEMIL and A. AWASTHI

Department of Chemical Engineering, Harcourt Butler Technological Institute, Kanpur, Uttar Pradesh, India

CONTENTS

7.1 Introduction 98
7.1.1 Lead 99
7.1.2 Cadmium 100
7.1.3 Chromium 100
7.2 Materials and Methods 101
7.2.1 Sample Collection and Preparation of Adsorbent 101
7.2.2 Stock Solutions 102
7.3 Result and Discussion 102
7.3.1 Cadmium 102
7.3.1.1 Effect of Contact Time 102
7.3.1.2 Effect of Adsorbent Dose 103
7.3.1.3 Effect of Initial Concentration 104
7.3.1.4 Effect of pH 105
7.3.1.5 Effect of Particle Size 106
7.3.2 Chromium 106
7.3.2.1 Effect of Contact Time 106

7.3.2.2 Effect of Adsorbent Dose 107
7.3.2.3 Effect of Initial Concentration........................... 108
7.3.2.4 Effect of pH.. 109
7.3.2.5 Effect of Particle Size..110
7.3.3 Lead...111
7.3.3.1 Effect of Contact Time.......................................111
7.3.3.2 Effect of Adsorbent Dose..................................112
7.3.3.3 Effect of Initial Concentration............................112
7.3.3.4 Effect of pH..113
7.3.3.5 Effect of Particle Size..114
7.4 Conclusion 115
Keywords 116
References...116

7.1 INTRODUCTION

Excessive release of heavy metals into the environment due to industrialization and urbanization has posed a great problem worldwide [1]. This fear has been heightened in recent times due to advancement in technology coupled with increasing industrial activities, both contributed to release of heavy metals into the environment [1–3]. The heavy metal pollution is the biggest problem of ground water. The most important toxic metals are Cadmium, chromium and Lead which originate from metal plating, mining activities, smelting, battery manufacture, tanneries, petroleum refining, paint manufacture, pesticides, pigment manufacture, printing and photographic industries, etc. [4, 5]. The release of large amount of heavy metals into the environment is done by anthropogenic activities out of which industrialization and irrigation are the main sources. Due to non-bio-degradability and persistence these heavy metals accumulate in food chain and possess a significant disaster to human health like as like headache, dizziness, nausea and vomiting, chest pain, tightness of chest, dry cough, shortness of breath, rapid respiration, nephritis and extreme weakness when their concentration is above the recommended limit [6]. Wastewater treatment methods, such as chemical precipitation,

ion-exchange, electroflotation, membrane separation, reverse osmosis, electrodialysis, solvent extraction, etc. Most of these methods are expensive so are not affordable for a developing country. Adsorption is the best purification and separation techniques of heavy metal from wastewater but in industries cost is very important parameter in comparison to efficiency of absorbent materials [7]. Activated carbon has been recognized as a highly effective adsorbent for the removal of heavy metal-ion from the concentrated and dilute metal bearing effluents. But the process has not been used by small and medium scale industries for the treatment of their metal bearing effluents, because of its high manufacturing cost. For this reason, the use of low cost materials as adsorbent for metal ion removal from the wastewater has been highlighted. According to Bailey et al. (1999), an adsorbent can be considered as cheap or low-cost if it is abundant in nature, requires little processing and is a byproduct of waste material from waste industry.

In recent years, a number of adsorptive materials such as aquatic plants, agricultural by-products and residues, industrial by-products, saw dust, clay, Zeolite, and micro organisms have been used for the removal of heavy metals from wastewaters. The utilization of unmodified or modified rice husk as a sorbent for the removal of pollutants. Unmodified rice husk has been evaluated for their ability to bind metal ions [8–10]. Various modifications on rice husk have been reported in order to enhance sorption capacities for metal ions and other pollutants [11–13], coconut husk [14–16].

The aim of this research is to determine the capacity compare the efficiency of three low cost adsorbents with removal of lead, chromium and Cadmium ions from aqueous solution. The research involved the modification of Rice husk, Treated Neem leaves and coconut husk with HCl, Sulphuric acid and investigation of some experimental conditions such as pH of solution, contact time, adsorbent loading and temperature as it relates to adsorption of the metal to the adsorbents. While isotherm studies were used to model the adsorption process.

7.1.1 LEAD

Lead is one such heavy metal with specific toxicity and cumulative effects. The chief sources of lead in water are the effluents of lead and lead processing industries. Lead is also used in storage batteries, insecticides, plastic water pipes, food, beverages, ointments and medicinal concoctions for flavoring and sweetening [17]. The permissible limit of Lead for drinking water is 0.005 mg/L (as total chromium) in EPA standard (EPA, 2007). Lead poisoning causes damage to liver, kidney and reduction in hemoglobin formation, mental retardation, infertility and abnormalities in pregnancy. Chronic lead poisoning may cause three general disease syndromes:

(a) Gastrointestinal disorders, constipation, abdominal pain, etc.
(b) Neuromuscular effects (lead lapsy) weakness, fatigue muscular atrophy.
(c) Central nervous system effects or CNS syndrome that may result to coma and death [18].

7.1.2 CADMIUM

Cadmium is a natural element in the earth's crust. It is usually found as a mineral combined with other elements such as oxygen (cadmium oxide), chlorine (cadmium chloride), or sulfur (cadmium sulfate, cadmium sulfide). It doesn't have a definite taste or odor. All soils and rocks, including coal and mineral fertilizers, have some cadmium in them. The cadmium that industry uses is extracted during the production of other metals like zinc, lead, and copper. Cadmium is an irritant to the respiratory tract and prolonged exposure to this pollutant can cause anemia and a yellow stain that gradually appears on the joints of the teeth. Cadmium and their salts are used in electroplating, paint pigments, plastics, silver cadmium batteries [19], smelting, cadmium nickel batteries, stabilizer, phosphate fertilizer, mining and alloy industries [20, 21]. The permissible limit of cadmium for drinking water is 0.01 mg/L (as total chromium) in EPA standard (EPA, 2007).

7.1.3 CHROMIUM

Chromium has three main forms chromium (0), chromium (III), and chromium (VI). Chromium (III) compounds are stable and occur naturally, in the environment. Chromium (0) does not occur naturally and chromium (VI) occurs only rarely. Chromium compounds have no taste or odor. Chromium (III) is an essential nutrient in our diet, but we need only a very small amount. Other forms of chromium are not needed by our bodies. Chromium is used in manufacturing chrome-steel or chrome-nickel-steel alloys (stainless steel) and other alloys, bricks in furnaces, and dyes and pigments [20], for greatly increasing resistance and durability of metals and chrome plating, leather tanning, and wood preserving. Manufacturing, disposal of products or chemicals containing chromium, or fossil fuel burning release chromium to the air, soil, and water. Particles settle from air in less than 10 days. It sticks strongly to soil particles; in water it sticks to dirt particles that fall to the bottom [19]; only a small amount dissolves. Small amounts move from soil to groundwater. The permissible limit of chromium for drinking water is 0.1 mg/L (as total chromium) in EPA standard (EPA, 2007).

7.2 MATERIALS AND METHODS

7.2.1 SAMPLE COLLECTION AND PREPARATION OF ADSORBENT

Rice husk was obtained from a local mill in India was pre-treated according to the method of treatment. The rice husk was screened and washed with de-ionized water to remove dirt and metallic impurities after which it was dried in the oven at about 105°C for 2 h. The dried rice husk was grounded and sieved in the mesh in the range between 250 μm and 150 μm in order to increase its surface area. The purpose of this work was to improve textural parameters of carbons obtained from rice husk 100 g of carbonized rice husks were soaked in 0.6 M of citric acid for 2 h at 20°C. Acid husk slurry is dried overnight at 50°C and the dried husks are heated to 120°C under aerobic conditions. The reacted product was washed repeatedly with distilled water (200 mL/g). Finally the cleaned rice husk was oven dried overnight at 100°C. It was reported that modified rice husk is a potentially useful material for the removal of Cr, Cd and Pb from aqueous solutions. The adsorbents were prepared as described by Ref. [21]. The adsorbents of coconut husk was cut into small pieces and blended. They were washed with distilled water to remove dirt and color and air-dried for 24 h to avoid thermal deactivation of the adsorbent surface, respectively. They were sieved to pass through a 2 mm stainless steel sieve and a portion of each of the adsorbent was stored in clean polyethylene containers (labeled as unmodified adsorbent) prior to analysis. For the modification of the adsorbents, about 400 g of each of the washed adsorbents were mixed with 600 mL of 0.1 mol dm^{-3} HCl. The mixture was heated at 120°C for 30 min with occasional stirring. The adsorbent was separated from water using Buckner funnel and a vacuum pump and washed with distilled water until the washings were free of color and the pH of wash solution was about 7. The washed adsorbents were air dried for 24 h and labeled as HCl-modified coconut husk.

7.2.2 STOCK SOLUTIONS

The stock solutions containing the 1000 mg/L concentration of lead, cadmium, and Chromium were prepared by dissolving lead nitrate [$Pb(NO_3)_2$], cadmium nitrate [$Cd(NO_3)_2$], potassium dichromate [$K_2Cr_2O_7$] according to weight of 1.6 g of $Pb(NO_3)_2$, 2.2124 g of $Cd(NO_3)_2$, and 2.8234 g of $K_2Cr_2O_7$, respectively in milli Q water. The salts of these chemicals were of analytical grade, procured from s.d. Fine-Chem Limited, Mumbai.

7.3 RESULT AND DISCUSSION

7.3.1 CADMIUM

Batch sorption studies were carried out with 100 mL Cd(II) solution and 1–5 gm adsorbent. The equilibrating mixtures were placed in stoppered glass bottles and agitated in temperature-controlled shaker for optimum time of contact. The contents were filtered through double what man no. 40 filter paper and then analyzed for the Cd(II) by using UV-visible spectrometer at λ_{max} = 326 nm wavelength. The pH of the sample solution was recorded initially and at the end of experiment using pH meter/pH paper.

7.3.1.1 Effect of Contact Time

The effect of contact time on adsorption of Cd(II) was investigated. 1.0 g each of the adsorbents was taken into 6 beakers containing 100 mL of metal solution at an initial concentration of 10 mg/L. The solutions were agitated at a fixed stirring speed of 180 rpm for varying time period ranging from 20–180 min at a fixed pH value of 6. The final concentration of solution was determined by UV-visible spectrometer, and the percentage Cd^{2+} was then determined. The change of adsorption of Cd(II) with time is presented in Figure 7.1. The adsorption increases with the increasing contact time and that mean the removal efficiency is slow after equilibrium was attained. The adsorption increases with the increasing contact time. And the equilibrium was attained after shaking for optimum time 90 min. Therefore, in each experiment the shaking time was set for 90 min.

7.3.1.2 Effect of Adsorbent Dose

Effect of adsorbent dose on percentage removal of metal ions was investigated by varying adsorbent dose in the range of 1–5 g. Separate masses of adsorbents (ranging from 0.5–3 g of modified and unmodified adsorbent) were each taken into 6 beakers containing 100 mL of metal solution of initial concentration of 10 mg/L. The solutions were agitated by orbital shaker speed of 180 rpm for a predetermined contact time of 90 min and a fixed pH value of 6.0. The final concentration of solutions was then

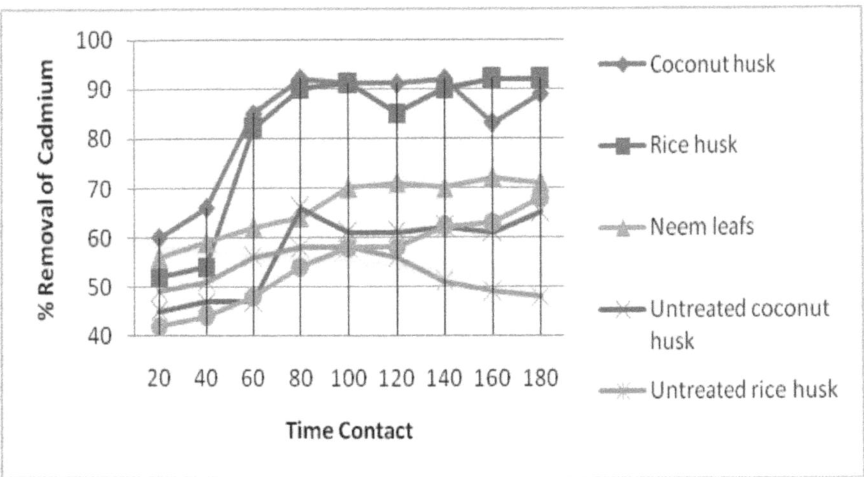

FIGURE 7.1 Effect of Contact Time, pH 6.0, RPM: 180, Time: 90 min, Temperature: 25°C, Initial Concentration: 10 mg/L, and Dose Amount: 3 g.

determined by UV-visible spectrometer and the percentage removal of Cd^{2+} determined. The removal of cadmium in the solution at pH 6 shows in Figure 7.2. The percentage adsorption increased from 77 to 87 for unmodified adsorbent and from 85 to 92.2% for modified adsorbent in time of 90 min 2–3 g/100 mL of adsorbent.

7.3.1.3 Effect of Initial Concentration

Effect of initial metal ion concentration on percentage removal of metal ions (adsorption) was investigated by varying concentration of Cd^{2+} in the range of 2 to 20 mg/L. 100 mL of separate concentrations of metal ions (ranging from 2–20 mg/L) were each taken into 6 beakers containing a fixed mass (1.0 g) of the adsorbents (modified and unmodified adsorbent). The solutions were agitated at a fixed speed of 180 rpm for a predetermined contact time of 120 min and fixed pH value of 6.0. The final concentration of each solution was determined by UV-visible spectrometer. The equilibrium sorption capacities of the sorbents obtained from experimental data at different initial cadmium concentration are showed in Figure 7.3. As seen from the results, the sorption capacities of the sorbent increased with increasing cadmium concentration while the adsorption

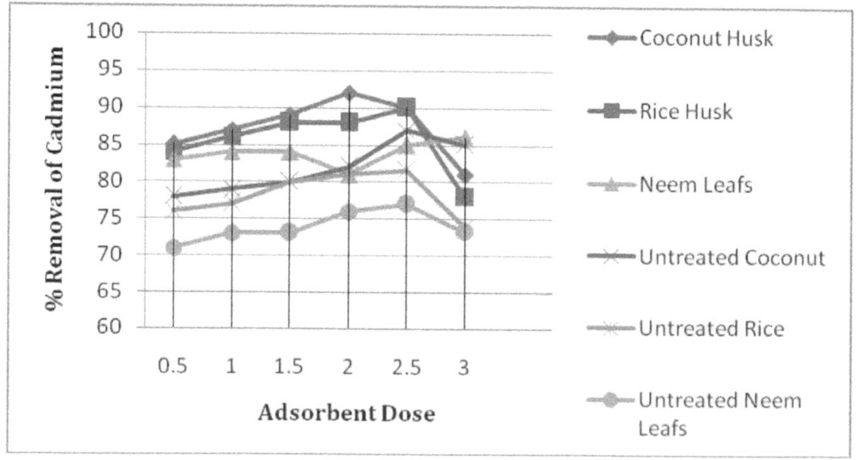

FIGURE 7.2 Effect of Adsorbent Dose, pH 6.0, RPM: 180, Time: 90 min, Temperature: 25°C, and Initial Concentration: 10 mg/L.

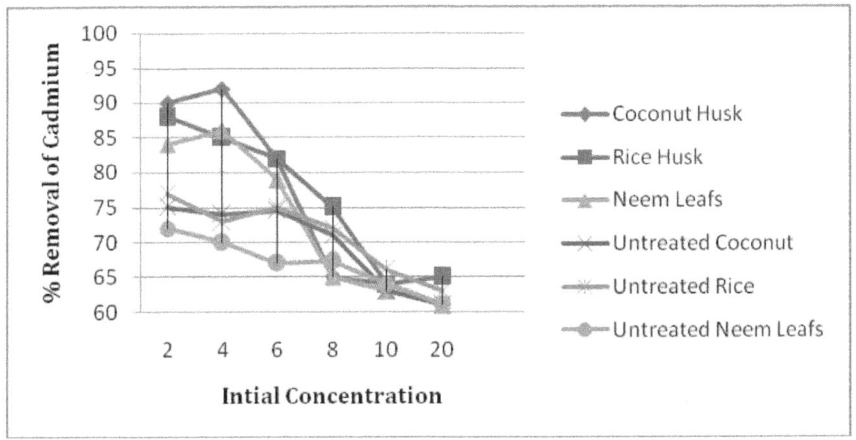

FIGURE 7.3 Effect of Initial Concentration, pH 6.0, RPM: 180, Time: 150 min, Temperature: 25°C, and Dose Amount: 3 g/100 mL.

yields of cadmium showed the opposite trend. Increasing the mass transfer driving force, and therefore, the rate at which cadmium ions pass from the bulk solution to the particle surface.

7.3.1.4 Effect of pH

The effect of pH on the adsorption of cadmium is presented in Figure 7.4. The pH of the aqueous solution is an important controlling parameter in the adsorption process and thus the role of H_2 ion concentration was examined from solutions at different pH converting a range of 2 to 7. Adsorption of cadmium at pH 2 was very low. It was observed that with the increase in the pH of the solution, the extent of removal increased to at most linearly between pH 2 to 6 attaining a maximum value of 640.27% around pH 6. Sorption had been found to enhance with increase in pH to reach optimum at pH 5–6, there was a decrease in the adsorption. This decrease may be due to the formation of soluble hydroxy complexes.

The hydrolysis of cations occurs by the replacement of metal ligands in the inner co- ordination sphere with the hydroxy groups. This replacement occurs after the removal of the outer hydration sphere of metal cat ions. Adsorption may not be related directly to the hydrolysis of the metal ion, but instead of the outer hydration sphere that precede hydrolysis.

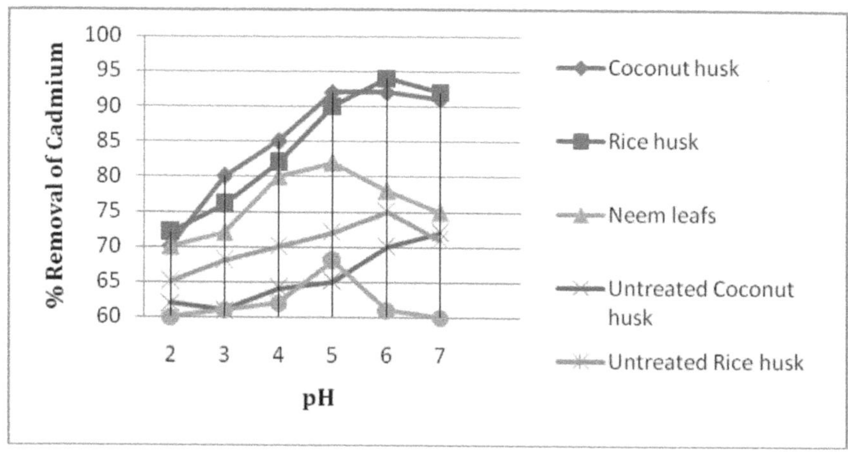

FIGURE 7.4 Effect of pH, RPM: 180, Time: 150 min, Temperature: 25°C, Dose Amount: 3 g/100 mL, and Initial concentration: 10 mg/L.

7.3.1.5 Effect of Particle Size

The batch adsorption experiments were carried out using the six particle size at fixed pH (6), adsorbent dose (2 g) and contact time (3 h). The selected particle mesh sizes were 60, 80, 100, 120 and 140. The percentage adsorption of cadmium was found to be 99.2, 98.5, 98.3, 97.7 and 96.5. Using the above-mentioned size, 60 were selected for adsorption studies due to the sufficient adsorption capacity and easiness of preparation.

7.3.2 CHROMIUM

Batch sorption studies were carried out with 100 mL Cr^{6+} solution and 1–5 gm adsorbent. The equilibrating mixtures were placed in Stoppard borosil glass bottles and agitated in temperature-controlled shaker for optimum time of contact. The contents were filtered through double what man no. 40 filter paper and then analyzed for the Cr^{6+} by using UV-visible spectrometer at λ_{max} = 324 nm wavelength. The pH of the sample solution was recorded initially and at the end of experiment using pH meter/pH paper.

7.3.2.1 Effect of Contact Time

The effect of contact time on adsorption of Cr^{6+} was investigated. 1.0 g each of the adsorbents was taken into 6 beakers containing 100 mL of metal solution at an initial concentration of 10 mg/L. The solutions were agitated at a fixed stirring speed of 200 rpm for varying time period ranging from 20–150 min at a fixed pH value of 2. The final concentration of solution was determined by UV-visible spectrometer, and the percentage Cr^{6+} was then determined. The change of adsorption of Cr^{6+} with time is presented in Figure 7.5. Removal was 67% for untreated adsorbent and 82% for treated adsorbent after 120 min. It is clear that, at the beginning % removal increased rapidly in few minutes, by increasing contact time, % removal increased lightly and slowly till reach maximum value and this can be explained on the basis that as initially a large number of vacant surface sites are available for adsorption of metal ions but with passage of time the surface sites become exhausted. These results indicate that the activated carbon and adsorbent have a very strong capacity for adsorption of Cr^{6+} ions in solutions.

7.3.2.2 Effect of Adsorbent Dose

Effect of adsorbent dose on percentage removal of metal ions was investigated by varying adsorbent dose in the range of 1 to 6 g. Separate masses of adsorbents (ranging from 1–6 g of modified and unmodified adsorbent) were each taken into 6 beakers containing 100 mL of metal solution of

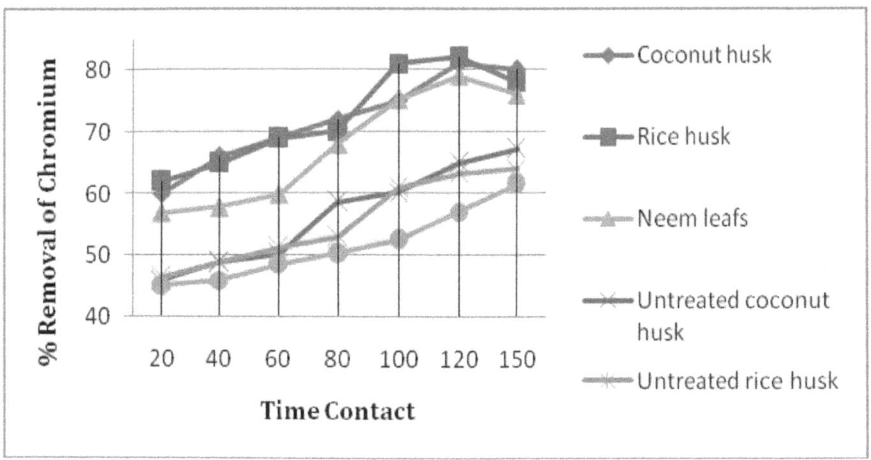

FIGURE 7.5 Effect of Contact Time, pH 2.0, Agitation: 200, Adsorbent Dose: 4 g, Temperature: 25°C, Initial Concentration: 10 ppm.

initial concentration of 10 mg/L. The solutions were agitated by orbital shaker inclulater of 200 rpm for a predetermined contact time of 20–150 min and a fixed pH value of 2.0. The final concentration of solutions was then determined by UV-visible spectrometer and the percentage removal of Cr^{6+} determined. The removal of chromium in the solution at pH 2 shows in Figure 7.6. The percentage adsorption increased from 45 to 62 for unmodified adsorbent and from 66.5 to 82.0% for modified adsorbent in time of 120 min 4 g/100 mL of adsorbent. Unit adsorption was, however, decreased with increasing in adsorbent dose. This is may be due to overlapping of adsorption sites as a result of overcrowding of adsorbent particles.

7.3.2.3 Effect of Initial Concentration

The initial heavy metal ions concentration is an important parameter in adsorption since a certain amount of adsorbent can adsorb a certain amount of heavy metal ions. Effect of initial metal ion concentration on percentage removal of metal ions (adsorption) was investigated by varying concentration of Cr^{6+} in the range of 2 to 20 mg/L. 100 mL of separate concentrations of metal ions (ranging from 2–20 mg/L) were each taken into 6 beakers containing a fixed mass (1.0 g) of the adsorbents (modified

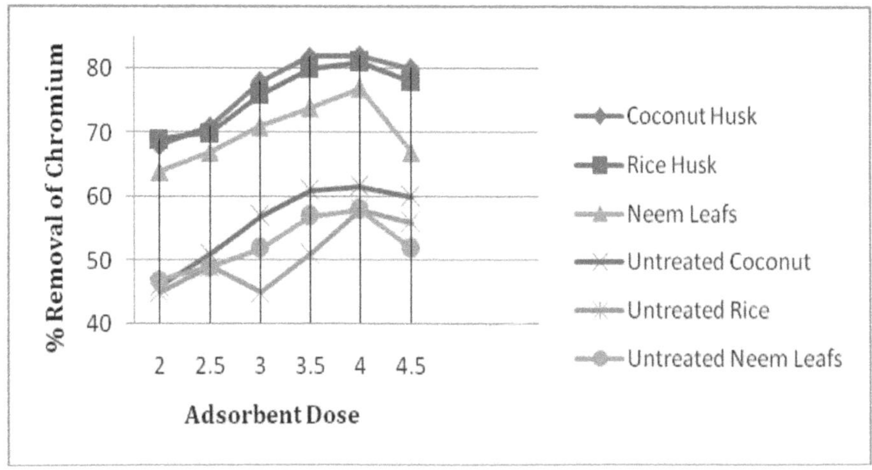

FIGURE 7.6 Effect of Adsorbent Dose, pH 2.0, Agitation: 200, Contact Time: 120 min, Temperature: 25°C, and Initial Concentration: 10 ppm.

and unmodified adsorbent). The solutions were agitated at a fixed speed of 200 rpm for a predetermined contact time of 120 min and fixed pH value of 2.0. The final concentration of each solution was determined by UV-visible spectrometer. As seen from the results, the sorption capacities of the sorbent decrease with increasing cadmium concentration. But the actual amount of Cr^{6+} ions adsorbed per unit mass of the adsorbent was increased with increasing in Cr^{6+} ions concentration in the test solution as shown in Figure 7.7. At low concentration the ratio of surface active sites to the total metal ions in the solution is high and hence all metal ions interact with the adsorbent and are removed quickly from the solution. However, the amount of metal ions adsorbed per unit weight of adsorbent, q, is higher at high concentration. According to these results, the initial Cr^{6+} ions concentration plays an important role in the adsorption capacities. Higher concentrations of metal ions were used to study the maximum adsorption capacity of adsorbent.

7.3.2.4 Effect of pH

The effect of pH on the adsorption of chromium is presented in Figure 7.8. The pH of the aqueous solution is an important controlling parameter in

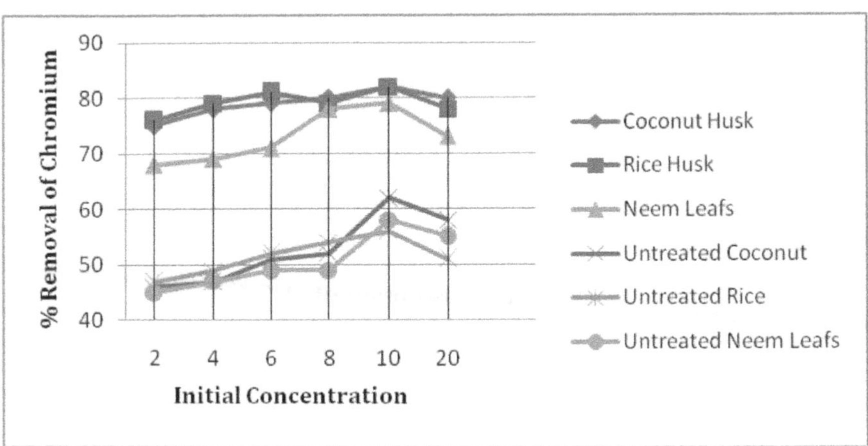

FIGURE 7.7 Effect of Initial Concentration, pH 2.0, Adsorbent Dose: 4.0 g, Agitation: 200, Contact Time: 120 min, Temperature: 25°C, Initial Concentration: 10 ppm, and Particle Sizes: 60 μm.

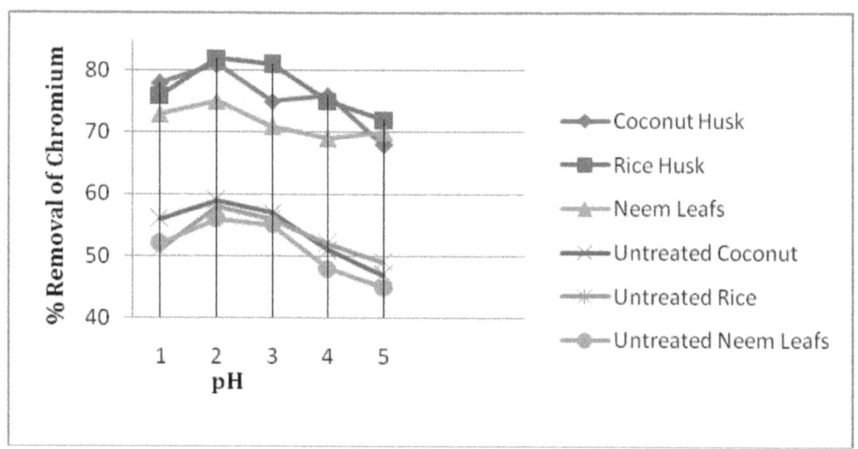

FIGURE 7.8 Effect of pH, Adsorbent Dose: 4.0 g, Agitation: 200, Contact Time: 120 min, Temperature: 25°C, and Initial Concentration: 10 ppm.

the adsorption process and thus the role of H_2 ion concentration was examined from solutions at different pH converting a range of 2–5. Hence, the influence of pH on the adsorption of Cr^{6+} ions onto modified and unmodified adsorbent was examined in the pH range of 1–5. These results were represented in Figure 7.8, which showed that the adsorption capacities of Cr^{6+} ions onto both adsorbents increased significantly, with decreasing pH value and the maximum removals of Cr^{6+} ions by both adsorbents for contact time (120 min) were carried out at pH (2.0). The improved removal of Cr^{6+} at low pH is probably due to reduction of hexavalent chromium to trivalent chromium ions.

7.3.2.5 Effect of Particle Size

The effect of Particle Size on the adsorption of chromium (Cr^{6+}) is shown in Figure 7.9. The batch adsorption experiments were carried out using the six-particle size at fixed pH (2), adsorbent dose (4 g) and contact time (2 h). The selected particle mesh sizes were 60, 80, 100, 120 and 140. The percentage adsorption of chromium was found to be 50.0–82%. Using the above-mentioned size, 60 were selected for adsorption studies due to the sufficient adsorption capacity and easiness of preparation.

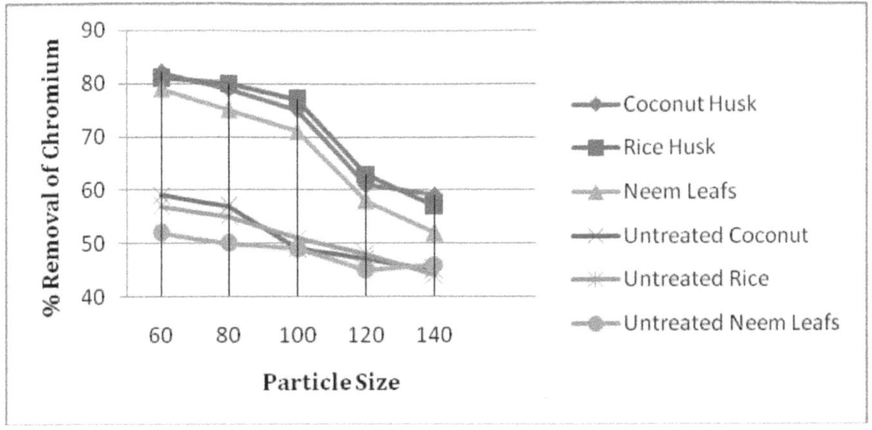

FIGURE 7.9 Effect of Particle Size, pH 2.0, Adsorbent Dose: 4.0 g, Agitation: 200, Contact Time: 120 min, Temperature: 25°C, and Initial Concentration: 10 ppm.

7.3.3 LEAD

In the determination of adsorption capacity of the adsorbent for Pb^{2+}, 100 mL of the working solution of Lead ions containing 10 mg/L of Pb^{2+} each was taken into six different borosil beakers and known masses (1, 2, 3, 4, 5 and 6 g, respectively) of the adsorbent was each added to the solution (specifically at the same and controlled initial pH of solution fixed at 6.0). The mixture was stirred at 180 revolutions per minute (rpm) for a predetermined time (for which equilibrium was attained). The supernatant of the mixture was then filtered through double Whatman filter paper (number 40). Determination of Lead ion concentration was done by UV-visible spectrometer at λ_{max} = 520 nm wavelength.

7.3.3.1 Effect of Contact Time

The effect of contact time on adsorption of Pb^{2+} was investigated. 0.5 g each of the adsorbents was taken into 6 beakers containing 100 mL of metal solution at an initial concentration of 10 mg/L. The solutions were agitated at a fixed stirring speed of 180 rpm for varying time period ranging from 20–180 min at a fixed pH value of 6. The final concentration of solution was determined by UV-visible spectrometer, and the percentage Pb^{2+} was then determined. The change of adsorption of Pb^{2+} with time is presented in Figure 7.10. The adsorption increases with the increasing contact time and that mean the removal efficiency is slow after equilibrium was attained.

7.3.3.2 Effect of Adsorbent Dose

Effect of adsorbent dose on percentage removal of metal ions was investigated by varying adsorbent dose in the range of 1.0 to 6.0 g. Separate masses of adsorbents (ranging from 1.0 g –6.0 g of modified and unmodified adsorbent) were each taken into 6 beakers containing 100 mL of metal solution of initial concentration of 10 mg/L. The solutions were agitated by orbital shaker inclulater speed of 180 rpm for a predetermined contact time of 120 min and a fixed pH value of 6.0. The final concentration of solutions was then determined by UV-visible spectrometer and the percentage removal of Pb^{2+} determined. The removal of lead in the solution at pH 6 shows in Figure 7.11. The percentage adsorption increased from 84 to 87 for unmodified adsorbent and from 88 to 92.5% for modified adsorbent in time of 90 min in the range of 4 g of adsorbent.

7.3.3.3 Effect of Initial Concentration

Effect of initial metal ion concentration on percentage removal of metal ions (adsorption) was investigated by varying concentration of Pb^{2+} in

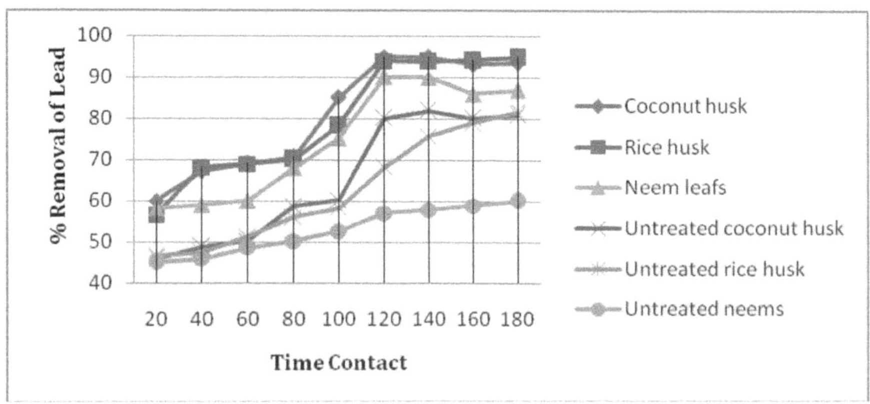

FIGURE 7.10 Effect of Contact Time, pH 6.0, RPM: 180, Dose: 1 g, and Temperature: 25°C.

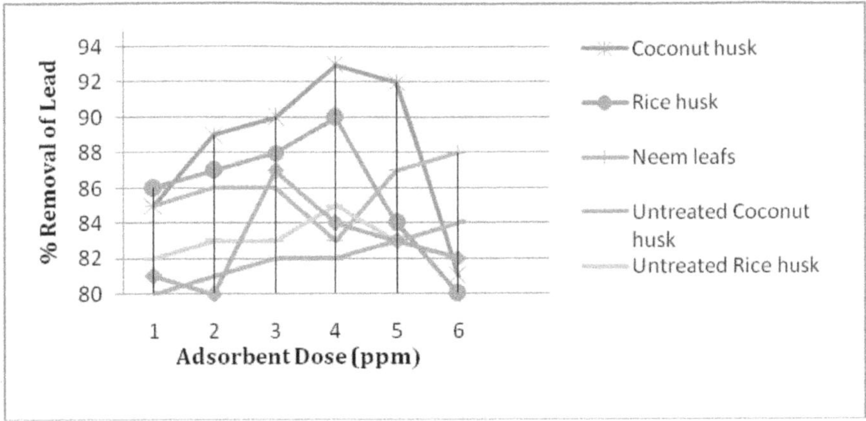

FIGURE 7.11 Effect of Adsorbent Dose, pH 6.0, RPM 180, Time: 120 min, and Temperature: 25°C.

the range of 2 to 20 mg/L. 100 mL of separate concentrations of metal ions (ranging from 2–20 mg/L) were each taken into 6 beakers containing a fixed mass (4.0 g) of the adsorbents (modified and unmodified adsorbent). The solutions were agitated at a fixed speed of 180 rpm for a predetermined contact time of 120 min and fixed pH value of 6.0. The final concentration of each solution was determined by UV-visible spectrometer. The equilibrium sorption capacities of the sorbents obtained from experimental data at different initial cadmium concentration are showed in Figure 7.12. As seen from the results, the sorption capacities of the sorbent increased with increasing lead concentration while the adsorption yields of lead showed the opposite trend increasing the mass transfer driving force.

7.3.3.4 Effect of pH

The effect of pH on the adsorption of lead is presented in Figure 7.13. The pH of the aqueous solution is an important controlling parameter in the adsorption process and thus the role of H_2 ion concentration was examined from solutions at different pH converting a range of 2 to 7. Adsorption of lead at pH 2 was very low. It was observed that with the increase in the pH of the solution, the extent of removal increased to at most linearly between pH 2–5 attaining a yield value of 876.14% around pH 6. Sorption

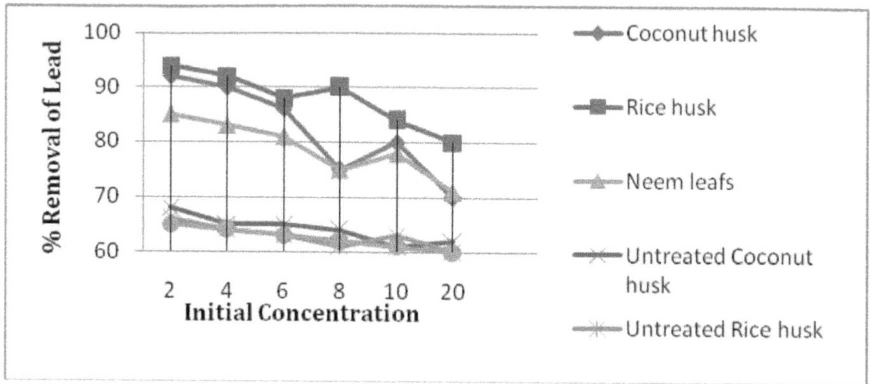

FIGURE 7.12 Effect of Initial Concentration, pH 6.0, RPM: 180, Time: 120 min, Temperature: 25°C, and Dose Amount: 4 g/100 mL.

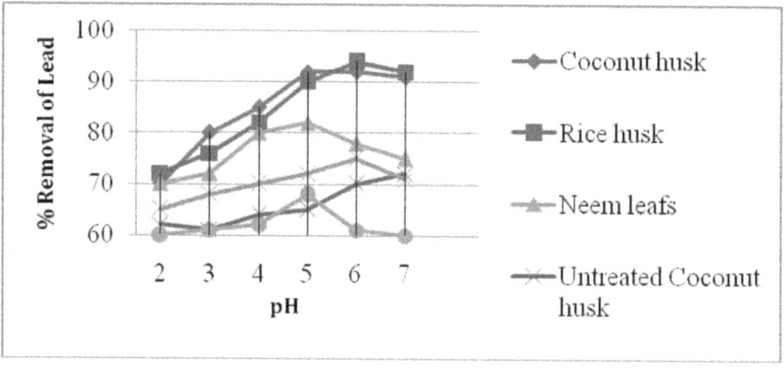

FIGURE 7.13 Effect of pH, RPM: 180, Time: 120 min, Temperature: 25°C, and Dose Amount: 4 g/100 mL.

had been found to enhance with increase in pH to reach optimum at pH 5–6, there was a decrease in the adsorption. This decrease may be due to the formation of soluble hydroxyl complexes.

7.3.3.5 Effect of Particle Size

The effect of particle Size on the adsorption of cadmium is shown in Figure 7.14. The batch adsorption experiments were carried out using the six-particle size at fixed pH (6), adsorbent dose (4 g) and contact time (2 h).

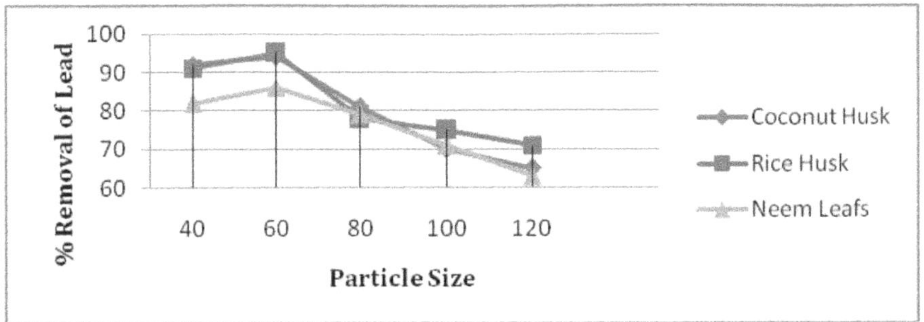

FIGURE 7.14 Effect of particle size, RPM: 180, Time: 120 min, Temperature: 25°C, Dose Amount: 4 g, Initial Concentration: 10 ppm, and pH 6.0.

The selected particle mesh sizes were 60, 80, 90, 100, and 120. The percentage adsorption of Lead was found to be 91.8, 82.2, 98.3, 97.1 and 96.7. Using the above-mentioned size, 60 were selected for adsorption studies due to the sufficient adsorption capacity and easiness of preparation.

7.4 CONCLUSION

The results from this study indicate that the studied of bio-adsorbent is suitable for use in the removal of metal ions for heavy metal solution. Adsorptive capacity and metal removal efficiency of the adsorbent studied varied significantly with each metal ion and with pH, contact time, dose, adsorbent particles size but with isotherm. Adsorption of cadmium increased with increase in pH while that of chromium exhibited an inverse relationship with increase in pH. The trend of lead adsorption varied with increasing pH. The efficiency of adsorbent per metal ion presented interesting results comparable to results from previous works. Chromium adsorption was generally poor exhibiting no efficiency for chromium removal but yielded maximum lead, chromium and cadmium removal of 95.01%, 78.26 and 63.13% at pH 6. The results show that the removal efficiency of each adsorbent is highly dependent on pH, and Metal ion removal occurred in the preferential order lead > cadmium > chromium, depicting strong contributions from the ionic radius of each metal ion. Finally, this works shows that locally available materials such as Coconut Husk, Rice Husk, Neem Leaves could easily be sourced to produce activated carbon which can be used as efficient adsorbents for Lead, chromium and cadmium ion removal from wastewater, representing an environmentally effective means of using these agricultural, bio residues.

KEYWORDS

- **Adsorbents**
- **Heavy Metals**
- **Particle Size**
- **Adsorbent Dose**

REFERENCES

1. Renuga, D. N., Manjusha, K., & Lalithia, P., Removal of Hex-avalent Chromium from Aqueous Solution Using an Eco-friendly Activated Carbon Adsorbent. *Advances in Applied Science Research* 2010, *1*, 247–254.
2. Sarin, V., Pant, K. K., & Mohan, D., Agricultural Waste Material as potential Adsorbent for Sequestering Heavy Metal ions from Aqueous Solutions. *Bioresources Technology* 2006, *99*, 6017–6027.
3. Samara, S. A. Toxicity, Heavy Metals, *e-Medicine* 2006, 1–2.
4. Kadirvelu, K., Thamaraiselvi, K., & Namasivayam, C., Removal of heavy metal from industrial wastewaters by adsorption onto activated carbon prepared from an agricultural solid waste. *Bioresour. Technol.* 2001, *76*, 63–65.
5. Kadirvelu, K., Thamaraiselvi, K., & Namasivayam, C., Adsorption of nickel(II) from aqueous solution onto activated carbon prepared from coir pith. *Sep. Purif. Technol.* 2001, *24*, 497–505.
6. Williams, C. J., Aderhold, D., & Edyvean, G. J., Comparison between biosorbents for the removal of metal ions from aqueous solutions. *Water Res.* 1998, *32*, 216–224.
7. Kannan, N., & Veeemaray, T., Detoxification of Toxic Metal Ions by Sorption onto Activated Carbon from Hevea Brasi-liensis Bark. *Global NEST Journal* 2010, *10*, 20–40.
8. Bailey, S. E., Olin, T. J., Bricka, R. M., & Adrian, D. D., A review of potentially low-cost sorbents for heavy metals. *Water Res.* 1999, *33*, 2469–2479.
9. Marshall, W. E., Champagne, E. T., & Evans, J. E., Use of rice milling by-product (hulls and bran) to remove metal ions from aqueous solution. *J. Environ. Sci. Health A* 1993, *28*, 1977–1992.

10. El-Said A. G., Badawy N. A., & Garamon, S. E., Adsorption of Cadmium (II) and Mercury (II) onto Natural Adsorbent Rice Husk Ash (RHA) from Aqueous Solutions: Study in Single and Binary System. *Journal of American Science* 2010, *6,* 400–409.
11. Roy, D., Greenlav, P. N., & Shane, B. S., Adsorption of heavy metals by green algae and ground rice husks. *J. Environ. Sci. Health A* 1993, *28*, 37–50.
12. Munaf, E., & Zein, R., The use of rice husk for removal of toxic metals from wastewater. *Environ. Technol.* 1997, *18*, 359–362.
13. Suemitsu, R., Uenishi, R., Akashi, I., & Nakanom, M., The use of dyestuff-treated rice hulls for removal of heavy metal from wastewater. *J. Appl. Polym. Sci.* 1986, *31*, 75–83.
14. Lee, C. K., Low, K. S., & Mah, S. J., Removal of a gold(III) complex by quaternized rice husk. *Adv. Environ. Res.* 1998, *2*, 351–359.
15. Low, K. S., & Lee, C. K., Quaternized rice husk as sorbent for reactive dyes. *Biores. Technol.* 1997, *61*, 121–125.
16. Chand, S., Aggarwal V. K., & Kumar, P., Removal of Hexavalent Chromium from the Wastewater by Adsorption. *Indian J Environ. Health*, 1994, *36*, 151–158.
17. Tan W. T., Ooi S. T., & Lee C. K., Removal of Chromium (VI) from Solution by Coconut Husk and palm Pressed Fiber. *Environmental Technology* 1993, *14,* 277–282.
18. Ho, Y. S., Bibliometric Analysis of Adsorption Technology in Environmental Science. *J. Environ. Prot. Sci.* 2007, *1*, 1–11.
19. Ponangi, S., Shyam, R. A., & Joshi, S. G., Trace pollutants in drinking water, *J. Indian Assoc. Environ. Manage.* 2000, *27*, 16–24.
20. Manahan, S., Environmental Chemistry, Book/colei California, USA, 1984.
21. Gao, H., Liu, Y., Zeng, G., Xu, W., Li, T., & Xia, W., Characterization of Cr(VI) removal from aqueous solutions by a surplus agricultural waste- Rice straw. *J. Hazard. Mater.* 2007, *150*, 446–452.

CHAPTER 8

ADSORPTION OF ANIONIC DYE ONTO TBAC-MODIFIED HALLOYSITE NANOTUBES

P. V. MANKAR,[1] S. A. GHODKE,[1] S. H. SONAWANE,[2] and S. MISHRA[3]

[1]*Chemical Engineering Department, Sinhgad College of Engineering, Pune, Maharashtra, India*

[2]*Chemical Engineering Department, National Institute of Technology, Warangal, Andhra Pradesh, India*

[3]*University Institute of Chemical Technology, North Maharashtra University, Jalgaon, Maharashtra, India*

CONTENTS

8.1 Introduction 120
8.2 Materials and Methods 122
8.2.1 Materials 122
8.2.2 Synthesis of TBAC Halloysite Modified Nanotubes 122
8.2.3 Adsorption of CR Dye 123
8.2.4 Characterization Techniques 124
8.3 Results and Discussion 124
8.3.1 Fourier Transform Infrared Ray (FTIR) of Adsorbent 124
8.3.2 Transmission Electron Microscope (TEM) 125
8.3.3 Scanning Electron Microscopy (SEM) 126
8.3.4 Thermo-Gravimetric Analyzes (TGA) 126

8.3.5 Effect of Loading 128
8.3.6 Effect of Initial Dye Concentration 129
8.3.7 Adsorption time 130
8.3.8 Effect of pH on the Adsorption Capacity and Removal Efficiency 131
8.3.9 Adsorption Isotherm 132
8.3.9.1 Langmuir Isotherm 133
8.3.9.2 Freundlich Isotherm 134
8.3.10 Adsorption Kinetics 134
8.3.10.1 Pseudo-First Order Kinetics 135
8.3.10.2 Pseudo-Second Order Kinetics 136
8.4 Conclusions 137
Keywords 138
References 138

8.1 INTRODUCTION

Synthetic dyes are used industrially in the field of textile, rubber, leather, printing, plastic etc. Treatment of effluents from such industries has received a considerable attention because of the carcinogenicity, toxicity to aquatic life and human life [1, 2]. The dyes include a broad spectrum of different chemical structures, primarily based on the substituted aromatic groups. Due to the complex chemical structure of these dyes, they are resistant to breakdown by chemical, physical and biological treatments [3]. The dye wastewater is found to have high salt content and low biodegradable potential, which proves limited use of conventional wastewater treatment processes [4, 5]. One of the most efficient methods for removal of organic dye pollutant is adsorption [6] and some of the research findings, activated carbon appears to be the best choice for dye adsorption [7]. However, activated carbon is expensive and difficult to regenerate after use. Hence, there is a need to produce relatively cheap adsorbents that can be applied to water pollution control. Therefore, in the recent times many studies have been reported focusing on the use of various low-cost adsorbents to replace activated carbon. Various studies reports use of bagasse fly ash [8], wood [9], maize cob [10], peat [11] as effective dye adsorbents.

Clay minerals (calcite, zeolite, bentonite and kaolinite) are found to be unconventional adsorbents for effective dye removal because of their high surface area, abundant resource and low-cost [12]. The ions and polar organic molecules can easily get adsorbed because of its lamellar structure providing higher specific surface areas [13]. Emna Errais et al. [14] reported that the natural clay exhibits a negative charge of structure, which allows it to adsorb positively charged dyes but induces a low adsorption capacity for anionic dyes. Thus literature mostly reports on cationic dye adsorption by clay and very few studies have been devoted to anionic dye adsorption onto natural clay or treated clay. Many studies report modification of mineral clays with the help of organic salts and metals so as to increase the adsorption of contaminants [12,15]. Halloysite in its dehydrated form is a 1:1 layered dioctahedral mineral from the kaolin group with a chemical composition of $Al_2Si_2O_5(OH)_4$. The layer is built from tetrahedral (silicic) and octahedral (aluminum) sheets linked by joint oxygen molecules. The interlayer water can lead to mismatch of the tetrahedral and octahedral layers, and affect the physicochemical properties of halloysite including organic intercalation and cation exchange capacity [16]. The main difference between halloysite and the other alumina-silicate minerals is its unique nanotubular structure. Halloysite occurs mainly in two different polymorphs, the hydrated form (basal distance around 10 Å) with the minimal formula $Al_2Si_2O_5(OH)_4.2H_2O$, and the dehydrated form (basal distance around 7 Å) with the minimal formula $Al_2Si_2O_5(OH)_4$, being identical to kaolinite. The hydrated form converts irreversibly into the dehydrated form when dried at temperatures below 100°C [17]. The outer surface of halloysite nanotubes contains SiO_4 tetrahedra, whereas the inner surface contains octahedral Al–OH functionalities [18].

In order to increase the performance of halloysite efforts have been made by intercalation [19], thermochemical modifications [20], acid treatments [18], etc. These methods were able to increase the adsorption characteristics of halloysite clays and are used for adsorption of dyes and metal ions [16]. It was also reported that, acid treatment and heat activation of halloysite for the adsorption purpose causes disaggregation and structural rearrangement of halloysite, which correspondingly changes the pore structure and surface properties and higher accuracy during experimental. Wang Jinhua reported the halloysite nanotubes modified with the cationic surfactant–hexadecyl trimethyl ammonium bromide (HDTMA-Br) to form a new adsorbent [21]. The adsorbent exhibited rapid adsorption rate and the adsorption capacity was found to be 90%. In this study we report the surface modification of halloysite nanotubes with a cationic adsorbent – tetrabutylammonium chloride (TBAC) to form a new adsorbent. Herein, we systematically investigated the surface modification of halloysite nanotubes with TBAC. The adsorbent was applied to remove the Congo Red dye from aqueous solution. The effect of adsorbent dose, initial concentration, adsorption time and pH on to dye adsorption was studied in detail. The data obtained was fitted using the adsorption isotherm models.

8.2 MATERIALS AND METHODS

8.2.1 MATERIALS

All the chemicals used in this study were of A.R. grade and were obtained from Sigma Aldrich. Deionized (DI) water was used for all experiments. Both HNTs and TBAC were used as supplied without any further purification. Congo red dye (CR) was obtained from Shree chemicals Pune, India. A stock solution of CR dye was prepared (100 mg/L) by dissolving a required amount of dye powder in deionized water (Millipore). The stock solution was diluted with deionized water to obtain the desired concentration ranging from 10 to 100 mg/L.

8.2.2 SYNTHESIS OF TBAC HALLOYSITE MODIFIED NANOTUBES

A two-step methodology was followed to synthesis of TBAC halloysite modified nanotubes [21]. Halloysite nanotubes (HNTs) were washed with 1 mol/L HCL under constant agitation at room temperature for 2 h and then kept solution for 24 h. Further the tubes were filtrated and wash with distilled water until a pH of 6 was obtained. The acidified HNTs were agitated for 20 h in a 1 mol/L NaCl solution at room temperature and then kept for 48 h in the NaCl solution. The product was then washed out and dried at 60°C for 4 hr. The product formed was Na-HNTs, this was to improve cation exchange capacity (CEC). Further, 4 g sample of Na-HNTs was equilibrated for 12 h using thermo magnetic stirrer (REMI) with 200 mL 0.014 M TBAC at 60°C finally the solid was washed out with DI water repeatedly. The modified HNTs were dried in the oven at 60°C and used in further experiments. All the dye removal experiment carried out at room temperature. Experiments were repeated in triplicate and the percentage deviation in results was found to be ±5%.

8.2.3 ADSORPTION OF CR DYE

The effect of parameters like loading, initial concentration, time, pH was studied in batch mode for both modified and unmodified HNTs. The concentration of CR in the experimental solution was determined from the calibration curve prepared by measuring the absorbance of different known concentrations of CR solutions at λ_{max} = 498 nm using a UV-vis spectrophotometer. Dye removal experiments with the TBAC modified HNTs and unmodified HNTs were performed in 100 mL Erlenmeyer flask by varying initial concentration of dye (Congo Red) solutions (50 mL) and adsorbent dose of modified and unmodified HNTs. Some experimental factors, such as loadings, adsorption time (0–180 min), initial concentration of dye ion (25–150 mg/L) and pH were chosen as the controlling parameters in the adsorption process. The amount of dye adsorbed at equilibrium (qe mg/g) and removal efficiency (% R) was calculated by using the following equations, respectively.

$$q_e = \frac{V(C_0 - C_e)}{M} \tag{2.1}$$

$$\%R = \frac{(C_0 - C_e)}{C_0} \times 100 \tag{2.2}$$

where, C_0 – Initial concentration; C_e – Equilibrium concentration; V – Volume in liter; M – Mass in g.

8.2.4 CHARACTERIZATION TECHNIQUES

Unmodified HNTs and modified HNTs were characterized by using FTIR, TEM, SEM, and TGA. Amount of dye adsorbed was analyzed using UV-visible spectroscopy.

Fourier transform infrared ray (FTIR) spectroscopic measurement was performed (Bruker Germany, 3000 Hyperion Microscope with Vertex 80 FTIR System) using KBr pellets Infrared spectra were recorded in the range 4000–400 cm^{-1} at Spectral resolution of FTIR 0.2 cm^{-1}. Transmission electron microscopy (TEM) image was taken on a (Philips CM 200) operating voltage 20–200 KV at resolution 2.4 Å. Scanning electron microscopy (SEM) was taken on (JEOL 6380A). Thermo-gravimetric analysis (TGA) analysis was performed using SHIMADZU model DTG-60H in temperature range 0–900°C. UV spectrophotometer CR dye concentration was determined by using double beam spectrophotometer (Thermofiesher 6.87 Spectrascan 2800). At ambient temperature the wavelength of maximum absorbance (λ_{max}) of CR dye was 498 nm.

8.3 RESULTS AND DISCUSSION

8.3.1 FOURIER TRANSFORM INFRARED RAY (FTIR) OF ADSORBENT

Infrared spectroscopy (IR) method was used to investigate whether the functional groups had been grafted onto the surface of nanotubes. Spectra were recorded with detector at 0.2 cm^{-1} resolution between 400 and 4000 cm^{-1} using KBr pellets. The results are shown in Figure 8.1. Before the modification Figure 8.1(a), the band at 910 cm^{-1} is assigned to bending vibration of Al–OH. Absorption bands at 3693 cm^{-1} and 3624 cm^{-1} are ascribed to –OH groups. Others bands at 1000–1100 cm^{-1} and 450–550 cm^{-1} are due to Si–O stretching vibration and Si–O bending vibration, 1680–1640 cm^{-1}–C=C– stretch respectively. In the spectra of modified halloysite Figure 8.1(b), two new peaks were observed at 2380 cm^{-1} and 2335 cm^{-1}. The significant peaks at 2380 cm^{-1} and 2335 cm^{-1} modified nanoclay, are ascribed to the asymmetric and symmetric vibration of methylene groups ($(CH_2)n$) of aliphatic carbon chain [22].

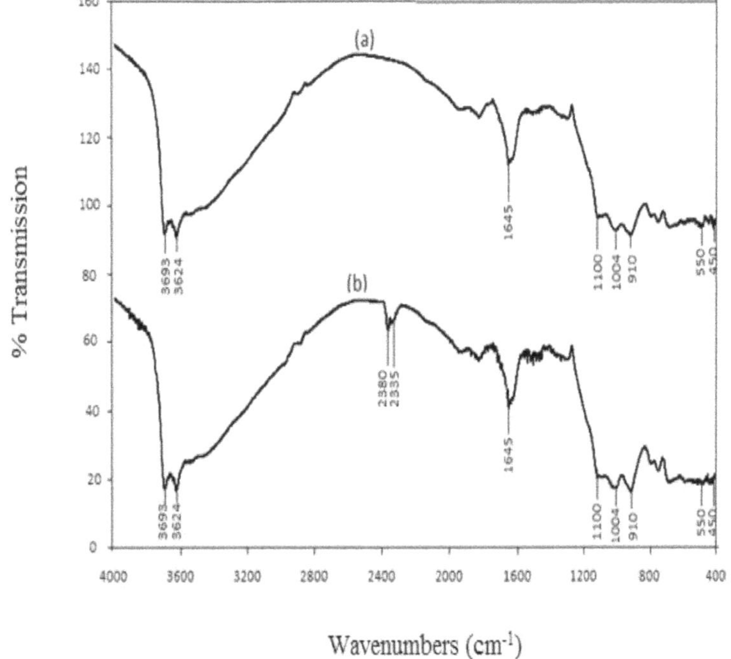

FIGURE 8.1 Infrared Spectra of original HNTs (a) and modified HNTs (b).

The peaks in case of TBAC modified HNTs are not pronounced due to less polarity of TBAC surfactant [23].

8.3.2 TRANSMISSION ELECTRON MICROSCOPE (TEM)

The transmission electron microscope (TEM) was used to observe morphological structure of unmodified and modified HNTs. Figure 8.2 displays TEM of unmodified and modified HNTs. The TEM images clearly depicts the halloysite particles have a typical cylindrical morphology with a transparent central area parallel to cylinder, indicating that the particles are hallow and open-ended. The diameter and length of unmodified HNTs Figure 8.2(a) are in the range from 10 to 50 nm and from 500 to 1000 nm respectively. In comparison with unmodified HNTs, tube walls of modified HNTs in Figure 8.2(b) are obviously thicker. The structure enables dye molecules to access and adsorb on the surface as well as inner lumen easily due to their open ends.

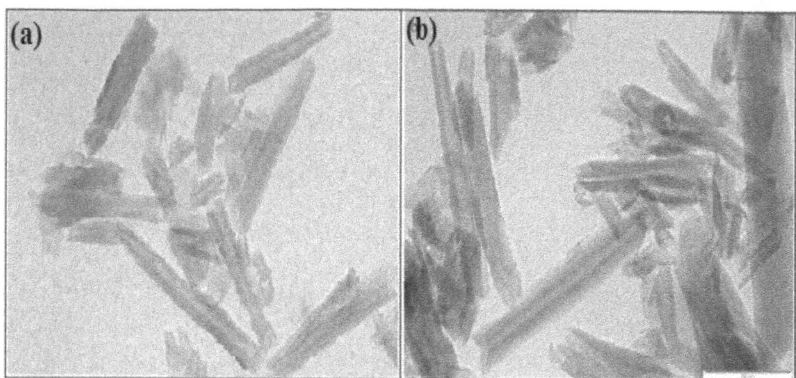

FIGURE 8.2 TEM images of unmodified HNTs (a) and modified HNTs (b).

8.3.3 SCANNING ELECTRON MICROSCOPY (SEM)

SEM images show Figure 8.3 surface morphology behavior of materials. Here SEM images of unmodified and modified HNTs clearly shows that the surface of HNTs, and its modification respectively. Figure 8.3(a) and Figure 8.3(c) of unmodified HNT at different resolution shows the irregular surface morphology of unmodified HNT as well the binding sites for TBAC. Figures 8.3(b) and 8.3(d) depicts the grafting of TBAC layers onto the HNTs surfaces. Also the images show uniformity of surface of modified HNT indicating adequate modification of HNT.

8.3.4 THERMO-GRAVIMETRIC ANALYZES (TGA)

Thermo-gravimetric analyzes (TGA) of original and modified HNTs were further carried out in N_2 from room temperature 0–900°C to test the grafted amount of the quaternary ammonium cations. The thermo-gravimetric curves that were obtained by measuring the weight loss of adsorbent from pyrolysis are shown in Figure 8.4. It can be observed that the mass-loss curve decreased continuously. During 105–900°C, the weight loss of adsorbent was found to be maximum. All the weight losses below 130°C are due to the loss of free water [24]. For the unmodified HNTs in Figure 8.4(a), the weight loss of 11.63% at 740°C is attributed to chemical

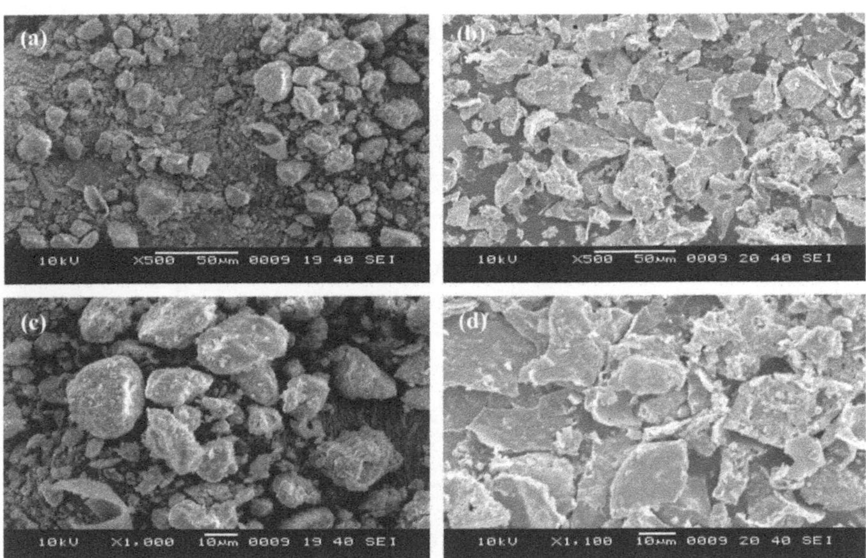

FIGURE 8.3 SEM images of unmodified HNTs (a), (c) and modified HNTs (b), (d).

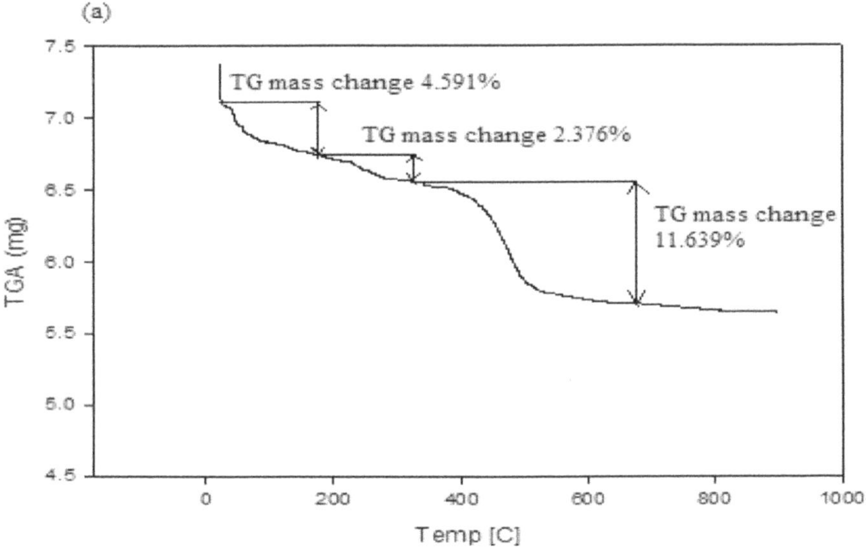

FIGURE 8.4 Continued

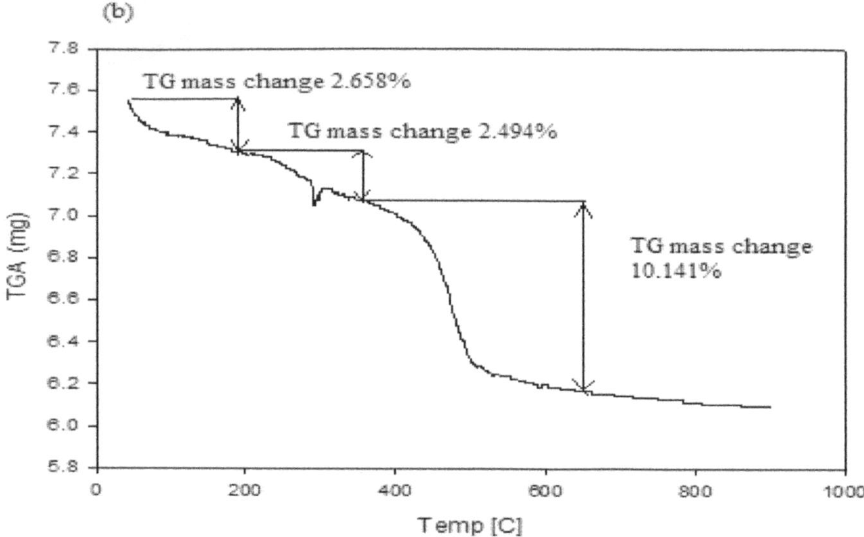

FIGURE 8.4 Thermogravimetric Analysis (TGA) of unmodified HNTs (a) and modified HNTs (b).

dehydration of the HNTs. For the modified HNTs in Figure 8.4(b), the weight loss of 10.141% at 600°C is due to chemical dehydration of the HNTs besides mass-loss mentioned above, there is obvious weight loss of 2.494% between 205°C and 350°C. Comparing Figure 8.4(b) with Figure 8.4(a), the weight loss between 205°C and 350°C can be attributed to thermal decomposition of the quaternary ammonium cations loaded on the HNTs.

8.3.5 EFFECT OF LOADING

Loading is an important parameter in the determination of adsorption capacity. The effect of the adsorbent dose was investigated by addition of various amounts of modified and unmodified HNTs in 50 mL of 50 mg/L CR dye aqueous solution at room temperature for 1 h. Also the effect of loading on the removal efficiency and adsorption capacity of modified HNTs and unmodified HNTs were studied.

From the result, it was observed that the removal efficiency shows an increasing trend, for example, 42.36% for 0.1 g loading and 78.18%

for 1.0 g loading in unmodified HNTs. For modified HNTs the removal efficiency shows increasing trend, for example, 49.92% for 0.1 g loading and 95.13% for 1.0 g loading; also the removal efficiency at the end of 1 his found to be more. This can be attributed to the increase in the adsorbent specific surface area and availability of more adsorption sites. Figure 8.5 depicts that the adsorption capacity trend decreases from 10.59 to 1.95 mg/g in unmodified HNTs whereas adsorption capacity trend decreases from 12.48 to 2.37 mg/g in modified HNTs with an increase in adsorbent dose from 0.1 g to 1.0 g. Consequently the adsorbent dose was maintained at 0.5 g, as thereafter even increase in adsorbent loading no significant increase in adsorption capacity was observed.

8.3.6 EFFECT OF INITIAL DYE CONCENTRATION

The adsorption capacity of modified HNTs and unmodified HNTs were determined by equilibrium adsorption studies, at different concentration of CR dye ranging from 25–150 mg/L with 0.5 g loading and the batch kept for continuous stirring for an hour at room temperature. In Figure 8.6

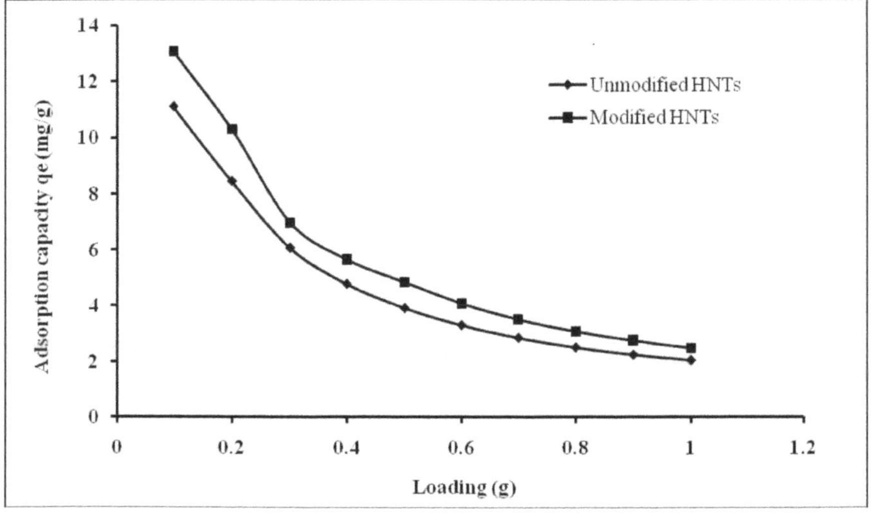

FIGURE 8.5 Effect of adsorbent dose on adsorption capacity qe (mg/g) of CR dye onto unmodified and modified HNTs.

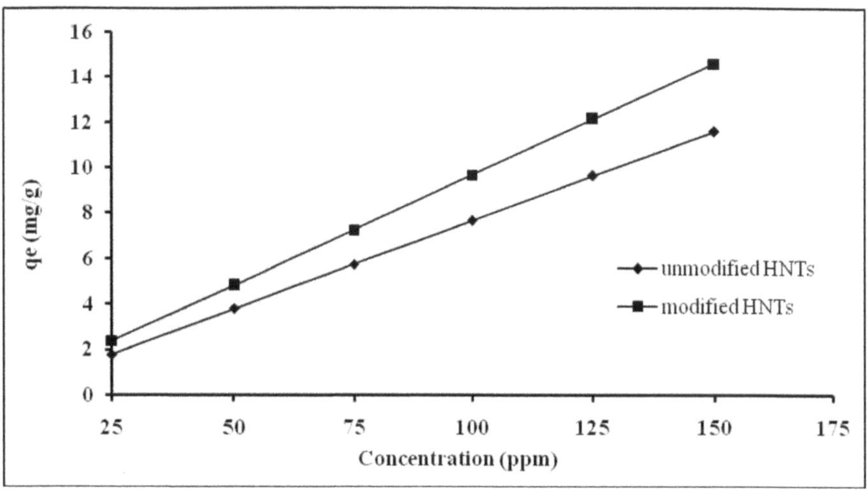

FIGURE 8.6 Effect of initial concentration of CR dye on adsorption capacity (qe mg/g) onto unmodified and modified HNTs.

results indicated that the adsorption capacity trend increased with increase in CR dye initial concentration and achieved maximum adsorption capacity finally. This may be due to the fact that more dye molecules are available for adsorption.

8.3.7 ADSORPTION TIME

The adsorption time is one of the important characteristics that define the efficiency of sorption.

In order to investigate the effect of adsorption time on modified and unmodified HNTs (0.5 g) were respectively added to a series of CR dye solution at a constant concentration of 50 mg/L at room temperature and the adsorption capacities were measured by different adsorption time from 5 to 180 min. The effect of adsorption time on CR dye adsorption capacity (qe mg/g) onto modified and unmodified HNTs is as shown in Figure 8.7. The observations indicated a sharp increase first 10 min and then gradually to reach adsorption equilibrium. The results indicated the availability of ample adsorption sites during initial phase of adsorption and then gradually the adsorption sites were saturated resulting in equilibrium state.

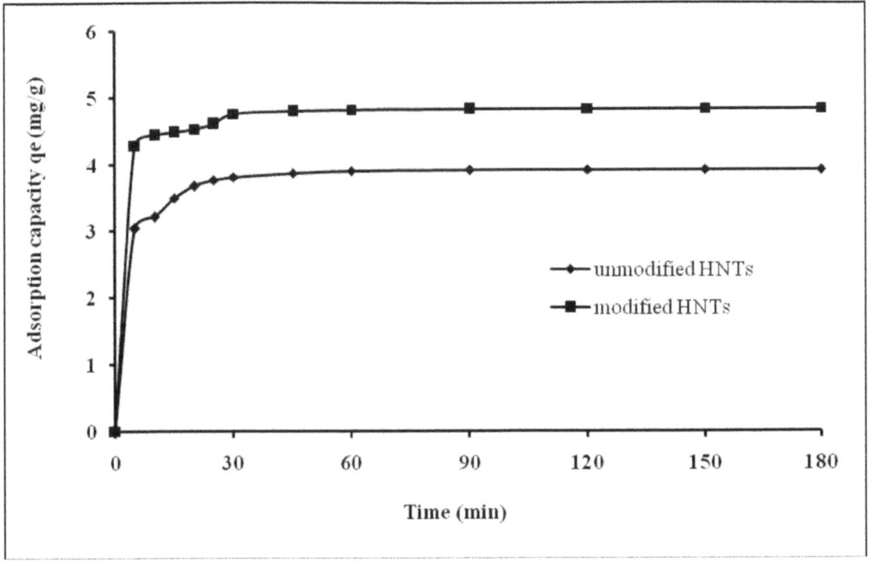

FIGURE 8.7 Effect of adsorption time on CR dye adsorption capacity (qe mg/g) onto unmodified and modified HNTs.

8.3.8 EFFECT OF PH ON THE ADSORPTION CAPACITY AND REMOVAL EFFICIENCY

The pH of the aqueous solution is an important controlling parameters that strongly affect the existed form of CR dye. Congo red is a diazo dye, and the initial pH influences the molecular form of Congo red in the aqueous solution [23]. It was reported that the dye solution changed its color from red to dark blue when pH was adjusted to 2, and the red color was different from the original red in the pH range 10–12 [25, 26]. Therefore, for the present experiment, the pH of the solution was kept between 3 to 11. In order to investigate the effect of pH values, modified and unmodified HNTs (0.5 g) were respectively added to CR dye solution with concentration of 50 mg/L at room temperature. The uptake of CR dye as a function of hydrogen ion concentration was examined over a pH range of 3–11 and is shown in Figure 8.8. It is evident that the adsorption capacity is highly pH dependent and the maximum adsorption is found at pH=3 but more in modified HNTs as compared to unmodified HNTs. When the pH value was maintained in the range of 3–11 the adsorption capacity was found to decreases with the increase in pH.

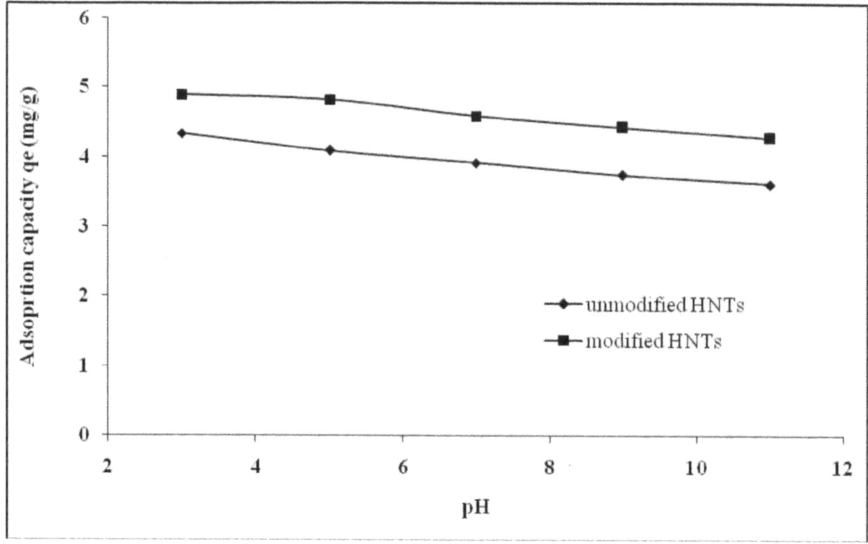

FIGURE 8.8 Effect of pH on CR dye adsorption capacity for unmodified and modified HNTs.

In aqueous solutions, CR dye anion is simple monovalent anion depending upon the pH of the solution. At low pH, the adsorbent surfaces become positively charged due to strong protonation, electrostatic force between the positively charged surface and the negatively charged CR dye, as well as the interaction between quaternary ammonium cations and CR dye in the internal nanotubes, will enhance the CR dye adsorption. However, when the pH value is less than 3, electrostatic interaction will reduce accordingly. With increase of pH, the degree of protonation of the surface reduces gradually and hence adsorption capacity decrease in the pH range of 3–6. Furthermore, the lower affinity of CR dye adsorption above pH 7 can be attributed to the strong competition between OH^- with CR dye since more OH^- anions are present. Similarly the removal efficiency of the CR dye decreases as the increases in pH of the respective solutions and again it proves the modified HNTs is more efficient than the unmodified HNTs.

8.3.9 ADSORPTION ISOTHERM

Equilibrium data in terms of adsorption isotherm is a basic requirement for the design of adsorption systems. The equilibrium removal of dyes was mathematically expressed in terms of Langmuir and Freundlich adsorption isotherms.

8.3.9.1 Langmuir Isotherm

The Langmuir equation is based on the assumption that maximum adsorption corresponds to saturated monolayer of the adsorbate molecule on the adsorbent surface [27]. In Langmuir equation,

$$\frac{C_e}{q_e} = \frac{1}{\beta Q_m} + \frac{C_e}{Q_m} \tag{1}$$

where, qe (mg/g) is the amount of adsorbed dye on the adsorbent surface, Ce (mg/L), is the equilibrium concentration of the dye in the solution, Q_m represents the maximum binding at the complete saturation of adsorbent binding sites, β is related to the energy of adsorption [28]. The values of β and Q_m can be obtained from slope and intercept of the linear plot of Ce v/s Ce/qe Figure 8.9, respectively. The values of Q_m and β are given in Table 8.1. It can be seen that the adsorption of the dye on modified HNTs followed the Langmuir isotherm.

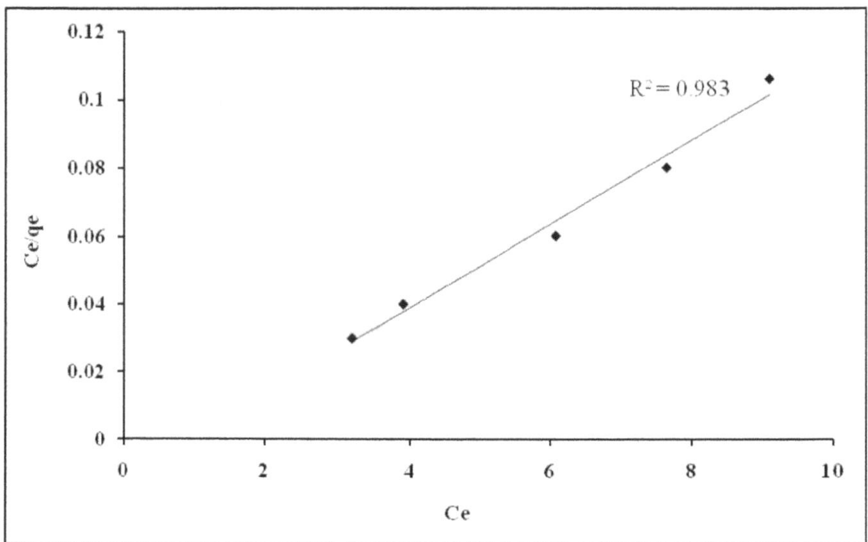

FIGURE 8.9 Langmuir sorption isotherms model.

TABLE 8.1 Langmuir and Freundlich Isotherm Constants of CR Dye Adsorption on Modified HNTs

	Langmuir isotherm	
Q_m	**β**	R^2
83.33	0.833	0.983
	Freundlich isotherm	
K_f	**n**	R^2
1.552	8.264	0.978

8.3.9.2 Freundlich Isotherm

Freundlich isotherm is an empirical equation employed to describe heterogeneous systems [27–29]. The Freundlich isotherm model proposes that sorption energy exponentially decreases on completion of the sorptional sites of adsorbent. The Freundlich isotherm is represented by equation

$$q_e = K_f C_e^{1/n} \tag{2}$$

where, K_f and n are the physical constants of the Freundlich adsorption isotherm. The values of n and K_f can be determined from the slope and intercept of the linear plot of ln qe vs ln Ce depicted in Figure 8.10.

The values of Freundlich constants are reported in Table 8.1. From the results it can be concluded that both the isotherms represent good adsorption capacity of the modified HNTs. Here the data best fitted in the Freundlich isotherm model.

8.3.10 ADSORPTION KINETICS

Sorption kinetics is determined to understand the mechanism of solute sorption onto a sorbent. The mechanism of solute sorption onto a sorbent can be expressed by various models. In this study, constants of sorption were determined using a pseudo-first and second-order equation of the Lagergen model.

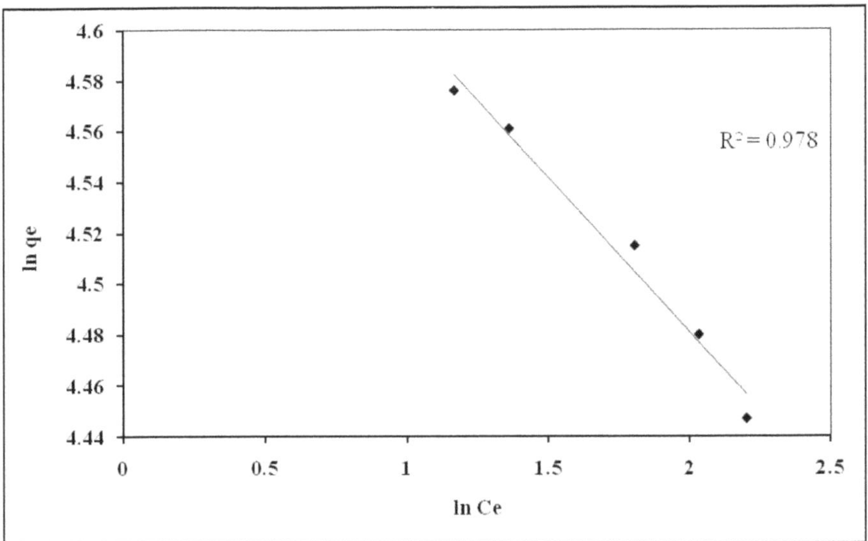

FIGURE 8.10 Freundlich sorption isotherm model.

8.3.10.1 Pseudo-First Order Kinetics

The Eq. (3) represents linear form of pseudo- first order is:

$$\frac{dq}{dt} = K \quad q_e - q \tag{3}$$

Where, qe and q are the amounts of adsorbed dye onto the modified HNTs at equilibrium and at time t, respectively. K_1represents the rate constant of first- order adsorption. The integrated form of Equation (3) is

$$\log(q_e - q) = \log q_e - \frac{t}{2.303} \tag{4}$$

The plot of log (qe – q) against t for the pseudo- first-order equation gives a linear relationship. Figure 8.11 depicts the plot of the linearized form of the pseudo- first-order equation. Values of K_1 and qe can be determined from the slope and intercept of this equation, respectively. Kinetic parameters and correlation coefficients of the first-order kinetic model are shown in Table 8.2.

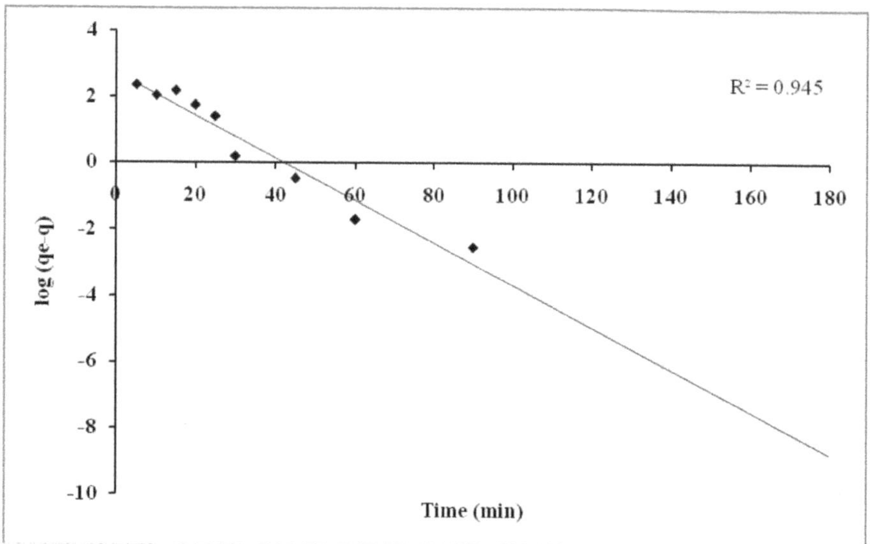

FIGURE 8.11 Adsorption kinetics of CR dye for pseudo-first-order.

TABLE 8.2 Constants for the First-Order and Second-Order Kinetics for the CR Dye Adsorption on the Modified HNTs

First-order kinetics		
K_1(min^{-1})	**qe (mg/g)**	**R^2**
0.108	1.82	0.945
Second-order kinetics		
K_2(g mg^{-1}/min) × 10^3	**qe (mg/g)**	**R^2**
8.1	111.1	0.99

8.3.10.2 Pseudo-Second Order Kinetics

For typical pseudo-second order adsorption kinetics, the kinetic rate equation is expressed as [28, 29]:

$$\frac{dq}{dt} = K_2(q_e - q)^2 \tag{5}$$

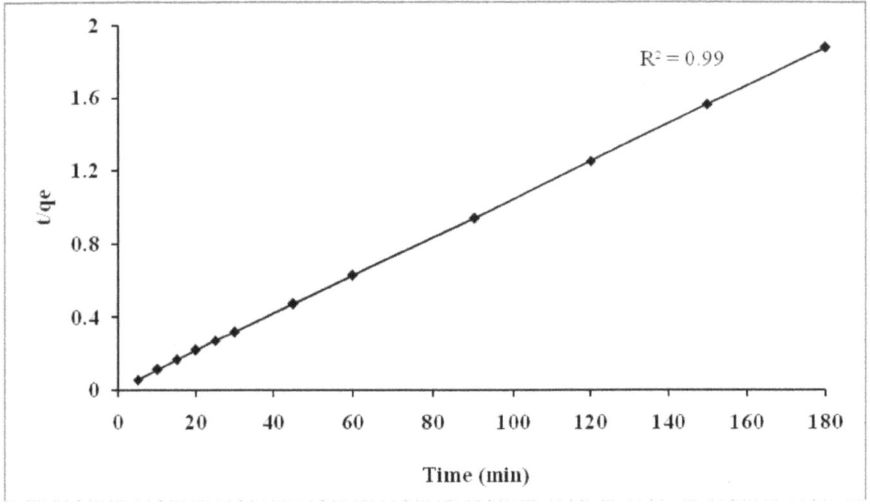

FIGURE 8.12 Adsorption kinetics of CR dye for pseudo-second-order.

where, K_2 is the rate constant of second-order adsorption. Integrating eq. (5) gives the simplified form represented as follows,

$$\frac{t}{q} = \frac{1}{K_2 q_e^2} + \frac{t}{q_e} \tag{6}$$

The plot of t/q vs t at various temperatures represented in Figure 8.12 shows a straight-line relationship between t/q and t. The values of K_1, K_2 (correlation coefficient) were calculated and shown in Table 8.2. From Table 8.2 the results of adsorption kinetics of dye show that the rates of sorption were found to conform to pseudo-second order kinetics.

8.4 CONCLUSIONS

This study attempt to investigate, the halloysite nanotubes (HNTs) modified with the surfactant of tetrabutyl ammonium chloride to form a new adsorbent by conventional method. Characterization of modified HNTs by FTIR, TEM and SEM showed that the surfactant was grafted onto the surface of HNTs. The thermal stability of material obtained from thermogravimetric analysis (TGA). The adsorption experimental results revealed that the adsorption capacity of the adsorbent (modified HNTs) decreased significantly by increasing pH. The variation in pH values has significant effect on adsorption and percentage removal of Congo Red (CR) dye. The removal efficiency was found to increases with increase in pH. The equilibrium data have been analyzed using Langmuir and Freundlich isotherm models and characteristic parameters for each have been determined. The experimental data fits with Langmuir adsorption isotherm. As well as pseudo-second order kinetics is well fitted for the adsorption data. Particularly, the experimental results show that the modified HNTs are evaluated as efficient adsorbent for anionic dyes with great adsorption rate as compared to unmodified HNTs.

KEYWORDS

- **Adsorption**
- **Anionic dye**
- **Halloysite nanotubes**
- **Modification**

REFERENCES

1. Liu, R., Zhang, B., Mei, D., Zhang, H., & Liu, J., Adsorption of methyl violet from aqueous solution by halloysite nanotubes. *Desalination.* 2011, *268,* 111–116.
2. Al-Degs, Y. S., El-Barghouthi, M. I., Khraisheh, M. A., Ahmad, M. N., & Allen, S. J., Effect of surface area, micropores, secondary micropores and mesopores volumes of activated carbons on reactive dyes adsorption from solution. *Sep. Purif. Technol.* 2004, *39,* 97–111.
3. Zhao, M., & Liu, P., Adsorption behavior of methylene blue on halloysite nanotubes. *Microporous Mesoporous Mater.* 2008, *112,* 419–424.
4. Errais, E., et al. Anionic RR120 dye adsorption onto raw clay: Surface properties and adsorption mechanism. *Colloids Surf., A.* 2012, *403,* 69–78.
5. Vimonses, V., Lei, S., Jin, B., Chow, C. W. K., Saint, C., Adsorption of Congo red by three Australian kaolins. *Appl. Clay Sci.* 2009, *43,* 465–472.
6. Crini, G., & Badot, P. M., Application of chitosan, a natural amino polysaccharide, for dye removal from aqueous solutions by adsorption processes using batch studies: A review of recent literature. *Prog. Polym. Sci.* 2008, *33,* 399.

7. Alteno, S., et al. Adsorption studies of Methylene Blue and phenol onto vetiver roots activated carbon prepared by chemical activation. *J. Hazard. Mater.* 2009, *165 (1–3),* 1029–1039.
8. Mall, I. D., Srivastava, V. C., & Agarwal, N. K., Removal of Orange-G and Methyl Violet dyes by adsorption onto bagasse fly ash-kinetic study and equilibrium isotherm analyzes. *Dyes Pigments.* 2006, *69(3)*, 210–223.
9. Jain, S., & Jayaram, R. V., Removal of basic dyes from aqueous solution by low-cost adsorbent: Wood apple shell (Feronia acidissima). *Desalination.* 2010, *250(3),* 921–927.
10. Tan, K. A., Morad, N., Teng, T. T., Norli, I., & Panneerselvam, P., Removal of Cationic Dye by Magnetic Nanoparticle (Fe_3O_4) Impregnated onto Activated Maize Cob Powder and Kinetic Study of Dye Waste Adsorption. *APCBEE Procedia.* 2012, *1*, 83–89.
11. Allen, S. J., Mckay, G., Porter, J. F., Adsorption isotherm models for basic dye adsorption by peat in single and binary component systems. *J. Colloid Interface Sci.* 2004, *280(2),* 322–333.
12. Sonawane, S. H., et al. Ultrasound assisted synthesis of polyacrylic acid–nanoclay nanocomposite and its application in sonosorption studies of malachite green dye. *Ultrason. Sonochem.* 2009, *16*, 351–355.
13. Gurses, A., et al. Determination of adsorptive properties of clay/water system: methylene blue sorption. *J. Colloid Interface Sci.* 2004, *269,* 310–314.
14. Errais, E., et al. Efficient anionic dye adsorption on natural untreated clay: Kinetic study and thermodynamic parameters. *Desalination.* 2011, *275,* 74–81.
15. Eren, E., & Afsin, B., Investigation of a basic dye adsorption from aqueous solution onto raw and pre-treated bentonite surfaces. *Dyes Pigments,* 2008, *76(1),* 220–225.
16. Wang, Q., Zhang, J., Zheng, Y., & Wang, A., Adsorption and release of ofloxacin from acid- and heat-treated halloysite. *Colloids Surf., B.* 2014, *113,* 51–58.
17. Nicolini, K. P., Regina, C., Fukamachi, B., Wypych, F., & Mangrich, A. S., Dehydrated halloysite intercalated mechanochemically with urea: Thermal behavior and structural aspects. *J. Colloid Interface Sci.* 2009, *338,* 474–479.
18. White, R. D., Bavykin, D. V., & Walsh, F. C., The stability of halloysite nanotubes in acidic and alkaline aqueous suspensions. *Nanotechnol* .2012, *23,* 0657051–2.
19. Horvath, E., Kristof, J., Kurdi, R., Mako, E., & Khunova, V., Study of urea intercalation into halloysite by thermoanalytical and spectroscopic techniques. *J. Therm. Anal. Calorim.* 2011, *105,* 53–59.
20. Yah, W. O., Takahara, A., & Lvov, Y. M., Selective Modification of Halloysite Lumen with Octadecyl phosphonic Acid: New Inorganic Tubular Micelle. *Am. Chem. Soc.* 2012, *134,* 1853−1859.
21. Jinhua, W., et al. Rapid adsorption of Cr(VI) on modified halloysite nanotubes. *Desalination.* 2010, *259*, 22–28.
22. Nicolini, K. P., Fukamachi, C. R., Wypych, F., & Mangrich, A. S., Dehydrated halloysite intercalated mechanochemically with urea: Thermal behavior and structural aspects. *J. Colloid Interface Sci.* 2009, *338,* 474–479.
23. Sonawane, S. H., et al. Combined effect of ultrasound and nanoclay on adsorption of phenol. *Ultrason. Sonochem.* 2008, *15*, 1033–1037.
24. Xie, Y., Qian, D., Wu, D., & Ma, X., Magnetic Halloysite nanotubes/iron oxide composites for the adsorption of dyes, *Chem. Eng. J.* 2011, *168*, 959–963.

25. Purkait, M. K., Maiti, A., & DasGupta, S., Removal of Congo red using activated carbon and its regeneration, *J. Hazard. Mater.* 2007, *145,* 287–295.
26. Lorenc-Grabowska, E., & Gryglewicz, G., Adsorption characteristics of Congo red on coal-based mesoporous activated carbon. *Dyes Pigments.* 2007, *74,* 34–40.
27. Mak, S., & Chem, D., Fast adsorption of methylene blue on polyacrylic acid-bound iron oxide magnetic nanoparticles. *Dyes pigments.* 2004, *61,* 93.
28. Li, S., Removal of crystal violet from aqueous solution by sorption into semi interpenetrated networks hydrogels constituted of poly(acrylic acid-acrylamide methacylate) and amylase. *Bioresour. Technol.* 2010, *101*, 2197–2202.
29. Emad, N., Qada, J., & Stephen, G., Adsorption of Methylene Blue onto activated carbon produced from steam activated bituminous coal: A study of equilibrium adsorption isotherm, *Chem. Eng. J.* 2006, *124,* 103.

CHAPTER 9

KINETIC STUDY OF ADSORPTION OF NICKEL ON GNS/δ-MnO_2

M. P. DEOSARKAR,[1] S. VARMA,[1] D. SARODE,[1] S. WAKALE,[1] and B. A. BHANVASE[2]

[1]*Chemical Engineering Department, Vishwakarma Institute of Technology, Pune, Maharashtra, India*

[2]*Chemical Engineering Department, Laxminarayan Institute of Technology, Nagpur, Maharashtra, India*

CONTENTS

9.1 Introduction.......... 142
9.2 Experimental Part.......... 144
9.2.1 Chemicals.......... 144
9.2.2 Experimental Setup and Procedure.......... 144
9.2.2.1 Synthesis of Adsorbent.......... 144
9.2.2.2 Batch Adsorption of Ni (II) [14].......... 145
9.3 Results and Discussion.......... 146
9.3.1 FTIR Analysis.......... 146
9.3.2 FT-Raman Analysis.......... 147
9.3.3 XRD Analysis.......... 147
9.3.4 SEM Analysis.......... 149
9.3.5 Effect of Different Parameters on Adsorption of Ni.......... 150
9.3.5.1 Effect of pH.......... 150
9.3.5.2 Effect of Temperature.......... 150

9.3.6 Adsorption Kinetics and Thermodynamics Study............ 151
9.3.6.1 Adsorption Isotherm.. 151
9.3.6.2 Adsorption Kinetics... 155
9.3.6.3 Thermodynamic Study 157
9.4 Conclusion 158
Keywords 159
References.. 159

9.1 INTRODUCTION

Now-a-days, heavy metal pollution is a global concern. Nickel being one of the noxious non-biodegradable and toxic heavy metal profoundly found in industrial wastewater. Its prominent sources are industrial exploration or processing operations like galvanization, smelting, ore mining, dyeing, battery manufacturing or protective finishing of metals [1]. Such dissolved nickel at certain concentration level can interfere with human body and can cause various types of cancer depending on the sources of entry into the human body also impairment of certain body functions like renal edema [1]. It can also produce gastrointestinal disorder when ingested through food or drinking water. For these reasons the American Conference of Governmental Industrial Hygenists (ACGIH) has identified nickel as possible carcinogen [1]. The permissible exposure limit for airborne nickel compounds is 1 milligram per cubic meter (mg/m^3) for an 8 TWA concentration, as prescribed by the OSHA. The Water Sanitation and Hygiene (WSH), the arm of WHO has established toxic PEL concentration of nickel as 1 mg/m^3 for Ni(II) in the form of insoluble compounds, soluble compounds of Ni(II) as 0.1 mg/m^3, nickel carbonyl as 0.05–0.12 mg/m^3 and nickel sulfide as 1.0 mg/m^3 [2].

For removal of Nickel from wastewater, methods used are following ion exchange techniques, chemical oxidation &reduction, chemical precipitation, membrane technology, filtration, electrochemical treatment or adsorption [3]. Each of the technique has its merits and demerits but what defines whether the method is applicable and most effective for certain separation is its suitability according to the different factors governing the conditions at which the effluent metals are released from the processing facilities such as pH, temperature, its intrinsic chemical properties and the cost of separation. Adsorption being the most cost effective method for dilutes, for example, wastewater containing trace amount of nickel, which is general case for the wastewater effluent streams from nickel processing industries [4]. It implies that adsorption is more suitable and can be easily implemented owing to its cost effective attributes. The primary requirement of an economical adsorption is selectivity, high adsorption capacity and lower operating cost. Some of the powerful adsorbents are NCS (Natural iron oxide Coated Sand) [5], ZrO-kaolinite [6], Nano-structured Hydrous Titanium (IV) oxide [7], peat [8], Multi-walled Carbon Nanotubes [9] or other natural materials with suitable adsorptive strength. Its now essential to develop a easily accessible, low cost, high adsorption capacity material for wastewater treatment that made remediate the environmental problems associated with nickel. In last few years several studies have been conducted to remove heavy metals Pb(II), Cu(II), Ni(II), Co(II), etc. from wastewater by using metal oxides such as $TiO_{2,}$ MnO_2, Al_2O_3, Fe_2O_3, and other composite metal oxides. Among them MnO_2 with different polymorphic phases (a, b, g and d-type) have stick out more attention particularly δ-MnO_2 with it chemical and physical properties. However, as an adsorbent it is susceptible to easy agglomeration, which yields poor dispersion and significant reduction in specific surface area [10]. When in conjunction with a strong carrier it has been observed that δ-MnO_2 exhibits superior adsorptive behavior [11].

Graphene Nanosheets (GNS) being honeycomb shaped array of SP^2 bonded carbon has a planar structure and a very high surface area. In contrast to conventional carbon materials such an activated carbon or carbon nanotubes, its specific surface area doesn't necessarily depends upon pores in solid state and thus makes it an excellent option for preparation of career matrix. The specific surface area of δ-MnO_2 can be conserved or rather can be enhanced by the application of GNS as carrier matrix. The metal nanocomposites of GNS has been synthesized using a microwave-assisted method and its adsorption characteristics along with effect of pH and temperature on the extent of adsorption were studied by carrying out a series of adsorption and desorption experiments in a laboratory. In this report the kinetic behavior of adsorption is extensively described where the experimental value of amount of adsorbed nickel derived from kinetic study is compared against the calculated value obtained from suitable adsorption isotherm.

9.2 EXPERIMENTAL PART

9.2.1 CHEMICALS

Graphite, Hydrazine solution, EDTA, $KMnO_4$, H_2SO_4, H_2O_2, HCl, Ethanol, De-ionized water was used to prepare all the solutions.

9.2.2 EXPERIMENTAL SETUP AND PROCEDURE

9.2.2.1 Synthesis of Adsorbent

A) GNS/δ-MnO_2

Graphite Oxide

For synthesis of nanosheets of graphene the chemical exfoliation method was used where elemental graphite was oxidized to form its oxide, which on reduction provides planar graphene sheets. The Hummers-Offeman method [12] was followed for synthesizing the said oxide in lab. About 10 g of lab grade graphite powder was dissolved in concentrated H_2SO_4 (230 mL, and dry ice bath) at very low temperature (<280–290 K) and potassium permanganate ($KMnO_4$, 30 g) was gradually added with continuous stirring and cooling using dry ice bath. Upon formation of a homogeneous suspension, dry ice bath was replaced by a water bath and the mixture was heated to 308 K for 30 min with gas release under continuous stirring. After that 460 mL of de-ionized water was added, this produced a rapid increase in solution temperature up to a maximum of 371 K. The process was allowed to continue for next 40 min for enhancing the degree of oxidation of the graphite oxide product. The resultant bright-yellow suspension was terminated by addition of more distilled water (140 mL) followed by hydrogen peroxide solution (H_2O_2, 30%, 30 mL). The solid product was separated by centrifugation at 3000 rpm and washed initially with 5% HCl until sulfate ions were no longer detectable with barium chloride. Solid product was then washed three times with acetone and air dried overnight at 338 K. After sonication for 30 min, the graphite oxide was transformed to graphene oxide.

Graphene (GNS) [11]
The synthesized graphene oxide powder (25 mg) was dissolved in 200 mL de-ionized water using magnetic stirring at 200 rpm for 10 min which yielded an inhomogeneous brown suspension which was when further treated with hydrazine solution (1:5, volume ratio of hydrazine to de-ionized water), a strong reductant, under ultrasonication (30 min) and yielded reduced graphene sheets after drying at 373 K.

GNS/δ-MnO_2 [13]
The GNS/MnO_2 was synthesized by a redox reaction assisted by microwave irradiation [2]. A homogeneous GNS suspension was prepared by dissolving 1.65 g of reduced graphene into 1 liter of de-ionized water. 100 mL of this suspension was subjected to ultrasonication for 60 min. Then potassium permanganate powder (0.9482 g) was added into said suspension, and stirred for 10 min. Subsequently, the consequential suspension was then heated using a household microwave oven for 5 min. The brownish black deposit obtained was subjected to filtration with distilled, de-ionized water and ethanol and was washed many times. It was then dried at 373 K for more than 8 h in a vacuum oven.

B) Dry MnO_2 [14]
Experiments to remove Ni (II) from wastewater are done by using 2 adsorbents one is GNS/δ-MnO_2 and other is dry δ-MnO_2. To compare adsorption phenomenon between 2 adsorbents δ-MnO_2 was synthesized by dispersing 18 g $KMnO_4$ in 200 mL deionized water under 80–90°C. Then 500 mL 1:1 (v/v) HCl was added from a fast dripping burette. The suspension obtained was centrifuged, and washed many times with distilled water. It was allowed to dry 373 K for 6 h in a vacuum oven. Black dry powder of MnO_2 is obtained.

9.2.2.2 Batch Adsorption of Ni (II) [14]

One-liter aqueous solution $NiCl_2 \cdot 6H_2O$ is prepared by dissolving 59.4275 gm. of $NiCl_2 \cdot 6H_2O$ in 1 L of water, concentration of this obtained solution is 0.5 N [2]. Two beakers were prepared pouring 250 mL of above solution in each one. 50 mg. of GNS/δ-MnO_2 was dissolved in one beaker and equal quantity of dry δ-MnO_2 is mixed in its counter-part to perform adsorption of Ni by using both the adsorbents separately and to evaluate their adsorption characteristics. Both the beakers were kept for stirring for almost 3 h to warrant equilibrium is attained [2]. The contents of beakers were subjected to centrifugation at 4000 rpm for 5 min. The solid deposit obtained was analyzed for determining amount of nickel adsorbed using UV–Visible Spectrophotometer at wavelength 530 nm [14].

9.3 RESULTS AND DISCUSSION

9.3.1 FTIR ANALYSIS

In order to confirm the reduction mechanisms forming the composites, FTIR analysis was carried out of GO, GNS and GNS/δ-MnO_2 whose spectra are as shown in Figure 9.1.

The presence of different types of oxygen functionalities in GO is confirmed at 3424 cm^{-1} (O-H stretching vibrations), at 1767 cm^{-1} (stretching

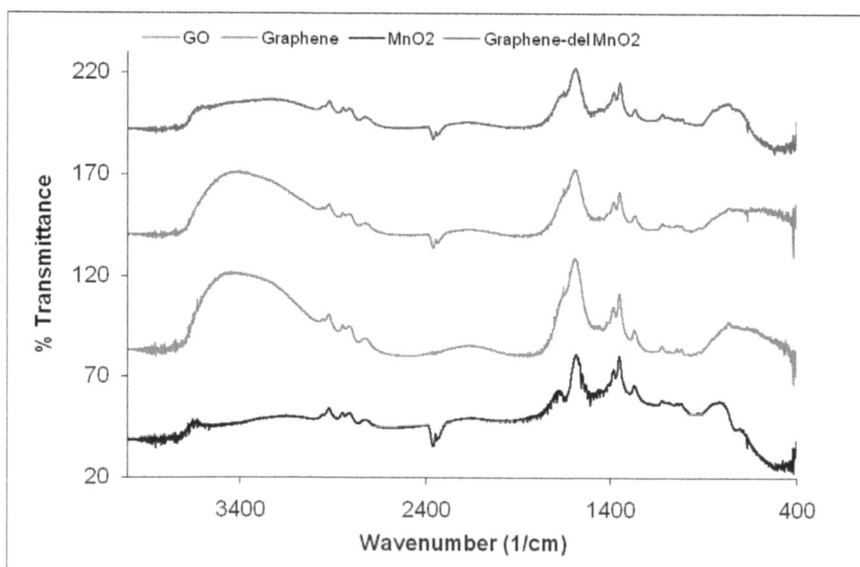

FIGURE 9.1 FTIR Spectra of GO, Graphene, Graphene/δ-MnO_2 and Dry MnO_2.

vibrations of C=O), at 1643 cm^{-1} (skeletal vibrations from unoxidized graphitic domains), at 1295 cm^{-1} (stretching vibrations of C-OH) and at 1032 cm^{-1} (C-O stretching vibrations). FTIR peak of reduced graphene or GNS shows that O-H stretching vibrations observed at 3424 cm^{-1} were significantly reduced to 2980 cm^{-1} due to deoxygenation, however, stretching vibrations from C=O at 1767 cm^{-1} were increased by very small extent (1084 cm^{-1}) while C-O stretching vibrations at 1067 cm^{-1} were damped. The dampening and subsequent shift of the peak shows the involvement of the O-H group in the reduction. Absorptions due to C=O group (1725 cm^{-1}) are decreased in intensity for obvious reasons whereas the absorptions at 1635 cm^{-1} (O-H groups) are nearly absent suggesting that the carboxyl group on the surface of GO have been reduced or modified by Mn particles. Furthermore, two new peaks located at 521 cm^{-1} observed in spectra of GNS/δ-MnO_2 can be attributed to Mn-O vibrations. While, The spectra of dry MnO_2 is almost horizontal with exception of number of vibrations at 525 cm^{-1}, 515 cm^{-1}, 495 cm^{-1}, and 485 cm^{-1} indicating presence of strong Mn-O vibrations.

9.3.2 FT-RAMAN ANALYSIS

The Figure 9.2 shows the Raman spectra of GNS and GNS/δ-MnO_2, which exhibit two typical peaks. The D-band located at 1301 cm^{-1} corresponds to the defects and disorder in organic structure of graphene layers which is most common in any exfoliated nanosheet. The ratio of intensities of the D-band and G-band (I_D/I_G) is approximately 1.34 for GNS and 1.36 for GNS/δ-MnO_2. This slight enhancement in I_D/I_G ratio of the metal nanocomposite of GNS indicates an almost even distribution of Mn-O groups on the planar GNS.

9.3.3 XRD ANALYSIS

The structural properties of synthesized metal nanocomposite were analyzed and compared against those of GNS by XRD analysis (Figure 9.3), which validates that monoclinic lamellar structure of MnO_2 is retained for both GNS/δ-MnO_2 and dry-MnO_2. The diffraction peak for GNS at 12^0 is similar to the standard GO peak. The characteristics peaks of MnO_2 at 12°, 18°, 27.5°, 29°, 31° and 41.24° corresponding to the (001), (002),

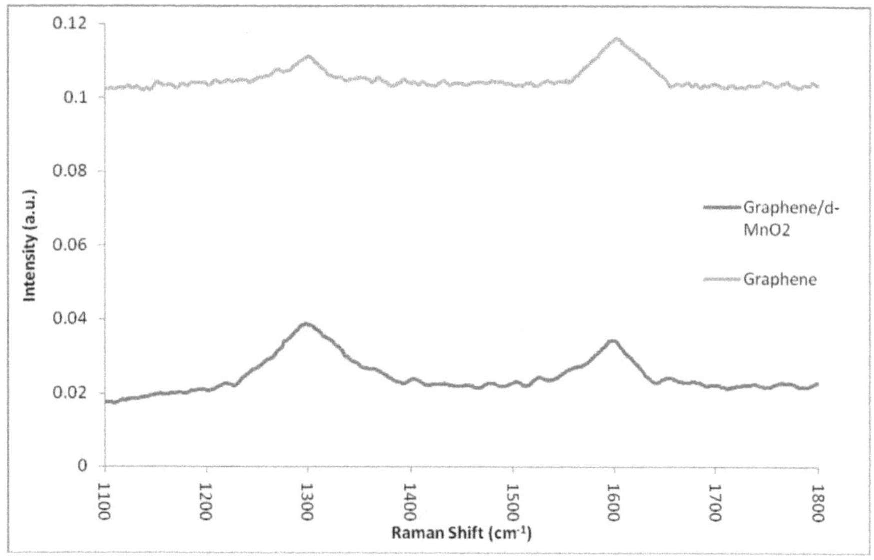

FIGURE 9.2 FT-Raman Spectra of Graphene and Graphene/δ-MnO_2.

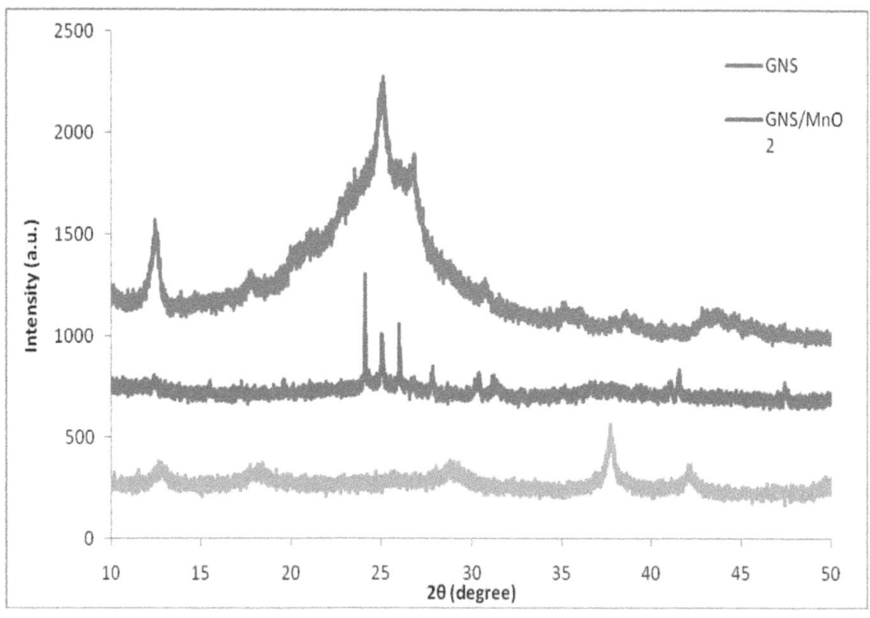

FIGURE 9.3 X-ray diffraction pattern of Graphene, Graphene/δ-MnO_2 and Dry MnO_2.

(100), (201) and (020) planes of the magnetic spinal structure were clearly evident. The low intensity and sharp diffraction of GNS/δ-MnO_2 were indexed to δ-type MnO_2. Furthermore, no evident peak corresponding to GO or graphite except at 12° strongly suggests that the bundle of sheets graphite or GO are exfoliated into GNS by a greater extent.

9.3.4 SEM ANALYSIS

SEM images (Figure 9.4) shows that amorphous MnO_2 has flower-like formation with a maximum diameter of 300 nm. The planer sheets of graphene are observed to have wrinkles and folds of edges. Its thickness ranges from 3 to 6 nm, which corresponds to an approximately 10–20 layer stacking of monoatomic graphene. The length of GNS stack is rather varying from 300–500 nm.

FIGURE 9.4 SEM Images of Graphene (a–c) and Graphene/δ-MnO_2 (d–f).

KMnO4 and Carbon (sp2) undergoes redox reaction in pH neutral solution is as follows:

$$4\ MnO_{4-} + 3C + H_2O \rightarrow 4\ MnO_2 + CO_3^{2-} + 2HCO_3$$

The reaction can be characterized by the carbon substrate acting as sacrificial reducing agent, which transforms soluble MnO_{4-} into insoluble MnO_2 (5–10 nm) flocks which are believed to get deposited on the surface of reducing agent itself forming the desired nonmaterial.

9.3.5 EFFECT OF DIFFERENT PARAMETERS ON ADSORPTION OF NI

9.3.5.1 Effect of pH

During the adsorption process, positively charged Ni(II) ions accumulate the active sites on the planer surface of the GNS/δ-MnO_2. At the same time, vacant sites can also get accumulated by the positively charged H^+ ions. So pH of the solution is deemed as a key parameter governing metal ion adsorption. It was studied that the adsorption of Ni(II) is strongly dependent on pH value independent of adsorbent (Figure 9.5). At the lower pH values, sorption capability for both samples is very small, which indicates the competition of an excess of H^+ ions with Ni(II) for available bonding sites is low.

At pH 5–8, the sorption is found to increases brusquely with increase in the pH value. The said effect can be described by taking into account the surface charge of GNS/δ-MnO_2& the extent of ionization. But, as the pH exceeds 8–8.5 value, the Ni(II) starts to precipitate and the extent of precipitation was observed to increase with pH value.

9.3.5.2 Effect of Temperature

To study the effect of temperature on the adsorption capacity, (c_e/q_e) vs (c_e) graphs at different temperatures are plotted referring the equation given Langmuir isotherms given below:

$$(c_e/q_e) = (1/q_m)*c_e + (K_L/q_m) \quad (1)$$

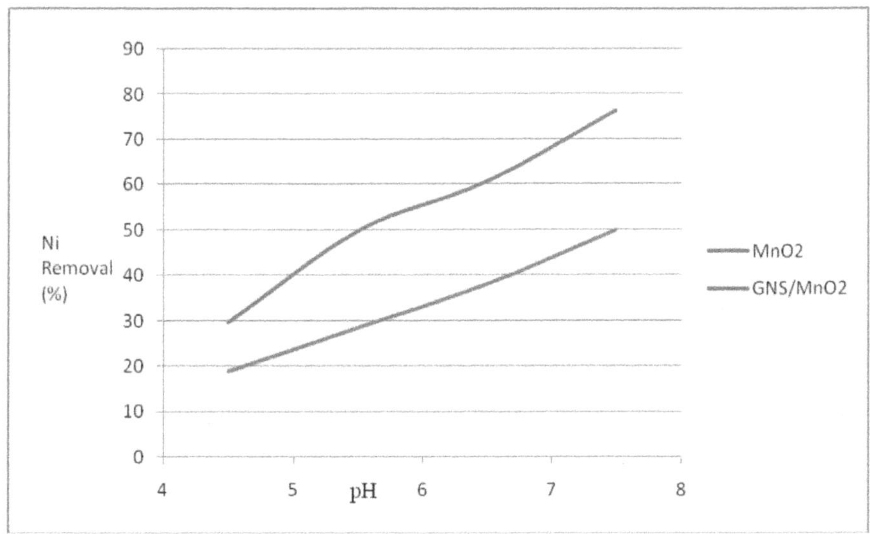

FIGURE 9.5 Effect of pH on the adsorption capacity.

where, c_e = concentration of Ni in the solution at equilibrium; q_e = amount of Ni adsorbed at equilibrium; q_m = maximum adsorption capacity; and K_L = Langmuir adsorption constant.

From the above graph (Figure 9.6), it is seen that the slope of the line decreases as the temperature of the solution goes on increasing. As the Maximum adsorption capacity is the reciprocal of the slope of the line, adsorption capacity increases as the temperature increases.

9.3.6 ADSORPTION KINETICS AND THERMODYNAMICS STUDY

9.3.6.1 Adsorption Isotherm

The adsorption isotherm is of primary significance in the design of adsorption based separation or purification system, which signifies the partition of Ni(II) ions between the adsorbent and substrate at equilibrium conditions as a function of increasing ions concentration. For studying the behavior of Adsorption with change in temperature, Langmuir isotherms are plotted for different temperatures where q_m and K_L constants

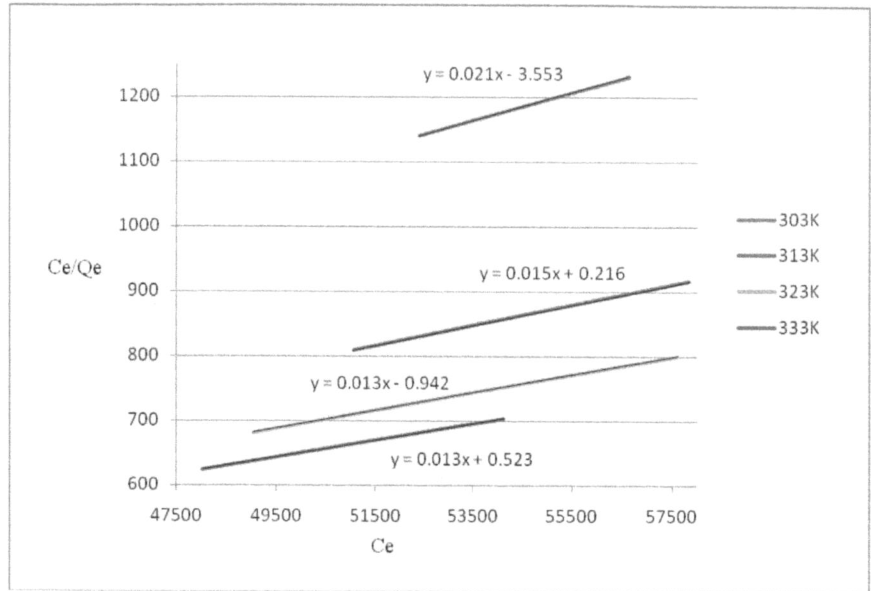

FIGURE 9.6 Effect of temperature on the adsorption capacity.

are calculated from the slope and intercept of the plot are presented in Table 9.1.

It is found that, at various temperatures adsorption of Ni(II) decreases by following order of GNS/δ-MnO_2 > MnO_2, for example, adsorption capacity of GNS/δ-MnO_2 is 1.5 times that of MnO_2. With increasing temperature from 303 K to 333 K, sorption capacity of GNS/δ-MnO_2 increase from 45.67 to 76.92 mg/g. High regression correlation coefficient (>0.99) states that Ni (II) adsorption on GNS/δ-MnO_2 is monolayer and forms a layer metal ion on the surface of adsorbent.

TABLE 9.1 Adsorption of Nickel on GNS/δ-MnO_2 and MnO_2: Parameters of Langmuir Isotherm

Temperature		303 K	313 K	323 K	333 K
GNS/δ-MnO_2	q_m(mg g^{-1})	45.97	63.13	72.01	76.92
	K_L (L mg^{-1})	0.31	0.458	0.628	0.641
MnO_2	q_m(mg g^{-1})	30.67	38.30	43.18	45.36
	K_L (L mg^{-1})	0.327	0.301	0.364	0.39

The equations formed by Langmuir adsorption isotherm results in linear plots when it is plotted at various temperatures. These are plots are given in Figures 9.7–9.10.

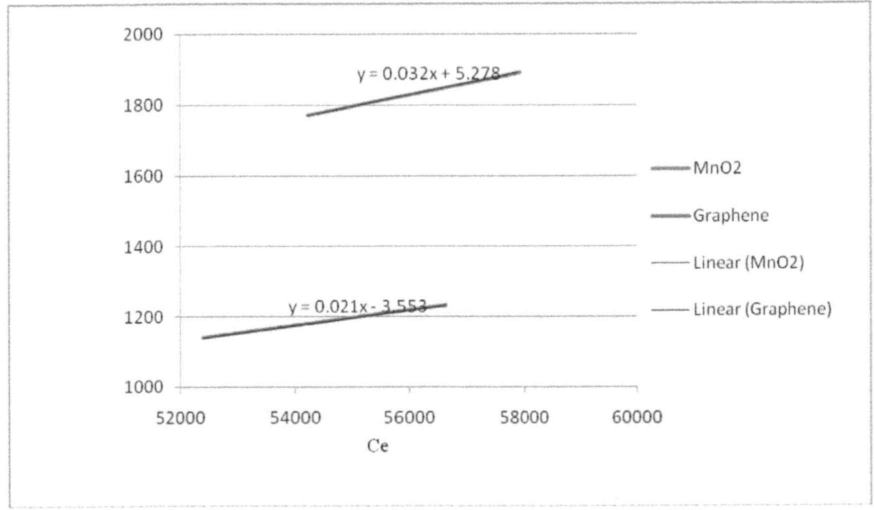

FIGURE 9.7 Langmuir isotherm for GNS/δ-MnO_2 composite and MnO_2 at 303 K.

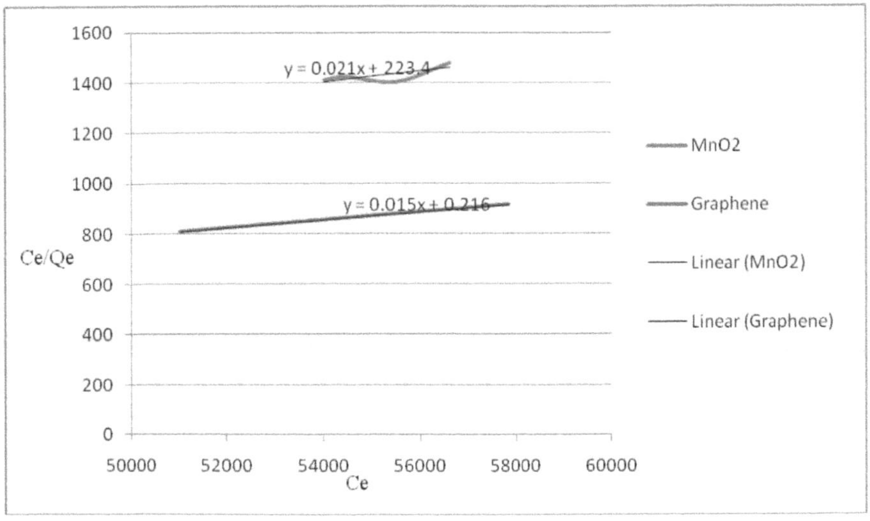

FIGURE 9.8 Langmuir isotherm for GNS/δ-MnO_2 composite and MnO_2 at 313 K.

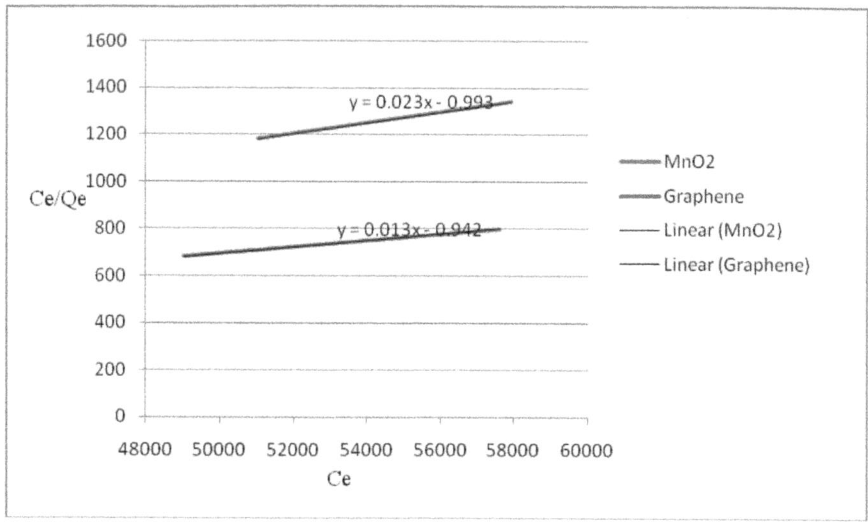

FIGURE 9.9 Langmuir isotherm for GNS/δ-MnO_2 composite and MnO_2 at 323 K.

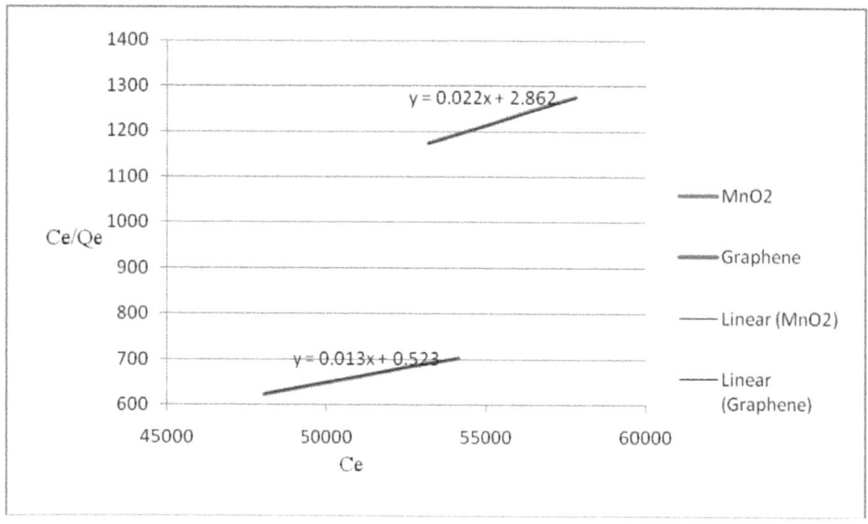

FIGURE 9.10 Langmuir isotherm for GNS/δ-MnO_2 composite and MnO_2 at 333 K.

9.3.6.2 Adsorption Kinetics

To study the effect of contact time on the rate of adsorption, q_t vs t (i.e., amount of Ni adsorbed vs time) is plotted. During the initial 60 mins, it is found that the rate of adsorption is very high. After that, rate of adsorption is gradually lowered indicating dynamic equilibrium is established. Based on the above plot, the shaking time of 3 h is considered as a suitable time for adsorption experiments. Pseudo-first order and pseudo-second order adsorption kinetics were studied to kinetic behavior of the adsorption. Based on the following two equations, graphs were plotted to calculate the regression correlation coefficients (R^2) in case of pseudo-first order (Eq. (2)) and pseudo-second order (Eq. (3)) adsorption kinetics.

$$\log (qe - qt) = \log qe - (k_1 t)/2.303 \quad (2)$$

$$(t/qt) = 1/(k_2 \times qe^2) + (1/qe) \times t \quad (3)$$

As seen from Figure 9.11 (t/qt vs t) and Table 9.2, pseudo-second-order kinetic model is well fitted for the adsorption due to the high ($R^2 > 0.99$) regression correlation coefficient as compared to pseudo-first order

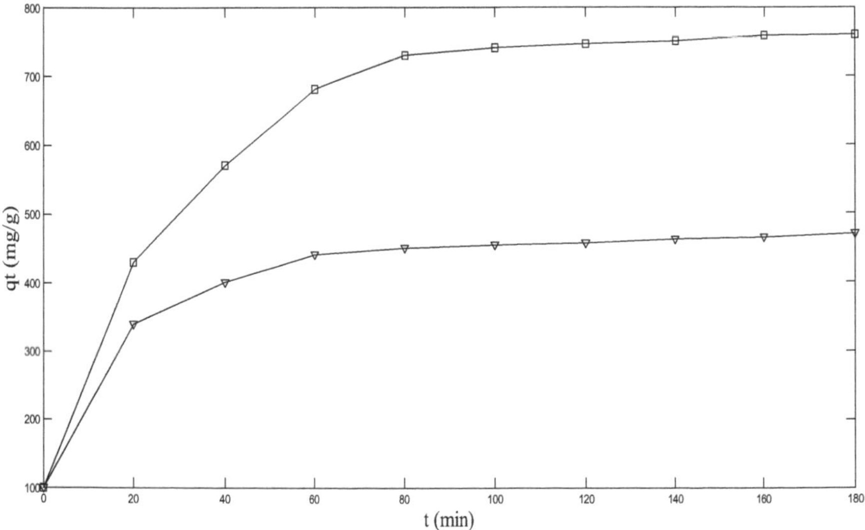

FIGURE 9.11 Amount of Nickel adsorbed onto GNS/δ-MnO_2 and MnO_2 VS time.

TABLE 9.2 Kinetics Constant of Adsorbed Ni (II)on GNS/δ-MnO_2 Composite and MnO_2 at 303 K

Sr. No.		1.	2.
Adsorbents		GNS/δ-MnO_2	MnO_2
(q_e)exp (mg g^{-1})		45.97	30.63
Pseudo 1st order kinetics	(q_e)cal (mg g^{-1})	29.37	16.57
	K_1 (min^{-1})	0.00085	0.00210
	R^2	0.8326	0.7934
Pseudo 2nd order kinetics	(q_e)cal (mg g^{-1})	46.45	28.98
	K_2 (g mg^{-1} min^{-1})	0.00893	0.0197
	R^2	0.9993	0.9998

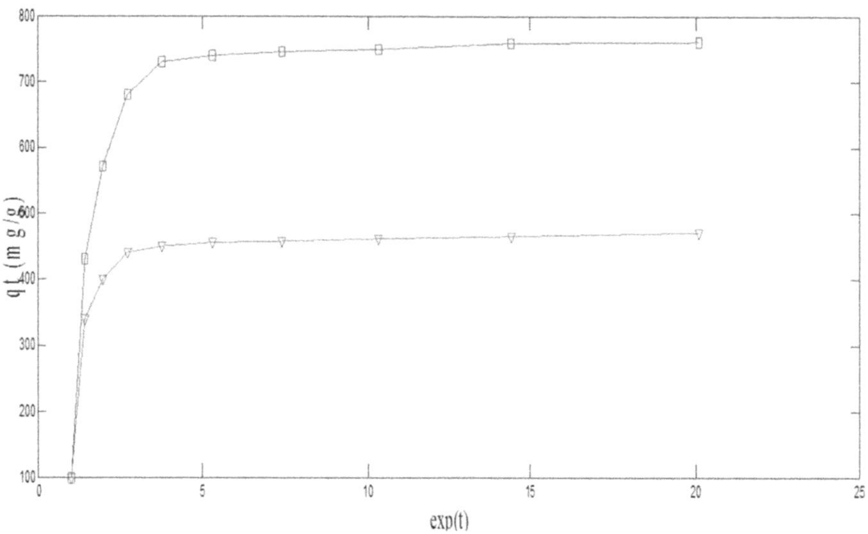

FIGURE 9.12 Graphical representation of pseudo-first order reaction.

kinetics. Also, the theoretical and experimental values of qe are approximately same for pseudo-second order kinetics.

By taking k_2 as the adsorption velocity, the constant k_2 (0.0197) for pure MnO_2 is higher than that of GNS/δ-MnO_2 (0.00893). After the equilibrium, 77.04% and 47.17% nickel is found to be adsorbed for GNS/δ-MnO_2

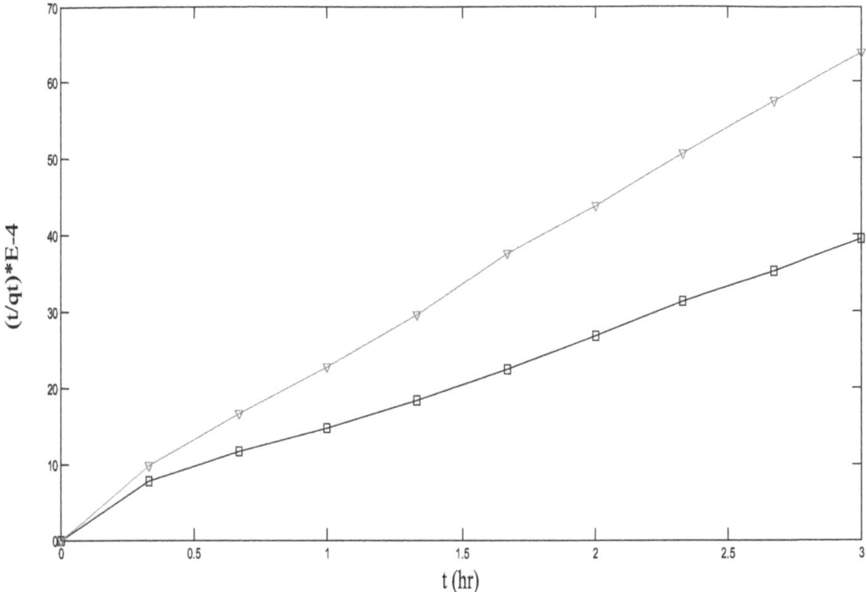

FIGURE 9.13 Graphical representation of pseudo-second order reaction.

and MnO_2 respectively concluding that the chemical sorption as a part of adsorption process.

9.3.6.3 Thermodynamic Study

To study the thermodynamic parameters, the graph of ln qm versus T^{-1} is plotted. The change in enthalpy and entropy (ΔH and ΔS) of the adsorption can be obtained from the slope and intercept of the plot (Figure 9.14). Table 9.3 shows the calculated values of these changes in the energy of Ni(II) adsorptionon GNS/δ-MnO_2 and MnO_2.

From the slope and intercept of the above plots, values of ΔH, ΔS and ΔG are calculated. ΔH and ΔS are found to be positive confirming the endothermic nature of reaction & increase in randomness at the solid–liquid interface during adsorption. While change in Gibbs energy is found to be –ve (–2.7 to −11.1 KJ mol^{-1}) shows the spontaneous nature of the adsorption. As the temperature increases ΔG values decreases confirms that the adsorption can be carried out at higher temperatures.

TABLE 9.3 Thermodynamic Parameters of Ni(II) Adsorption Onto GNS/δ-MnO_2 and MnO_2 and MnO_2

Thermodynamic parameters		ln (q_m)	ΔG (KJ mol^{-1})	ΔS (KJ mol^{-1} K^{-1})	ΔH (KJ mol^{-1})
GNS/δ-MnO_2	303 K	3.8258	−9.6377	79.31	14.23
	313 K	4.1452	−10.7869		
	323 K	4.2768	−11.48		
	333 K	4.3428	−12.0233		
MnO_2	303 K	3.4233	−8.6237	64.89	10.94
	313 K	3.6455	−9.4866		
	323 K	3.7654	−10.1116		
	333 K	3.8146	−10.5609		

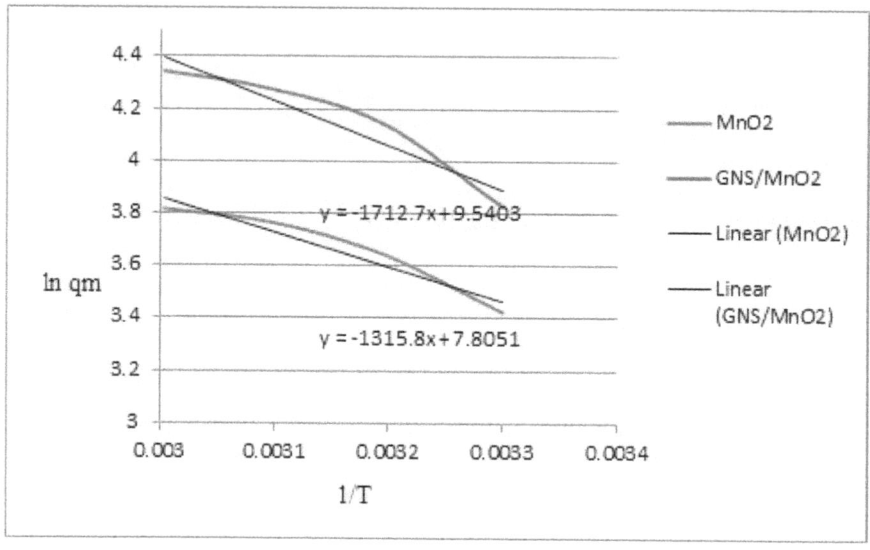

FIGURE 9.14 Plot of ln qm versus T−1 for GNS/δ-MnO_2 and MnO_2.

9.4 CONCLUSION

Microwave irradiation method was used to prepare GNS/δ-MnO_2 which was later used as adsorbent for the adsorption of Nickel from the aqueous solution. It is found that the adsorption capacity of Ni(II) for GNS/δ-MnO_2 is 1.5 time higher than that of MnO_2, which mainly comes from MnO_2

nanoparticles (5–10 mm) growth on the surfaces of GNS. According to linear behavior of Langmuir isotherm, pseudo second order kinetic is found to be well fitted for the Ni(II) adsorption. As a result of thermodynamic study, it is found that the reaction taking place is endothermic (+ve values of ΔH and ΔS) and more disturbance at the solid-liquid interface is observed. Whereas, the negative values of ΔG dictate the spontaneous nature adsorption process. The results presented in this work indicate that GNS/δ-MnO_2 composite as a promising adsorbent has great potential for the removal of metal ions from wastewater.

KEYWORDS

- **Adsorption**
- **Chemical exfoliation**
- **GNS/δ-MnO_2**
- **Graphene**
- **Metal nanocomposites**
- **Nickel**

REFERENCES

1. Compel, M., & Nikel, G., Nickel: a review of its sources & environmental toxicology. *Pol. J. of Environ. Stud.* 2006, *15*, 375–382.
2. Ren, Y., Yana, N., Wena, Q., Fana, Z., Weira, T., Zhanga, M., & Mab, J., Graphene/δ-MnO_2 composite as adsorbent for the removal of nickel ions from wastewater. *Chem. Eng. J.* 2011, *175,* 1–7.
3. Kudesia, V. P., Water pollution – toxicology of metals, Science Report SC050021.
4. Radenvoic, A., Malina, J., & Strkalj, A., Removal of Ni(II) from aqueous solution by low cost adsorbents. *Hol. Approach. Environ.* 2011, *1*, 109–120.
5. Boujelben, N., & Bouzid, J. Adsorption of nickel and copper onto natural iron coated sand from aqueous solutions: study in single and binary systems. *J. Hazard. Mater.* 2009, *163*, 376–382.
6. Bhattacharya, K. G., & Gupta, S. S., Adsorption of Fe(III), Co(II) and Ni(II) on ZrO-Kaolinite and ZrO-montmorillonite surfaces in aqueous medium. *Colloid. Surf. A: Physiochem. Eng. Aspects* 2008, *317*, 71–79.
7. Debnath, S., & Ghosh, U., Nanostructured hydrous titanium (IV) oxide: synthesis, characterization and Ni (II) adsorption behavior. *Chem. Eng. J.* 2009, *2–3*, 480–491

8. Sen Gupta, B., & Curran, M., Adsorption characteristics of Cu and Ni on Irish peat moss. *J. Environ. Manag.* 2009, *90*, 954–960.
9. Yang, S., & Hu, J., Adsorption of Ni (II) on oxidized multi-walled carbon-nanotubes: effect of contact time, pH, foreign ions and PAA. *Chem. Eng. J.* 2009, *166,* 109–116.
10. Tripathy, S. S., Adsorption of CO_2^+, Ni^{2+}, Cu^{2+} and Zn^{2+} from 0.5 M NaCl and major ion sea water on amixture of δ-MnO_2 & amorphous FeOOH. *J. Colloid. Interface Sci.* 2005, *284*, 30–38.
11. Zhu, Y., & Murali, S., Graphene and graphene oxide: synthesis, properties and applications. *Adv. Mettallurg.* 2010, *22*, 1–19.
12. Hummers, W. S., & Offeman, R. E., Preparation of graphitic oxide. *J. Amer. Chem. Soc.* 1958, *80*, 1339–1341.
13. Hou, C., One-step synthesis of magnetically functionalized reduced graphite sheets and their use in hydrogels. *Carbon* 2011, *49*, 47–53.
14. Varma, S., Sarode, D., Wakale, S., Bhanvase, B. A., & Deosarkar, M. P., Removal of nickel from wastewater using graphene nanocomposite. *Int. J. Chem. Phy. Sci.* 2013, *2*, 2319–6602.

CHAPTER 10

ULTRASOUND ASSISTED SYNTHESIS OF HYDROGELS AND ITS EFFECTS ON WATER/DYE INTAKE

R. S. CHANDEKAR,[1] K. PUSHPARAJ,[1] G. K. PILLAI,[1] M. ZHOU,[2] S. H. SONAWANE,[3] M. P. DEOSARKAR,[1] B. A. BHANVASE,[4] and M. ASHOKKUMAR[2]

[1]*Department of Chemical Engineering, Vishwakarma Institute of Technology, Pune–411037, Maharashtra, India*

[2]*School of Chemistry, University of Melbourne, VIC 3010, Australia*

[3]*Chemical Engineering Department, National Institute of Technology, Warangal, Telangana–506004, India*

[4]*Chemical Engineering Department, Laxminarayan Institute of Technology, Nagpur–440033, Maharashtra, India*

CONTENTS

10.1 Introduction 162
10.2 Experimental Work 163
    10.2.1 Materials 163
    10.2.2 Ultrasound Assisted Synthesis of PAA and Its Nano Composite Hydrogel 164
    10.2.3 Characterization Techniques 165
    10.2.4 Water Uptake of Hydrogel and Nano Composites 165
    10.2.5 Dye Adsorption Experiments 166

10.3 Results and Discussions 166
10.3.1 Water Swelling Behavior of Hydrogel and Nano Composite in Absence and Presence of Ultrasound 168
10.3.2 Comparative Study of PAA and PAA/MMT Hydrogel for Dye Adsorption 171
10.3.3 Adsorption Isotherm Models 173
10.4 Conclusion 175
Keywords 176
References 176

10.1 INTRODUCTION

The treatment of industrial wastewater has always been a key aspect of research due to the increasing awareness about the environmental cleanliness and water scarcity. Continuous attempts are being made in developing new wastewater treatment techniques, such as chemical precipitation, adsorption, ion exchange, filtration, photocatalysis, wet air oxidation, electrochemical treatment, and reverse osmosis along with conventional methods for removing heavy metals, dyes and organic pollutants from wastewater [1–8]. The combinations of two or more techniques such as sonophotocatalysis and sonosorption have been reported to be more efficient in the removal of contaminants from wastewater [9, 10]. The use of ultrasound promotes degradation and also increases the rate of mass transfer in sorption process [11–14]. The phenomenon of cavitation, the formation, growth and collapse of bubbles, generates free radicals due to localized extreme temperature and pressure conditions. In aqueous medium, water molecules are dissociated into $OH^{\bullet}$ and $H^{\bullet}$ radicals inside the collapsing bubble. These radicals can be used to decompose organic pollutant molecules [5].

Hydrogel possesses three dimensional cross-linked polymer structures and can be used for a number of engineering applications, such as adsorption, drug delivery devices, protein separation, artificial muscles, sensors, actuators and in membrane separation technique. Swelling in water is an inherent property of the hydrogels [15–18]. The equilibrium degree of swelling involving the volume change is an important and inherent property of a hydrogel for the separation of chemicals. The volume change, caused by water sorption with respect to time, is strongly responsive to different factors such as temperature, pH, ionic strength, degree of crosslinking, etc. Faster response time for hydrogel to reach equilibrium makes it more useful in many important industrial applications, such as adsorbent filled membranes. The hydrogel dispersed with nano clay materials forms a new class of nano composite material with increased elasticity, permeability and adsorption capability. The addition of organically modified clay into hydrogel has two distinct advantages: (i) it will act as a good adsorbent for dye molecules; and (ii) it will act as a cross-linker for the hydrogel network. Various techniques are being used for the preparation of hydrogels. Among them, the radiation cross-linking method has more advantage over others, especially with respect to a clean environment and higher production rate [19]. Rokita et al. [20] used ultrasound assisted cavitation technique for the synthesis of 3D polymer network consisting of covalently linked polymer chains, polyethylene glycol, diacrylate polyethylene glycol dimethacrylate and vinylpyrrolidone. Nanoclay materials are known to be inexpensive good adsorbent materials that have been used in wastewater treatment due to their unique octahedral and exfoliated structures. It is found to be difficult to use these clay materials in free form; hence clay is incorporated into polymer to form a nano composite. This nanocomposite hydrogel could be useful in number of applications such as drug delivery, pharmaceutical products, waster treatment, adsorbent, etc. [21–24].

In this study, polyacrylic acid and its nano composite hydrogels were synthesized using ultrasound technique. The MMT clay was introduced and exfoliated into the PAA so as to increase the absorption/adsorption capacity of the hydrogel. The effects of ultrasound on mass transport, rate of water diffusion and dye uptake were explored.

10.2 EXPERIMENTAL WORK

10.2.1 MATERIALS

Acrylic acid (AA) of purity 99%, procured from M/s Loba Chemie., India was used as the monomer for synthesis of hydrogel. Montmorillonite clay (M/s Sigma Aldrich) was used as a filler and a cross-linking agent.

Ammonium persulfate (APS) (M/s CDH) was used as an initiator. All the above reactants were used as supplied without further purification.

10.2.2 ULTRASOUND ASSISTED SYNTHESIS OF PAA AND ITS NANO COMPOSITE HYDROGEL

Figure 10.1 shows experimental setup for hydrogel synthesis. As shown in Figure 10.1, ultrasound probe (Hielscher 22 kHz, 120 W) was used for polymerization process. Polymerization reaction was carried out at 60°C by heating the reaction mixture using a hot water bath. The initiator was added dropwise as function of time (semibatch mode).

As shown in Figure 10.2(a) and (b), the synthesis of the polymer hydrogel and polymer/clay nano composite was carried out using ultrasound assisted polymerization process. Acrylic acid (36 g) along with 100 mL demineralized water was added to a reactor and the mixture was sonicated for five minutes using an ultrasound to generate a homogenous mixture. The temperature of the reactor was maintained at 60°C by using hot water circulation from a water bath. Initiator solution (2.28 g in 20 mL water) was added drop wise into the reactor. Due to the effective dissociation of the initiator by sonication, the formation of polymer takes place very fast

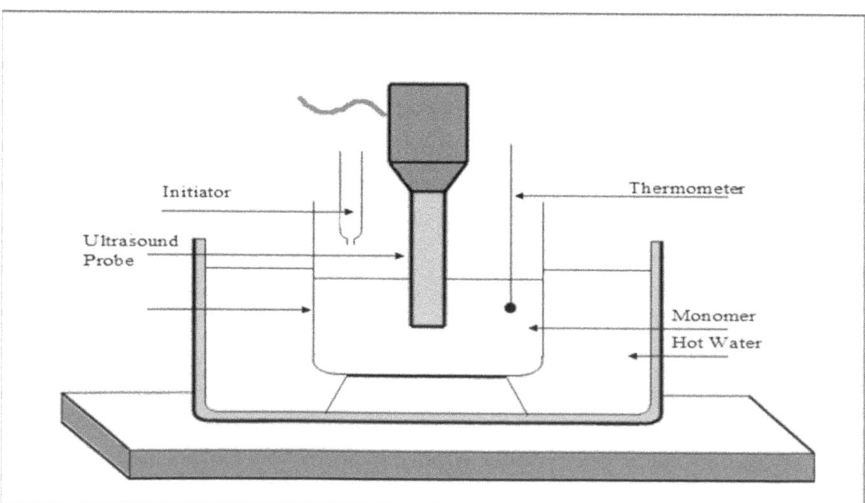

FIGURE 10.1 Experimental setup for hydrogel synthesis.

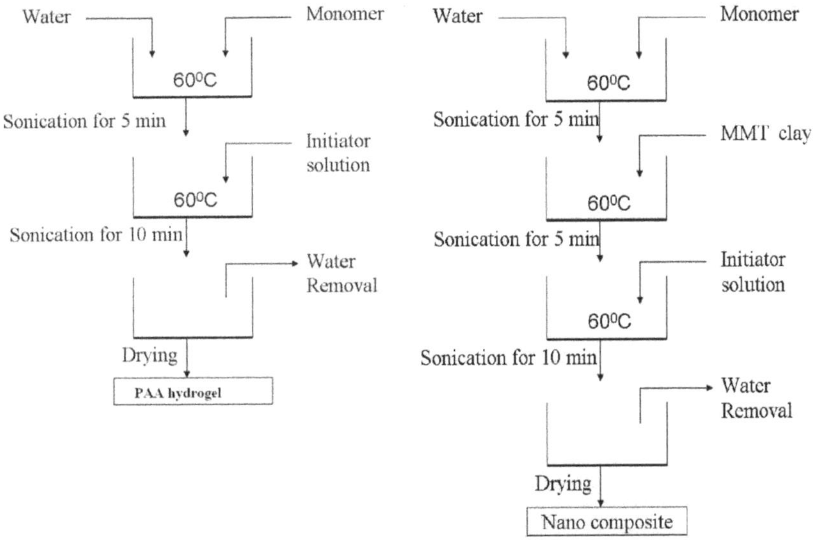

FIGURE 10.2 (a) Synthesis procedure of PAA hydrogel under ultrasound environment; (b) Synthesis procedure of PAA/MMT hydrogel under ultrasound environment.

resulting in a viscous gel after the completion of polymerization. Excess water was removed from the hydrogel and hydrogel was dried at 80°C. For the nano composite preparation, a similar procedure as described above was used in the presence of 2 g of MMT clay. The procedure of the addition of monomers/initiator and MMT clay is shown in Figure 10.2(b).

10.2.3 CHARACTERIZATION TECHNIQUES

UV-visible spectrophotometer (SHIMADZU 160A model) was used to determine the concentration of MB dye. The wavelength of maximum absorbance (λ_{max}) of MB dye was found to be 662.5 nm. Demineralized water was used as a reference.

10.2.4 WATER UPTAKE OF HYDROGEL AND NANO COMPOSITES

To determine water absorption property of hydrogel, a predetermined quantity of hydrogel was added to a reactor containing 200 mL water and kept for 6 days at room temperature to attain equilibrium. The weight of the hydrogel was noted at definite time intervals. To understand the diffusion effect of ultrasound onto the absorption property of the hydrogel and nano composite, similar experiments were performed under sonication environment for 3 h to attain equilibrium. The swelling ratio of hydrogel and nano composites was determined using the following equation:

$$\text{Swelling ratio} = \frac{M_t}{M_d} \quad (1)$$

where M_t and M_d are the weight of the wet gel after swelling and the dry gel, respectively. The equilibrium degree of swelling (EDS) was calculated by using the above equation by replacing M_t with $M_{e,}$ which is the equilibrium weight of swollen gel.

10.2.5 DYE ADSORPTION EXPERIMENTS

A comparative study of adsorption of methylene blue (MB) dye was carried out in the presence of hydrogel and nano composite at different concentrations of MB dye. A specific quantity of the hydrogel was added to the reactor containing a known concentration of the (200, 400, 800 or 1000 ppm) MB dye solutions and kept undisturbed 6 days. The concentration of MB dye solution and weight of the hydrogel were recorded at finite time intervals. Experiments in the presence of ultrasound were carried out by using different concentrations for 2 hrs. In order to understand the loading capacities of the hydrogel and the nano composite, experiments were carried out using different quantities of the hydrogel and the nano composite (2, 4, 6 and 8 g) and using different dye concentrations (200 and 400 ppm) in the presence and absence of sonication.

10.3 RESULTS AND DISCUSSIONS

Acrylic acid was polymerized using free radical polymerization to form the hydrogel in the presence of ultrasound and in the absence and presence of MMT. MMT clay is modified clay by quaternary ammonium salts, which help for cross-linking of polymer chains. The monomer present in the reaction medium diffuses inside the clay galleries and generates the cross-linking between the polymer chains. The presence of C-C double bonds in the quaternary ammonium salts attached to the clay are involved in the cross-linking process. Further as monomer to polymer conversion increases, exfoliation of clay into the polymer occurs. Figure 10.3(a) and (b) shows the TEM images of hydrogel and nanocomposite hydrogel, Figure 10.3(a) shows porous structure containing most of white part represents the hydrogel. While in Figure 10.3(b) shows nanoclay layered

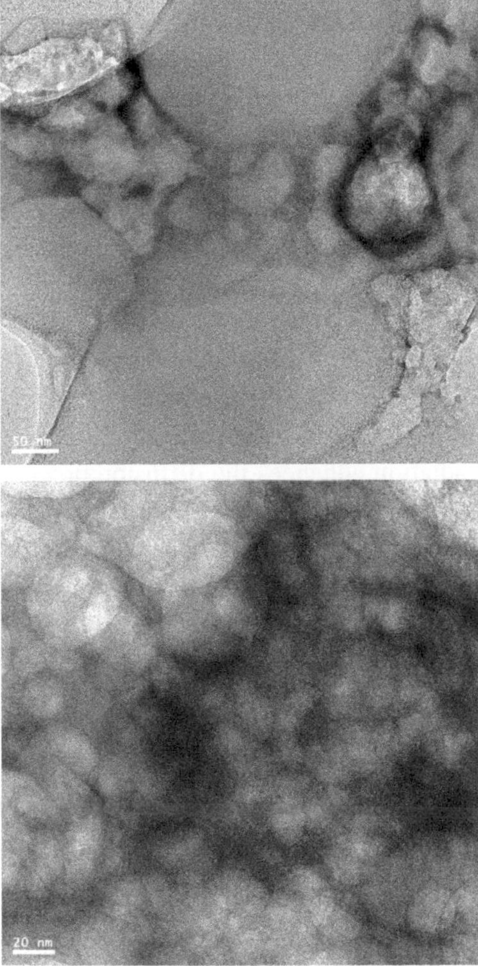

FIGURE 10.3 (a) TEM image of hydrogel; (b) TEM image for nanocomposite hydrogel.

structure in which hydrogel contains tactoids of MMT clay in which clay platelets are not been fully separated from each other.

10.3.1 WATER SWELLING BEHAVIOR OF HYDROGEL AND NANO COMPOSITE IN ABSENCE AND PRESENCE OF ULTRASOUND

Figure 10.4(a) and (b) show the swelling behavior of the hydrogel and nano composite in the absence and presence of ultrasound. As shown in Figure 10.4(a), PAA and PAA/MMT hydrogel shows very small difference in the water absorption in absence of ultrasound. While the presence of ultrasound leads to an increase in 60% water absorption for PAA hydrogel and 120% absorption in the case of PAA/MMT clay. To attain the equilibrium of water uptake, the hydrogel and nano composite takes about 6 days in absence of ultrasound with 80% swelling, while the nano composite shows 120% swelling in 6 days. The reason for the higher swelling property of the nano composite can be attributed to exfoliation of platelets of clay in the hydrogel. A significant reduction in the time from 8640 to 200 min was observed due to ultrasonic irradiation. It can also be seen that the hydrogel shows 75% swelling, whereas the nano composite shows 130% swelling in the presence of ultrasound. Higher swelling indicates that ultrasonic irradiation enhances the diffusion process by breaking the barrier film between the liquid and solid interface.

The water uptake data was further analyzed using the following equation for determining the transport coefficient for water uptake [19, 23, 25].

$$\frac{M_t}{M} = kt^n \tag{2}$$

where M_t and M represent the amount of water uptake at times t and infinity, respectively. 'k' is a characteristic constant of the hydrogel and 'n' is a characteristic exponent of the mode of transport of thc penetrating molecule. The plots of ln 't' as a function of ln (M_t/M) are shown in Figure 10.5 for both hydrogel and nano composite and the coefficient 'k' and exponent 'n' were determined from the graphs. The values of 'k' and 'n' were found to be 20.1 and 0.6, respectively.

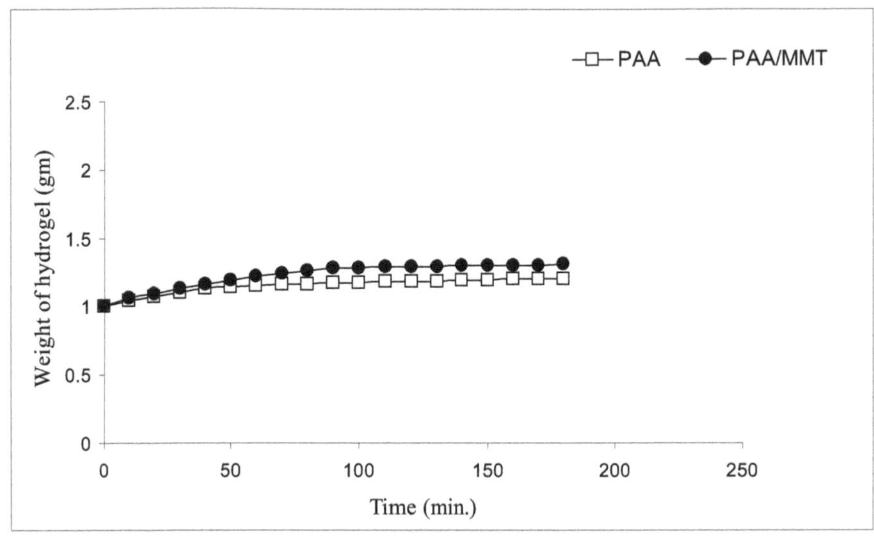

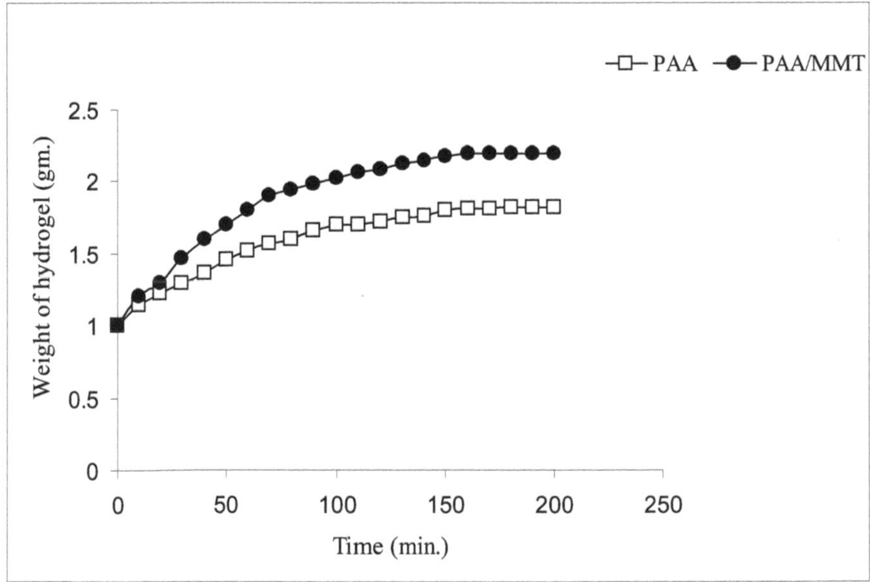

FIGURE 10.4 (a) Water uptake capacity of hydrogel and nano composite in absence of ultrasound; (b) Water uptake capacity of hydrogel and nano composite in absence of ultrasound.

It is observed from Figure 10.6 that the data of % sorption versus time$^{0.6}$ shows a linear behavior for both cases of hydrogel and nano composite. It is also found that the % water absorption is more in the case of nano

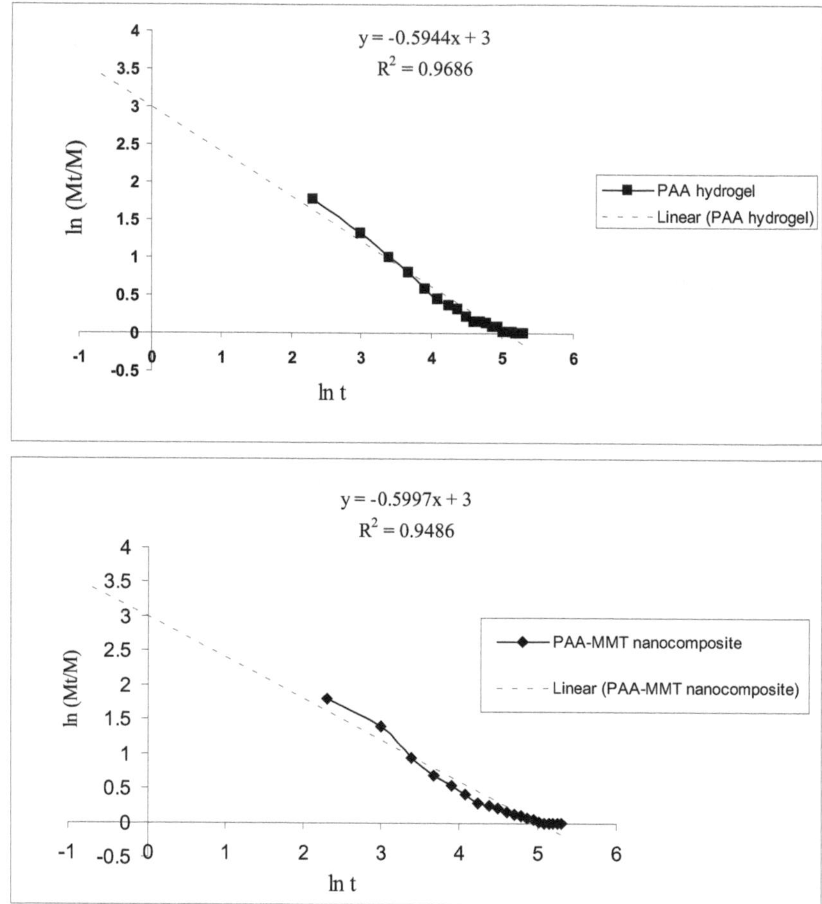

FIGURE 10.5 (a) ln (M_t/M) as a function ln 't' for water diffusion in presence of hydrogel and ultrasound; (b) ln (M_t/M) as function ln 't' for water diffusion in presence of PAA nano composite and ultrasound.

composite as compared to hydrogel. From Figure 10.6, it is evident that both hydrogel and nano composite obey the Fick's law of diffusion.

To calculate the diffusion coefficient, thin films of both hydrogel and nano composite of approximate thickness (100 μ) were prepared. The diffusion coefficients at room temperature were calculated using Eq. (3).

$$\frac{M_t}{M} = \frac{4}{d}\sqrt{\frac{Dt}{\pi}} \tag{3}$$

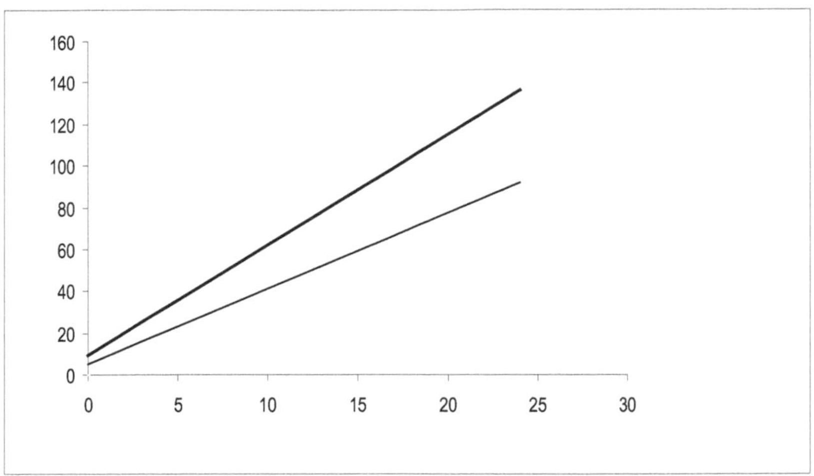

FIGURE 10.6 Water uptake as a function of $t^{0.6}$ for (a) hydrogel and (*b*) nano composite.

where M_t is the amount absorbed at time *t*, *M* is the amount absorbed at thermodynamic equilibrium, 'd' is thickness of thin film hydrogel and composite and D is the diffusion coefficient. The diffusion values for the both hydrogel and nano composite were found to be 2.2×10^{-9} and 2.4×10^{-9} cm^2/s, respectively, as shown in Table 10.1. The 'S' shape water absorption isotherms are shown in Figure 10.7 for the PAA hydrogel as well as PAA-nano composite. The 'S' shape curve for nano composite is slightly shifted towards left indicating that the hydrogel nano composite shows a favorable isotherm for water uptake.

10.3.2 COMPARATIVE STUDY OF PAA AND PAA/MMT HYDROGEL FOR DYE ADSORPTION

Tables 10.2(a) and (b) show the effect of different dye concentrations (800 to 1000 ppm) on the dye adsorption capacity of the hydrogel in the

TABLE 10.1 Diffusion Coefficient of Hydrogel and Nano Composite in Presence of Ultrasound for Water

Type	Diffusion coefficient (cm^2/s) $\times 10^9$
Hydrogel	2.2167
Nano composite	2.367

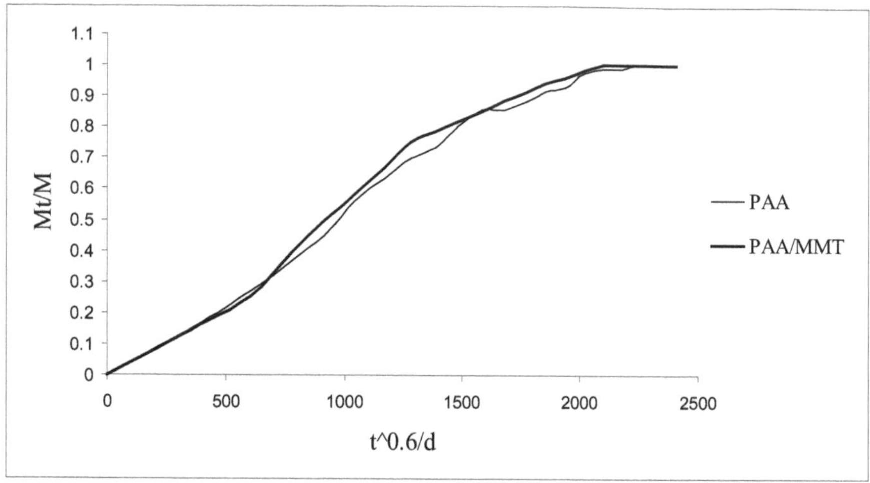

FIGURE 10.7 Water sorption isotherm of hydrogel and nano composite at room temperature in presence of ultrasound.

absence and presence of ultrasound for PAA and PAA/MMT hydrogel. In Table 10.2(a) it is shown without ultrasound that PAA and PAA/MMT take about 6 days for the establishment of adsorption equilibrium. It is also found that the % adsorption decreases with an increase in the dye concentration. About 50% adsorption was found in case of PAA hydrogel and PAA/MMT without ultrasound. Whereas the % adsorption was found

TABLE 10.2(A) Effect of Different Concentration Dye on PAA and PAA/MMT in Absence of Ultrasound

	PAA in absence of ultrasound		**PAA/MMT in absence of ultrasound**	
Time (days)	**Dye Concentration (ppm)**	**Dye Concentration (ppm)**	**Dye Concentration (ppm)**	**Dye Concentration (ppm)**
0	800	1000	800	1000
1	701	936	633	724
2	621	867	595	663
3	508	774	504	600
4	473	665	450	534
5	458	651	443	507
6	456	648	442	502

TABLE 10.2(B) Effect of Different Concentration Dye Adsorption on PAA and PAA/MMT in the Presence of Ultrasound

	PAA in presence of ultrasound		PAA/MMT in presence of ultrasound	
Time (minutes)	**Dye Concentration (ppm)**	**Dye Concentration (ppm)**	**Dye Concentration (ppm)**	**Dye Concentration (ppm)**
0	800	1000	800	1000
20	691	873	547	640
40	557	736	369	454
60	440	568	207	295
80	344	460	160	244
100	289	368	157	219
120	272	348	157	221

to be 80%, when PAA and PAA/MMT clay was used in presence of ultrasound. Also, it was observed that the rate of swelling of nano composite was more compared to the hydrogel; this might be because the addition of MMT clay to PAA produces more surface area which increased the water uptake capacity. Moreover, the hydrophilic nature of the clay produces forces of attraction for the OH molecules.

10.3.3 ADSORPTION ISOTHERM MODELS

The well-known Langmuir isotherm represents as shown in Eq. (4), it gives possibility of monolayer coverage of dye molecule, which shows monolayer possibility of adsorption process [24].

$$C_e/q_e = 1/ab + C_e/a \tag{4}$$

where C_e is the equilibrium concentration (mg/l), q_e is the amount of adsorbate present per unit mass of adsorbent at equilibrium (mg/g), 'a' is the theoretical maximum adsorption capacity (mg/g) and constant 'b' is related to the affinity to the binding sites (l/mg). The Freundlich isotherm is represented as shown in Eq. (4), it shows the possibility of multilayer adsorption based on affinity parameter 'n' The Freundlich intensity parameter, 1/n indicates deviation from linearity. If 1/n = 1, the adsorption sites are homogeneous and there is no interaction between the adsorbed sites. If 1/n is less than 1, the adsorption is favorable [24]

$$\ln q_e = \ln K + 1/n\, C_e \quad (5)$$

To understand the effect of PAA and nano composite hydrogel, the Freundlich and Langmuir isotherm curves were plotted as shown in Figures 10.8 and 10.9, respectively. The Freundlich isotherm model suggests that sorption energy exponentially decreases on completion of sorption of hydrogel adsorbent. The linear plot of ln 'q' versus ln 'Ce' proves that hydrogel sorption of MB dye on nano composite fits very well with Freundlich isotherm model. Freundlich parameters were determined from the values of slope (intensity parameter) and intercept (capacity constant). The results are shown in Figure 10.8 and Table 10.3. The fit of the Freundlich model with the experimental data is very close agreement in the entire range of concentration of MB dye.

The linear plot of 'Ce/q' versus 'C_e' shows the validity with the Langmuir isotherm model for hydrogel adsorption system (Figure 10.9). From the values of parameters and R^2 (0.87), it is observed that hydrogel sorption

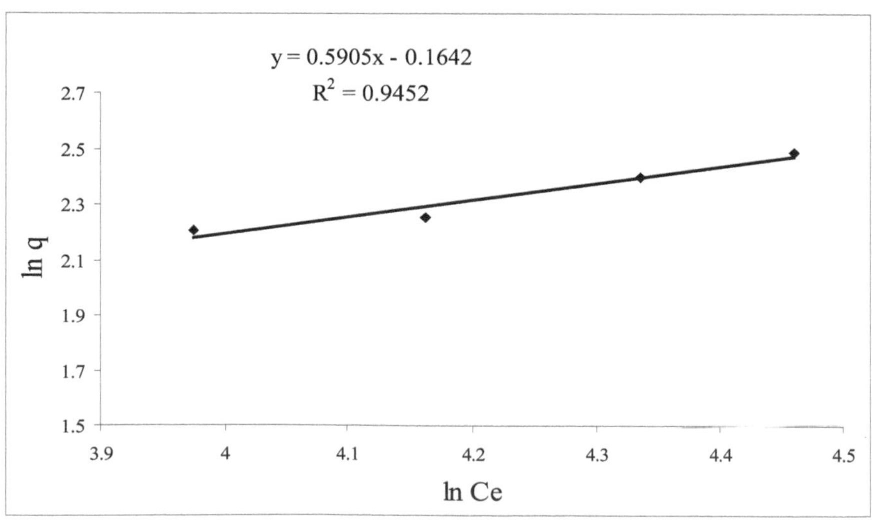

FIGURE 10.8 Freundlich adsorption isotherm for MB dye adsorption onto PAA/MMT nano composite (Concentration = 200 ppm, Temperature = 30°C).

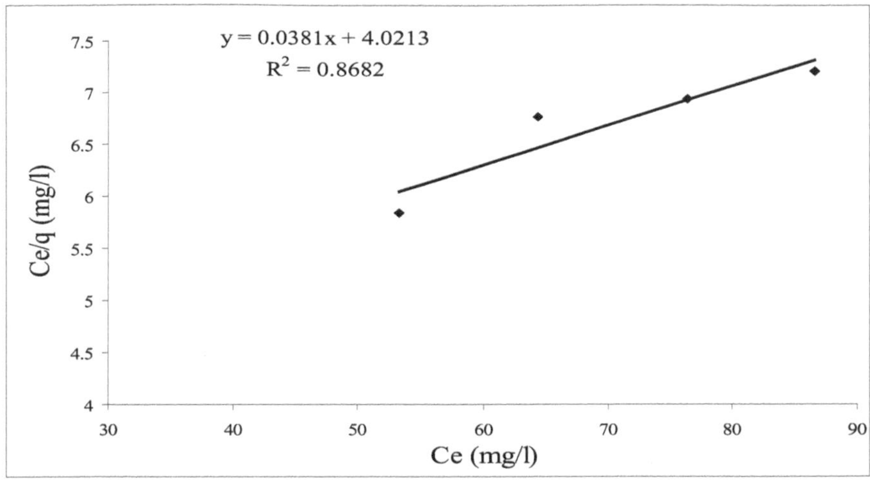

FIGURE 10.9 Langmuir adsorption isotherm for MB dye adsorption onto PAA/MMT Nano composite (Concentration = 200 ppm, Temperature = 30°C).

TABLE 10.3 Parameters of Freundlich and Langmuir Adsorption Isotherm Models

Freundlich Parameters	**Slope**	**Intercept**	**Correlation coefficient**
	$1/n = 0.59$	$\ln K_f = -0.16$	$R^2 = 0.95$
Langmuir Parameters	**Affinity Factor**	**Mono. Capacity**	**Correlation coefficient**
	$b = 0.04$/mg	Qo = 4.02 mg/g	$R^2 = 0.87$

data of MB dye on nano composite did not fit very well to Langmuir isotherm model. This could be due to strong interaction between the MB dye molecules and nano composite hydrogel, which can be attributed to chemisorption. The Freundlich model fits very well as compared to the Langmuir model, suggesting the formation of thick boundary layer (multilayer).

10.4 CONCLUSION

PAA hydrogels and PAA/MMT nano composites were prepared using acoustic cavitation assisted polymerization technique. The prepared hydrogels and nano composites were tested for water uptake properties. It has been observed that the nano composite significantly reduces the diffusion barrier between liquid film and solid hydrogel. Due to ultrasonic irradiation, the PAA hydrogel and nano composite show very fast water absorption capacity and reduction in time from 6 days to 2 h. It is also seen that dye uptake is increased when ultrasound and nano composite were used. The results indicate at higher concentration of MB dye, the adsorption of dye decreases. Freundlich Isotherm shows the multilayer adsorption of dye onto the MMT nano composite indicates that the prepared hydrogel can be very much useful for the industrial wastewater treatment plant applications.

KEYWORDS

- **Methylene blue**
- **PAA/MMT hydrogel**
- **Polymerization**
- **Ultrasound**
- **Water swelling**

REFERENCES

1. Saquib, M. M., & Muneer, M., *Dyes and Pigments* 2003, *56*, 37–49.
2. Yapar, S., & Yilmaz, M., *Adsorption* 2004, *10*, 287–298
3. Patel, H. A., Somani, R. S., Bajaj, H. C., & Jasra, R. V., *Bull. Mater. Sci.* 2006, *29*, 133–145.
4. Modirshahla, N., & Behnajady, M. A., *Dyes and Pigments* 2006, *70*, 54–59.
5. Torres, R A., Pe'trier, C., Combet, E., Carrier, M., & Pulgarin, C., *Ultrason. Sonochem.* 2008, *15*, 605–611.
6. Daneshvar, N., Ashassi-Sorkhabi, H., & Tizpar, A., *Sep. Purif. Technol.* 2003, *31*, 153–162.
7. Petrier, C., Jiang, Y., & Lamy, M.-F., *Environ. Sci. Technol.* 1998, *32,* 1316–1318.
8. Madhavan, J., Grieser, F., & Ashokkumar, M., *Ultrason. Sonochem.* 2010, *17*, 338–343.
9. Wang, J., Hiang, Y., Zhang, Z., Zhao, G., Zhan, G., Ma, T., & Sun, W., *Desalination* 2007, *216*, 196–208.
10. Kubo, M., Matsuoka, K., Takahashi, A., Shibasaki-Kitakawa, N., & Yonemoto, T., *Ultrasonics Sonochemistry* 2005, *12*, 263–269.

11. Entezari, M. H., & Al-Hoseini, Z. S., *Ultrason. Sonochem.* 2007, *14*, 599–604.
12. Iida, Y., Kozuka, T., Tuziuti, T., & Yasui, K., *Ultrasonics* 2004, *42*, 635–639.
13. Breitbach, M., & Bathen, D., *Ultrason. Sonochem.* 2001, *8*, 277–283.
14. Juang, R. S., Lin, S. H., & Cheng, C. H., *Ultrason. Sonochem.* 2006, *13*, 251–260.
15. Weian, Z., Weib, L., & Yue'e, F., *Materials Letters* 2005, *59*, 2876–2880.
16. Kurumada, K., Yamada, Y., Igarashi, K., Pan, G., & Umeda, N., *J. Chem. Eng. Jpn.* 2005, *38*, 657.
17. Yan, X., & Gemeinhart, R. A., *J. Controlled Release* 2005, *106*, 198.
18. Nossal, R., *Macromolecules* 1985, *18*, 49–54.
19. Hoffman, A. S., *Radiation Physics and Chemistry* 1981, *18*, 323.
20. Rokita, B., Rosiak, J. M., & Ulanski, P., *Macromolecules* 2009, *42,* 3269–3274
21. Cass, P., Knower, W., Pereeia, E., Holmes, N. P., & Hughes, T., *Ultrason. Sonochem.* 2010, *17*, 326–332.
22. Ekici, S., Işıkver, Y., & Saraydın, D., *Polymer Bulletin* 2006, *57*, 231–241.
23. Xia, X., Yih, J., D'Souza, A. D., & Hu, Z., *Polymer* 2003, *44*, 3389.
24. Sonawane, S. H., Chaudhari, P. L., Ghodke, S. A., Parande, M. G., Bhandari, V. M., Mishra, S., & Kulkarni, R. D., *Ultrason. Sonochem.* 2009, *16*, 351–355.
25. Kasgöz, H., & Durmus, A., *Poly. Adv. Tech.* 2008, *19*, 838–845.
26. Adnadjevic, B., & Jovanovic, J., *J. Appl. Poly. Sci.* 2007, 107, 3579–3587.

PART III

MATERIALS AND APPLICATIONS

CHAPTER 11

ULTRASONICALLY CREATED RECTANGULAR SHAPED ZINC PHOSPHATE NANOPIGMENT

S. E. KAREKAR,[1] A. J. JADHAV,[1] C. R. HOLKAR,[1] N. L. JADHAV,[1] D. V. PINJARI,[1] A. B. PANDIT,[1] B. A. BHANVASE,[2] and S. H. SONAWANE[3]

[1]*Chemical Engineering Department, Institute of Chemical Technology, Matunga, Mumbai 400019, India*
E-mail: dv.pinjari@ictmumbai.edu.in, dpinjari@gmail.com;
Tel: +91-22-3361-2032; Fax: +91-22-33611020

[2]*Chemical Engineering Department, Laxminarayan Institute of Technology, Nagpur, 440001, India*

[3]*Chemical Engineering Department, National Institute of Technology, Warangal, 506004, India*

CONTENTS

11.1 Introduction 182
11.2 Experimental Part 184
 11.2.1 Materials and Methods 184
 11.2.2 Synthesis of ZP by Ultrasound Method 184
 11.2.3 Effect of pH on the yield of Zinc Phosphate 186
 11.2.4 Conductivity of ZP with Respect to Sonication Time 186
 11.2.5 Pretreatment of Mild Steel Panels Prior to Paint Application 186

11.2.6 Paint Preparation (2 Pack System) 187
11.2.7 Application of Paint on the Surface of Mild Steel Panels 188
11.2.8 Characterizations 188
11.3 Results and Discussion 189
11.3.1 Particle Size, XRD, FTIR and SEM Analysis of Zinc Phosphate Nanoparticles 189
11.3.1.1 Particle Size Analysis 189
11.3.1.2 XRD Analysis 190
11.3.1.3 FTIR Analysis 191
11.3.1.4 SEM Analysis 192
11.3.2 Corrosion Rate Analysis 192
11.3.3 Dried Film Properties of ZP-2K Epoxy System 193
11.3.4 Corrosion Mechanism, and Role of Zinc Phosphate and Binder Resin in Corrosion Inhibition 194
11.4 Conclusions 195
Acknowledgments 195
Keywords 195
References 196

11.1 INTRODUCTION

The corrosion inhibition efficiency of the various coatings on the metallic surfaces can be enhanced by the incorporation of corrosion inhibiting pigments (organic or inorganic) in the protective coatings such as barrier coatings, sacrificial coatings, and inhibitive pigmented and un-pigmented lacquer coatings [1, 2]. Commonly, anticorrosive coatings are generally selected based on the binders or pigments used in a coating formulation. Corrosion inhibition is possible by barrier, sacrificial mechanism like zinc and aluminum galvanization [3, 4] or cathodic as well as anodic protection method (passivation technique). By using nano particles, dense barrier coating can be formed which may help to form an obstacle for the passage of corrosive species. Application of corrosion inhibitors such as organic [5] or inorganic with hybrid [6] polymers as epoxy binders have various industrial applications. Dense barrier coatings can be applied easily and render low penetration to the corrosive species and hence, is the most accepted mode for corrosion inhibition. Epoxy resin has been used as a carrier medium in the preparation of corrosion inhibiting coatings as the former consists of non-reactive regularly spaced ether linkage along the hydrocarbon chains, esterifiable hydroxyl group spaced inside the chain, aromatic nature of the backbone and terminal reactive polar group. These features help to improve the chemical resistance against severe corrosive conditions by strongly adhering to the substrates [7]. The corrosion inhibiting ability of anti-corrosion coatings also depend on the properties of the resin materials used in the coating formulation. Selection of epoxy resin is a key factor in efficient corrosion inhibition of the final coating formulation. Epoxy contains aromatic groups in the chain, which provides resistance to the migration of species corrosive and humid atmosphere [8]. Polyamine and polyamide are the commercially used hardeners for 2K epoxy coating, for example, it needs to be mixed with a hardener, catalyst or an activator. Amine based hardener's are more durable and chemical resistant than amide based hardeners but most have a tendency to 'blush' in moist conditions. Blushing produces a waxy surface layer on the surface of the epoxy resin, these results into a reaction with the hardener and moisture in the air and potentially toxic chemicals within the hardener can also be released in the same manner. Thus, amines based hardener's suffer with these potential shortcomings, however, amide based hardeners are more surface tolerant and less altered by moisture. Due to this reason polyamide based hardener has been selected for 2K epoxy coating system in this work. Also epoxy-polyamide 2K system shows better flexibility, resistance to water and has a longer pot life [9].

In the present study, epoxy-polyamide has been selected as a binder and nano zinc phosphate as an anticorrosive pigment. Zinc phosphate has been widely used in coating industry, medical, electrical applications etc. It exhibits corrosion-inhibiting behavior due to its low solubility in water/biological environment [10, 11]. Pigments such as lead and chromates are not chosen in the surface coating due to its inherent toxicity and are now being replaced by the ecofriendly pigments. Such pigments include phosphate based pigments among which zinc phosphate, $Zn_3(PO_4)_2$, has been found to have a wide range of application due to its non-toxic nature, excellent anticorrosive properties and can be readily used as a replacement pigment in the coatings [12–15].

On the other hand, all these methods were not found to be useful at an industrial level for the production because of the high costs and complex technology. Nevertheless main drawback of these methods was related to the proper dispersion of zinc phosphate nanoparticles in practical applications. Ultrasonically assisted synthesis has been proved to be a useful tool for the intensification of the synthesis of nanoparticles [16]. The chemical effects of ultrasonic irradiation arise from acoustic cavitation, which can be described as the formation, growth and implosive collapse of bubbles in a liquid medium resulting in the generation of high temperature and pressure pulse and intense turbulence associated with microscopic liquid circulation currents [17, 18]. These extreme conditions of high temperature, pressure and local intense micro mixing helps in the formation of nanoparticles with near-uniform size distribution [18, 19].

11.2 EXPERIMENTAL PART

11.2.1 MATERIALS AND METHODS

Epoxy resin (B.P: >400°F, Viscosity: 1000 (cP), Specific Gravity: 1.15, 75%)

NVM with 500 EEW), Polyamide hardener (B.P: > 350.01°F, Viscosity: >600 (cP), Specific Gravity: 0.96, 70%NVM with 166 AHEW) and zinc rich epoxy primer (B.P: >150°F, Viscosity: 1000 (cP), Specific Gravity: 3.9) were procured from Sonal engineering and plastic fabricators Ltd Alibaug, Maharashtra, India. Chemicals used for zinc phosphate nanopigment synthesis consist of Zinc chloride, which was procured from Thomas baker and phosphoric acid, ammonia solution were procured from S.D. Fine Chemicals Ltd, Mumbai. Mild steel panels of two different sizes for corrosion analysis were used as received from the supplier.

11.2.2 SYNTHESIS OF ZP BY ULTRASOUND METHOD

During the synthesis of zinc phosphate, initially aqueous solutions of zinc chloride and potassium dihydrogen phosphate were prepared separately by adding, 106 g of zinc chloride in 50 mL distilled water and 71 g of potassium dihydrogen phosphate in 30 mL distilled water respectively. The solution of potassium dihydrogen phosphate was then added drop wise to the solution of zinc chloride under sonication within a time span of 10 min, which results into the original clear solutions turning turbid. During the course of the reaction the pH of solution was adjusted to 3.5 by ammonia solution under acoustic cavitation by using Ultrasonic Horn (Hielscher Ultrasonics GmbH, 22 KHz Frequency, 240 W powers) at 50% amplitude, which results into the formation of dense white precipitate. The optimum value of pH was adjusted here to achieve maximum ionization of zinc chloride to convert it into zinc phosphate by reacting it with potassium dihydrogen phosphate. If pH goes higher than 3.5 then precipitate of zinc phosphate gets slowly dissolved and at lower pH than 3.5 dense white precipitate of zinc phosphate doesn't forms. Above synthesis method can be explained by the reaction put forth below (R1). The above-reaction was completed in 20 min under sonication at room temperature (25 ± 2°C). In order to separate the formed product after the completion of the reaction, the reaction mixture was kept in water bath at 150°C for 10 min. The separated product was washed thrice using distilled water to remove the byproducts and dried at 100°C for 3 h in oven. $Zn_3(PO_4)_2$ (ZP) powder obtained was (whitish in nature), cooled, checked for yield (98%) and characterized by XRD, TGA, FTIR, and SEM analysis.

Reaction: R1

$$4ZnCl_2 + H_2O \rightarrow 4Zn^{2+} + 8\,Cl^- \quad \text{(i)}$$

$$2KH_2PO_4 + H_2O \rightarrow 2K^{2+} + 4H^+ + 2PO_4^{2-} \quad \text{(ii)}$$

$$2NH_4OH + H_2O \rightarrow 2NH_4^+ + 2OH^- \quad \text{(iii)}$$

$$4Zn^{2+} + 8\,Cl^- + 2K^{2+} + 4H^+ + 2PO_4^{2-} + 2NH_4^+ + 2OH^- \rightarrow Zn_3(PO_4)_2 + 2KCl + 4HCl + 2NH_4Cl + Zn(OH)_2 \quad \text{(iv)}$$

Overall Reaction:

$$4ZnCl_2 + 2KH_2PO_4 + 2NH_4OH \rightarrow Zn_3(PO_4)_2 + 2KCl + 4HCl + 2NH_4Cl + Zn(OH)_2\,(R\text{–}1) \quad \text{(v)}$$

11.2.3 EFFECT OF PH ON THE YIELD OF ZINC PHOSPHATE

Figure 11.1 shows effect of pH on the % Yield of zinc phosphate. It is observed that %yield of zinc phoshpate goes on increasing and reaches 98% of the theoretical at pH 3.5 which is considered to be as on optimum pH for the synthesis of the zinc phosphate.

11.2.4 CONDUCTIVITY OF ZP WITH RESPECT TO SONICATION TIME

Figure 11.2 depicts the variation in the conductivity with time during sonochemical synthesis of ZP. In case of sonochemical synthesis of ZP, conductivity goes on increasing steadily and finally reaches a value of 1.6 μS/cm because of the release of Zn^{2+} ions from zinc chloride and combination of these ions with PO_4^{3-} ions of potassium dihydrogen phosphate.

11.2.5 PRETREATMENT OF MILD STEEL PANELS PRIOR TO PAINT APPLICATION

Bare mild Steel material is known for its tendency to undergo rapid oxidation when exposed to atmosphere. The result of oxidation leads to the

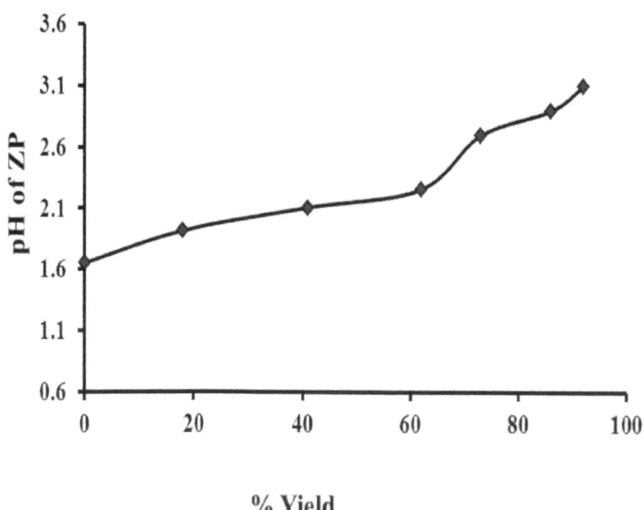

FIGURE 11.1 Effect of pH on % Yield of Zinc Phosphate nanopigment.

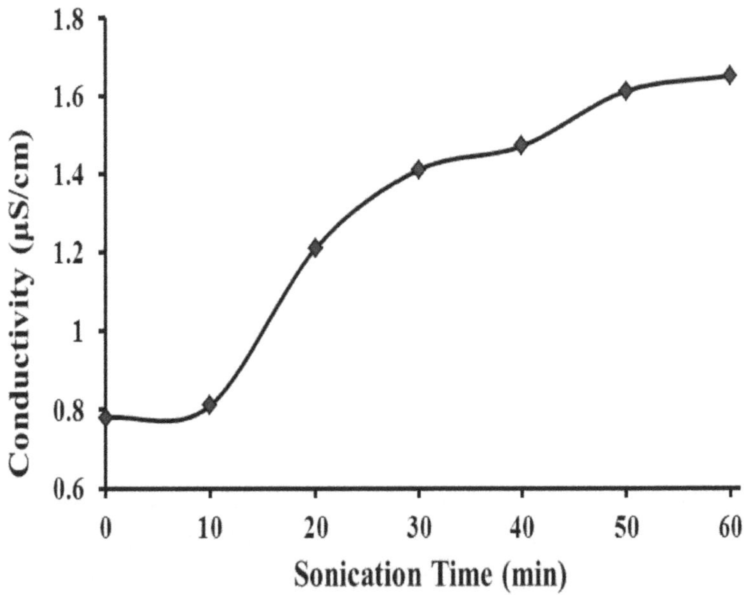

FIGURE 11.2 Conductivity variation of Zinc Phosphate nanopigment.

formation of oxide layer on the surface, which is called as rust. In order to remove this oxide layer the surface of mild steel needs to be cleaned before the application of paint. Cleaning of panels include rubbing with a sand paper and then treatment with a solvent (i.e., analytical grade acetone), lastly wiping with cotton. Before the application of the actual formulated paint, panels were coated with a thin layer of zinc rich epoxy primer and dried for a day. Primer was used to provide proper adherence of formulated paint on mild steel panels.

11.2.6 PAINT PREPARATION (2 PACK SYSTEM)

2K-epoxy coating system of epoxy resin and zinc phosphate pigment was prepared with a pigment muller. Firstly, the muller base was cleaned with a solvent to ensure proper sanitation of muller and then varying composition of zinc phosphate pigment and resin mixture (0–16% w/v value of zinc phosphate and resin) were prepared separately while muller operating at 2500 RPM with it. Finally, before the application of the actual paint, resin was mixed with a hardener, for example, base and hardener ratio for the 2K system was maintained at 1:1 (by weight) and the real integration of base and the hardener has been achieved under constant stirring using mechanical overhead stirrer. This ratio was calculated on the basis of epoxy value of the epoxy resin and hydroxyl value of polyamide hardener. Studies have been performed at varying loading of zinc phosphate pigments (0 to 16% W/V) in the resin paint formulations with a purpose of finding the optimum loading for the desired mechanical properties and required corrosion inhibition.

11.2.7 APPLICATION OF PAINT ON THE SURFACE OF MILD STEEL PANELS

After the application of primer on the surface panels and preparation of 2K-Epoxy system and (0–16% W/V composition) zinc phosphate in epoxy resin, panels were coated by prepared paint with a fine brush. Uniform thickness of coating on the surface of the panels was achieved to obtain accuracy in the result of the corrosion inhibition. All the panels were kept for drying for a week at room temperature before further evaluation of corrosion inhibition.

11.2.8 CHARACTERIZATIONS

The particle size distribution of nano zinc phosphate was carried out by Malvern Zetasizer Instrument (Malvern Instruments, Malvern, UK). To find out corrosion inhibition we did characterizations such as dip test over the entire range of zinc phosphate to resin ratio and salt spray analysis (ASTM B 117) for 2K-epoxy system at 12% W/V concentration zinc phosphate to resin ratio. Dip test was used to calculate the corrosion rate in cm/year for each one of the prepared samples by using the corrosion rate calculations. This test is based on weight loss of mild steel samples between initial weight (before the dip test) and final weight (after dip test). Further, Fourier transform infrared (FTIR) spectroscopic analysis was carried out for zinc phosphate nanoparticles using SHIMADZU 8400S analyzer in the region of 4000–8500 cm^{-1}. X-ray diffraction (XRD) pattern of zinc phosphate nanoparticles was recorded by using powder

X-ray diffractometer (Rigaku Mini-Flox, USA/Philips PW 1800). The Cu-Ka radiation was selected for the analysis. The morphological analysis of zinc phosphate nanoparticles was investigated by using Scanning Electron Microscopy (SEM), (PHILIPS, CM200, 20–200 KV, magnification 1,000,000X). Mechanical and anti-corrosion properties Cross cut adhesion, Pull of adhesion, Resistance to neutral salt spray, Resistance to water condensation, Resistance to cyclic corrosion were tested according to ASTM standards.

11.3 RESULTS AND DISCUSSION

11.3.1 PARTICLE SIZE, XRD, FTIR AND SEM ANALYSIS OF ZINC PHOSPHATE NANOPARTICLES

11.3.1.1 Particle Size Analysis

Figure 11.3 shows particle size distribution of ZP synthesized by ultrasound method described in this work. The average particle size of $Zn_3(PO_4)_2$

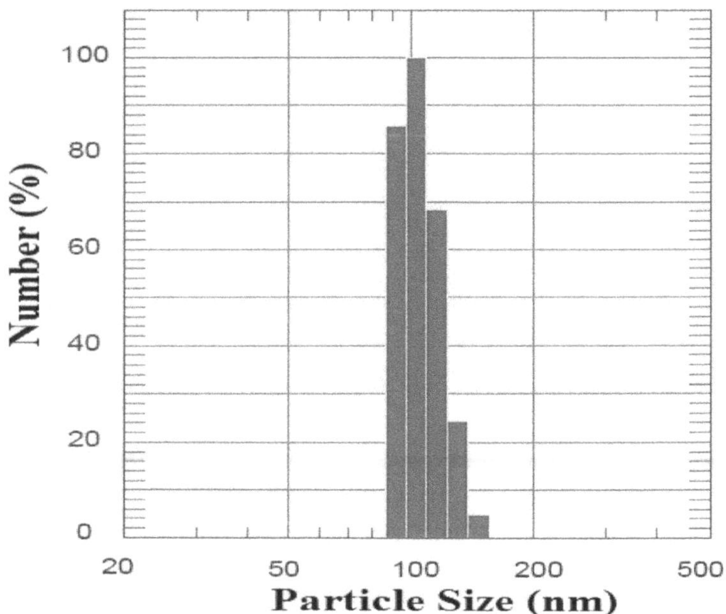

FIGURE 11.3 Particle size distribution of Zinc Phosphate nanopigment.

nanoparticles prepared by ultrasound method is found to be 119.8 nm, measured by Malvern Zetasizer Instrument (Malvern Instruments, Malvern, UK).This smaller particle size obtained probably due to an efficient micromixing and due to physical effects of the ultrasonic irradiation, which results into faster reaction to form ZP nanoparticles. The probable reason for this observation is possibly due to the facilitated reaction due to ultrasonic irradiation, which does not, provides enough time for growth and agglomeration of the particles.

11.3.1.2 XRD Analysis

Figure 11.4 depicts the XRD analysis of ZP nanoparticles (A) and ZP-2K Epoxy clear coat (B). The XRD pattern of ZP nanoparticles shows the diffraction peaks at 19.2, 22.1, 28.2, 31.1, 34.1, 34.4, 40, and 47° (2θ). It is

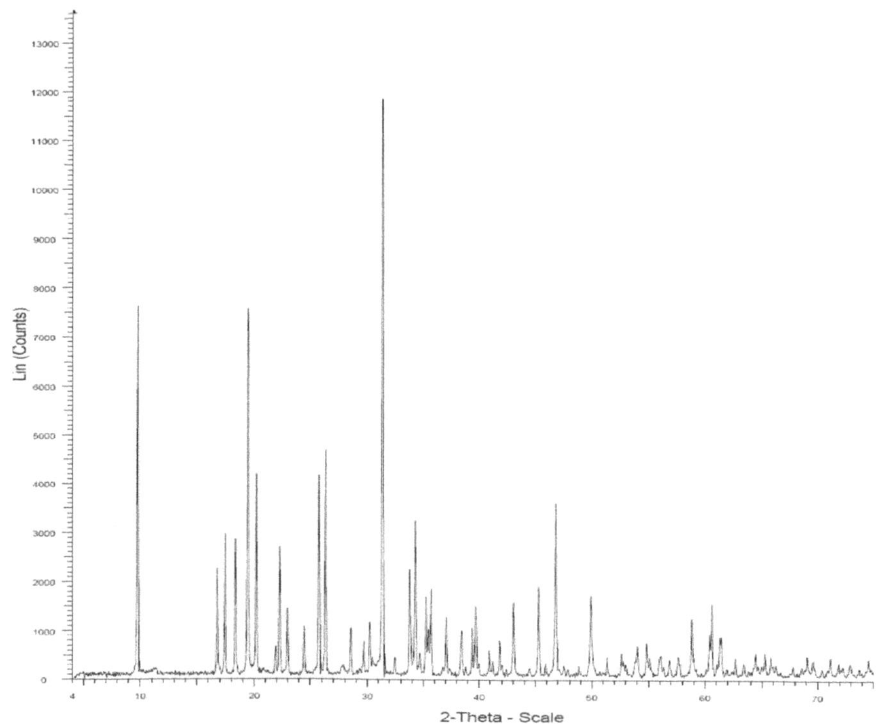

FIGURE 11.4 XRD analysis of Zinc Phosphate nanopigment.

found that the synthesized ZP nanoparticles are semi crystalline in nature [20] having monoclinic structure and observed phase of ZP nanoparticles is scheelite phase [21]. Ultrasound is known to have a major impact on the crystal structure and on the kinetics of crystallization. Ultrasound probe produces cavitation in the system, which leads to the production of free radicals, which results into improved crystal structure [22]. Further the crystallite size of ZP at $2\theta = 28.2°$ estimated from Debye Scherrer formula is 98.02 nm.

11.3.1.3 FTIR Analysis

The infrared spectra of the zinc phosphate nanoparticles in the range 4000–800 cm^{-1} has been shown in Figure 11.5. FTIR spectra of the zinc phosphate nanoparticles show stretching and bending modes. The O-H stretching broad band at 3207 cm^{-1} and 1695 cm^{-1} can be observed, implying the existence of crystal water in the synthesized product. The prominent peaks at 989, 1020 and 1160 cm^{-1} were the complex stretching vibrations of PO_4^{3-} group as reported [23–26]. The characteristic peaks at 1060 and 1120 cm^{-1} were the anti-symmetric stretching and symmetric

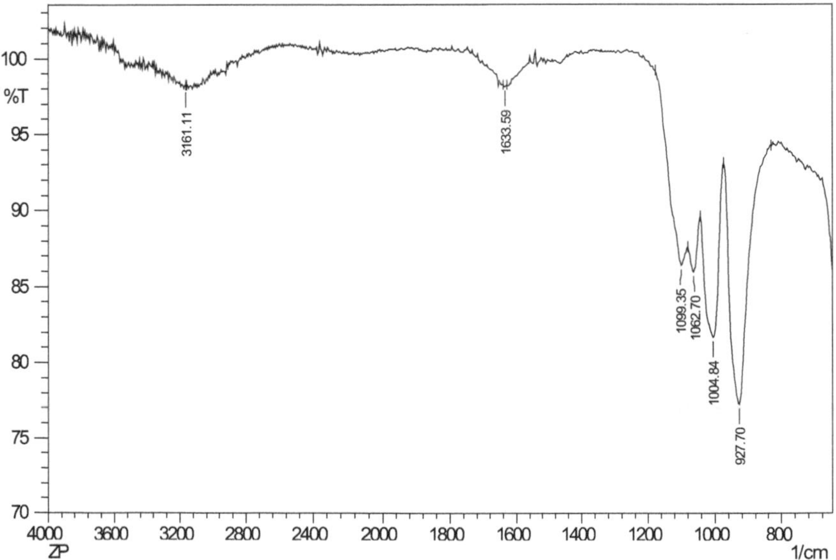

FIGURE 11.5 FTIR spectra of Zinc Phosphate nanopigment.

stretching of PO_4^{3-}, respectively. The P-O bending vibrations at 987 cm^{-1} were also observed.

11.3.1.4 SEM Analysis

Figure 11.6 shows the SEM analysis of ZP nanoparticles. The scanning electron microscopic analysis was performed on zinc phosphate nanoparticles. It can be seen that ZP nanoparticles shows distributed particles with lower agglomeration, which is attributed to efficient micromixing and cavitational effects generated by ultrasonic irradiations [27]. SEM analysis clearly shows the uniformly sized rod plates like structures of zinc phosphate nanoparticles.

11.3.2 CORROSION RATE ANALYSIS

Figure 11.7 shows the corrosion rate per year at different wt % loading of ZP nanoparticles in resin coatings. ZP nanoparticles were dispersed in the resin with the help of pigment muller and were then coated on MS

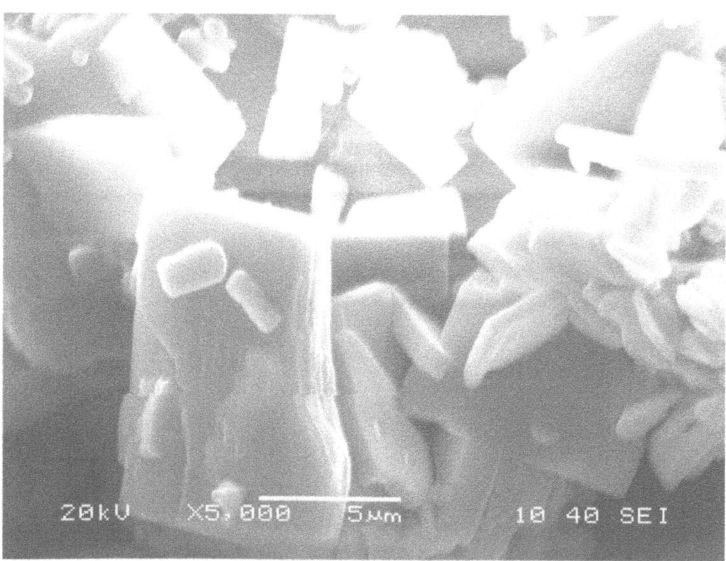

FIGURE 11.6 SEM Image of Zinc Phosphate nanopigment.

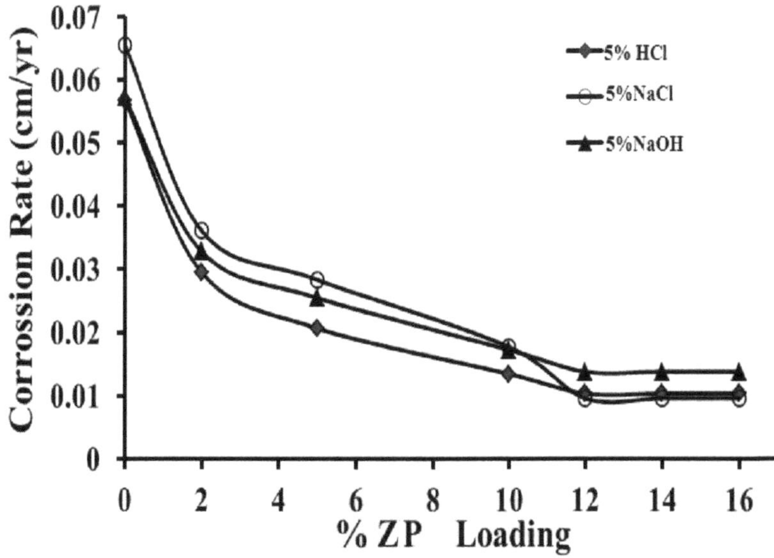

FIGURE 11.7 Corrosion rate analysis of Zinc Phosphate nanopigment in acid, basic and salt solution.

plates for corrosion rate analysis. Coated MS plates were further dipped in HCl, NaCl and NaOH (5 wt % each) solutions for a period of 750 h. The weight loss was measured by gravimetric analysis for the estimation of corrosion rate using dip test. Corrosion rate for 0% loading of ZP nanoparticles (Primer and 2K epoxy coat) was 0.056 cm/yr, 0.057 cm/yr and 0.066 cm/yr in case of 5% HCl, NaOH and NaCl solution respectively. From the Figure 11.7, it can be concluded that the corrosion rate was maximum for NaCl solution and minimum for HCl solution. The corrosion rate was found to vary at different percent loading of ZP nanoparticles in all the cases. 2% loading of zinc phosphate itself reduces the corrosion rate by 50% and then gradually further goes down with increase in the loading, till about 12% and then no further reduction was observed.

11.3.3 DRIED FILM PROPERTIES OF ZP-2K EPOXY SYSTEM

Table 11.1 shows the improvement in the resistance to neutral salt spray, water condensation and cyclic corrosion with an incorporation of zinc phosphate nanoparticles in the base resin. It was found that, resistance to

TABLE 11.1 Properties Evaluations of Dried Films for ZP-2K Epoxy Coating

Properties	Standards	ZP-2K Epoxy system (12%W/V)
Dry Film Thickness in microns	ASTM B499	320
Cross cut adhesion	ASTM D 3359	4B
Pull of adhesion (MPa)	ASTM D 4541	7
Resistance to neutral salt spray (hr pass)	ASTM B 117	440
Resistance to water condensation (hr pass)	ASTM D 2247	220
Resistance to cyclic corrosion (hr pass)	ASTM D 5894	720

neutral salt spray was for 440 hrs, water condensation was observed for 220 h, while resistance to cyclic corrosion was observed only till 56 h. This is due to the combination of epoxy and polyamide. Cyclic corrosion resistance is found to be less because epoxy has very poor stability against UV light due to aromatic rings in its backbone structure and double bond in the ring and gets easily affected by UV rays.

11.3.4 CORROSION MECHANISM, AND ROLE OF ZINC PHOSPHATE AND BINDER RESIN IN CORROSION INHIBITION

The corrosion mechanism of zinc rich coating and zinc phosphate based coating are distinguished here. Iron passivation occurs due to the presence of a zinc phosphate layer and the O_2 barrier effectiveness increases due to the coherent mass of epoxy zinc phosphate nano composite [28]. The inhibition of corrosion can possibly occur as zinc phosphate gets adsorbed on the metal surface, so the electrolytic corrosion cell formation is restricted. Schaefer et al. explains the improvement of electrochemical action of zinc-rich paints by the addition of nanoparticulate zinc [29]. When metal surface is coated with zinc rich coating, zinc metal Zn^{2+} gets converted to nonreactive complex by combining with O_2 and the moisture. The consumption of zinc metal continues this way until zinc gets completely exhausted [30]. Above mechanism of the coating is called as sacrificial coating. Schmidt et al. explains the action of corrosion inhibition due to sacrificial coatings in marine environment with respect to time [31]. Therefore, corrosion of ferrous metal panel gets delayed due to the sacrificial tendency of zinc against ferrous.

11.4 CONCLUSIONS

In the present study, nano zinc phosphate was prepared using ultrasound. The coating of 2K epoxy and the matrix of nano zinc phosphate along with polyamide were made over mild steel substrate. Evaluation for salt spray resistance, resistance to water condensation and cyclic corrosion test shows remarkable improvement in achieving corrosion inhibition. It was observed that when nano zinc phosphate was added to 2K epoxy polyamide coatings, the coatings show significant improvement in anti-corrosion properties. This improvement was based on the mechanism of adsorption where interface with the metallic substrate happened, also it filled the gaps of metal surface and promoted the formation of adherent oxide layer that prohibited the corrosion of underlying substrate by passivation.

ACKNOWLEDGMENTS

The author is thankful to Institute of Chemical Technology, Matunga, Mumbai for providing facility to carry out work, also author is grateful of CII, SERB and SUYOG NIRMITI for providing sufficient financial support to carry out research work.

KEYWORDS

- **Mild steel**
- **Paint coatings**
- **Passive films**
- **SEM**
- **Weight loss**
- **XRD**

REFERENCES

1. Fauvet, P., Balbaud, F., Robin, R., Tran, Q. T., Mugnier, A., & Espinoux, D. Corrosion Mechanisms of Austenitic Stainless Steels in Nitric Media used in Reprocessing Plant, *J. Nucl. Mater.* 2008, *375*, 52.
2. Glass, R. S., Overturf, G. E., Van Konynenburg, R. A., & McCright, R. D. Gamma Radiation Effects on Corrosion—I. Electrochemical Mechanisms for the Aqueous Corrosion Processes of Austenitic Stainless Steels Relevant to Nuclear Waste Disposal in Tuff, *Corros. Sci.* 1986, *26*, 577.
3. Thébaulta, F., Vuillemina, B., Oltra, R., Ogle, K., & Allely, C. Investigation of Self-Healing Mechanism on Galvanized Steels Cut Edges by Coupling SVET and Numerical Modeling, *Electrochim. Acta.* 2008, *53*, 5226.
4. Mekeridis, E. D., Kartsonakis, I. A., & Kordas, G. C. Multilayer Organic–Inorganic Coating Incorporating TiO_2 Nanocontainers Loaded with Inhibitors for Corrosion Protection of AA2024-T3, *Prog. Org. Coat.* 2012, *73*, 142.
5. Poznyak, S. K., Tedim, J., Rodrigues, L. M., Salak, A. N., Zheludkevich, M. L., Dick, L. F. P., & Ferreira, M. G. S. Novel Inorganic Host Layered Double Hydroxides Intercalated with guest Organic Inhibitors for Anti Corrosion Applications, *ACS Appl. Mater. Interfaces*. 2009, *1*, 2353.
6. Neto P. L., Araújo A. P., Araújo W. S., & Correia A. N. Study of the Anticorrosive Behaviour of Epoxy Binders Containing Non-Toxic Inorganic Corrosion Inhibitor Pigments, *Prog. Org. Coat.* 2008, *62*, 344.
7. Anandakumar S., Denchev Z., & Alagar M. Synthesis and Thermal Characterization of Phosphorus Containing Siliconized Epoxy Resins, *Eur. Polym. J.* 2006, *42*, 2419.
8. Johannes, K. F. *Epoxy Resins*, 2nd ed.; Reactive Polymers Fundamentals and Applications, 2013, 95–153.
9. Schweitzer, P. A. *Paint and Coatings: Applications and Corrosion Resistance*; CRC Press, 2006, 528–529.
10. Romagnoli, B. D. A. R., Vetere, V. F., Hernandez, L. S. Study of the Anticorrosive Properties of Zinc Phosphate in Vinyl Paints, *Prog. Org. Coat.* 1998, *33*, 28–35.
11. Czarnecka, B., Nicholson, J. W. Ionrelease, Dissolution and Buffering by Zinc Phosphate Dental Coments, *J. Mater. Sci-Mater M.* 2003, *14*, 601–604.
12. de Lima-Neto, P., de Araújo, A. P., Araújo, W. S., Correia, A. N. Study of the anticorrosive behavior of epoxy binders containing non-toxic inorganic corrosion inhibitor pigments, *Prog. Org. Coat.* 2008, *62*, 344–350.
13. Abd El-Ghaffar, M. A., Youssef, E. A. M., & Ahmed, N. M. High performance anticorrosive paint formulations based on phosphate pigments, *Pigm. Resin Technol.* 2004, *33*, 226–237.
14. Delamo, B., Romagnoli, R., & Vetere, V. F. Steel corrosion protection by means of alkyd paints pigmented with calcium acid phosphate, *Ind. Eng. Chem. Res.* 1999, *38*, 2310–2314.
15. Zeng, R., Lan, Z., Kong, L., Huang, Y., & Cui, H. Characterization of calcium-modified zinc phosphate conversion coatings and their influences on corrosion resistance of AZ31 alloy, *Surf. Coat. Technol.* 2011, *205*, 3347–3355.
16. Sonawane, S. H., Shirsath, S. R., Khanna, P. K., Pawar, S., Mahajan, C. M., Paithankar, V., Shinde, V., & Kapadnis, C. V. An innovative method for effective

micro-mixing of CO_2 gas during synthesis of nano-calcite crystal using sonochemical carbonization, *Chem. Eng. J.* 2008, *143*, 308–313.
17. Gogate, P. R., Tayal, R. K., & Pandit, A. B. Cavitation: a technology on the horizon, *Corr. Sci.* 2006, *91*, 35–46.
18. Mahulkar, A. V., Riedel, C., Gogate, P. R., Neis, U., & Pandit, A. B. Effect of dissolved gas on efficacy of sonochemical reactors for microbial cell disruption: experimental and numerical analysis, *Ultrason. Sonochem.* 2009, *16*, 635–643.
19. Pinjari, D. V., & Pandit, A. B. Cavitation milling of natural cellulose to nanofibrils, *Ultrason. Sonochem.* 2010, *17*, 845–852.
20. Ryu, J. H., Yoon, J. W., Lim, C. S., Oh, W. C., & Shim, K. B. Microwave-Assisted Synthesis of $CaMoO_4$ Nano-Powders by a Citrate Complex Method and its Photoluminescence Property, *J. Alloys Compd.* 2005, *390*, 245.
21. Raj, A. M. E. S., Mallika, C., Swaminathan, K., Sreedharan, O. M., & Nagaraja, K. S. Zinc (II) oxide-zinc (II) molybdate composite humidity sensor, Sensor Actuat B-Chem. 2002, *81*, 229–236.
22. Patrick, M., Blindt, R., & Janssen, J. The effect of ultrasonic intensity on the crystal structure of palm oil, *Ultrason Sonochem.* 2004, *11*, 251–255.
23. Zhang, M., Liu, J. K., Miao, R., Li, G. M., & Du, Y. J. Preparation and Characterization of Fluorescence Probe from Assembly Hydroxyapatite Nanocomposite, *Nanoscale Res. Lett.* 2010, *5*, 675–679.
24. Li, Y. B., & Weng, W. J. In vitro synthesis and characterization of amorphous calcium phosphates with various Ca/P atomic ratios, *J. Mater. Sci.: Mater. Med.* 2007, *18*, 2303–2308.
25. Wang, K. W., Zhou, L. Z., Sun, Y., Wu, G. J., Gu, H. C., Duan, Y. R., Chen, F., & Zhu, Y. J. Calcium phosphate/PLGA-mPEG hybrid porous nanospheres: A promising vector with ultrahigh gene loading and transfection efficiency, *J. Mater. Chem.* 2010, *20*, 1161–1166.
26. Pawlig, O., & Trettin, R. Synthesis and characterization of α-hopeite, $Zn_3(PO_4)_2\cdot 4H_2O$, *Mater. Res. Bull.* 1999, 1959–1966.
27. Bhanvase, B. A., Kutbuddin, Y., Borse, R. N., Selokar, N. R., Pinjari, D. V., Gogate, P. R., Sonawane, S. H., & Pandit, A. B. Ultrasound assisted synthesis of calcium zinc phosphate pigment and its application in nanocontainer for active anticorrosion coatings, *Chem. Eng. J.* 2013, *231*, 345–354.
28. Valanezhad, A., Tsuru, K., Maruta, M., Kawachi, G., Matsuya, S., & Ishikawa, K. Zinc phosphate coating on 316 L-type stainless steel using hydrothermal treatment. *Surf. Coat. Technol.* 2010, *205*, 2538–2541.
29. Schaefer, K., & Miszczyk, A. Improvement of electrochemical action of zinc-rich paints by addition of nanoparticulate zinc, *Corro. Sci.* 2013, *66*, 380–391.
30. Attanasio, S. A., & Latanision, R. M. Corrosion of rapidly solidified neodymium-iron-boron (Nd-Fe-B) permanent magnets and protection via sacrificial zinc coatings, *Mat. Sci. Eng. A-Struct.* 1995, *198*, 25–34.
31. Schmidt, D. P., Shaw, B. A., Sikora, E., Shaw, W. W., Laliberte, L. H. Corrosion protection assessment of sacrificial coating systems as a function of exposure time in a marine environment, *Prog. Org. Coat.* 2006, *57*, 352–364.

CHAPTER 12

BIOSYNTHESIS OF SILVER NANOPARTICLES USING *Raphanus sativus* EXTRACT

P. D. JOLHE,[1] B. A. BHANVASE,[2] V. S. PATIL,[1] and S. H. SONAWANE[3]

[1]*University Institute of Chemical Technology, North Maharashtra University, Jalgaon, Maharashtra, India*

[2]*Department of Chemical Engineering, Laxminarayan Institute of Technology, Nagpur–440033, Maharashtra, India*

[3]*Department of Chemical Engineering, National Institute of Technology Warangal, Telangana–506004, India*

CONTENTS

12.1 Introduction 200
12.2 Materials and Methods 202
  12.2.1 Materials 202
  12.2.2 Preparation of *Raphanus sativus* Bio-extract 202
  12.2.3 Synthesis and Characterization of Silver Nanoparticles 203
12.3 Results and Discussions 204
  12.3.1 XRD Analysis of Silver Nanoparticles 204
  12.3.2 Morphological Analysis of Silver Nanoparticles 205
  12.3.3 Effect of Reaction Time in Batch Reactor 207

12.3.4 Effect of *Raphanus sativus* Bio-Extract on Silver Nanoparticles Formation 207
12.3.5 Effect of $AgNO_3$ Concentration 210
12.3.6 Effect of Surfactant (SDS) Loading 212
12.4 Conclusion 212
Keywords 214
References 214

12.1 INTRODUCTION

In the last few years there has been an increasing impact on materials and surface science by the research on the synthesis and modification of nanoparticles [1]. Noble metal nanoparticles such as gold, silver and platinum are widely applied in medicinal applications and nanotechnology is emerging as a rapidly growing field with its application in science and technology [2]. Nanoparticles are clusters of atoms in the size range of 1–100 nm. The metallic nanoparticles are most promising as they show good antibacterial properties due to their large surface area to volume ratio, which is coming up as the current interest in the researchers due to the growing microbial resistance against metal ions, antibiotics and the development of resistant strains [3].

Nanoparticles have been used as a physical approach to alter and improve the effective properties of some types of the synthetic chemical pesticides or in the production of bio-pesticides directly [4]. Metal nanoparticles embedded paints synthesized using vegetable oils have been found to have good anti-microbial activity [5]. It is well known, that silver is an effective antibacterial agent and possesses a strong antibacterial activity against bacteria, viruses and fungi, the mechanism and the manner of action still being unknown [6]. The toxicity of nanomaterials is influenced by the structural features such as size, shape, composition and the surface chemistry. There is a growing need to develop an environmentally friendly process for the synthesis of nanoparticles that does not employ toxic chemicals [7]. The development of reliable green process for the synthesis of silver nanoparticles is an important aspect of current nanotechnology research. Recently biosynthesis of silver nanoparticles (AgNPs) was achieved by a novel, simple green chemistry procedure using citrus sinensis peel extract as a reducing and a capping agent, Parthenium leaf, extracts of Ananas comosus, leaf broth of Ocimum sanctum, aqueous leaves extracts of Euphorbia prostrate, Pomegranate peel extract, Cassia auriculata flower, extract of Lantana camara fruit, culture of supernatant of Pseudomonas aeruginosa, *Myrica esculenta* leaf extract, switchgrass (Panicum virgatum) extract, plant leaf extracts (Lantana camara) [8–17]. Biosynthesis of nanoparticles is a kind of bottom up approach where the main reaction occurring is reduction/oxidation. The microbial enzymes or the plant phytochemicals with anti oxidant or reducing properties are usually responsible for reduction of metal compounds into their respective nanoparticles. Chemical synthesis methods lead to presence of some toxic chemical absorbed on the surface that may have adverse effect in the medical applications. Biosynthesis provides advancement over chemical and physical method as it is cost effective, environment friendly, easily scaled up for large scale synthesis and in this method there is no need to use high pressure, energy, temperature and toxic chemicals [18, 19] Reducing the particle size of materials is an efficient and reliable tool for improving their biocompatibility. In fact, nanotechnology helps in overcoming the limitations of size nanomaterials can be modified for better efficiency to facilitate their applications in different fields such as bioscience and medicine [20] Chemical reduction also has obvious disadvantages. For instance, most of the reactants used in the reaction system are toxic chemical agents, which have potential risks for environment and health. As above-mentioned, the protective agent on the surface of nanoparticles may have influence on the characteristics of nanoparticles. The development of dependable, environmentally benign process for the synthesis of nanoscale materials is an important aspect of nanotechnology. Pharmaceutical companies and researchers in this field are searching for new antibacterial agents due to the outbreak of the infectious diseases caused by different pathogenic bacteria and the development of antibiotic resistance [21]. The current energy intensive and toxic chemicals employing chemical methods for synthesizing nanoparticles are becoming outdated, expensive and inefficient and highlight the growing need for the uses of bio-compatible, non-toxic, cost-effective and eco-friendly methods for production of silver nanoparticles [22]. For silver, this precious metal was originally used as an effective antimicrobial agent and as a disinfectant, as it was relatively free of adverse effects. However, with the development of modern antibiotics for the treatment of infectious diseases, the use of silver agents in the clinical setting had been restricted mainly to topical silver sulfadiazine cream in the treatment of burn wounds [23]. Working with nanomaterials has allowed a better understanding of molecular biology. As a consequence, there is the potential of providing novel methods for the treatment of diseases which were previously difficult to target due to size restrictions. For biomedical applications, the synthesis of bio-functional nanoparticles is very important, and it has recently drawn the attention of numerous research groups, making this area constantly evolve [24].

Considering the growing technological demand for eco-friendly and development of reliable process for the synthesis of silver nanoparticles, the present work was undertaken. Therefore, in the present work, biosynthesis of silver nanoparticles was achieved by a novel, simple green chemistry procedure using *Raphanus sativus* bio-extract as a reducing and a capping agent in a batch reactor was carried out.

12.2 MATERIALS AND METHODS

12.2.1 MATERIALS

Analytical grade silver nitrate ($AgNO_3$, 99%) and sodium dodecyl sulfate (SDS, $NaC1_2H_25SO_4$, 99%) were obtained from Merck Specialties Pvt. Ltd., Mumbai and Thomas Baker (Chemical) Ltd., respectively, and used as received. Millipore deionized water (resistivity > 1 MΩcm) was used in all experimentation. The bio-extract of *Raphanus sativus* was used as a reducing agent during preparation of Ag colloidal nanoparticles.

12.2.2 PREPARATION OF Raphanus sativus BIO-EXTRACT

Fresh *Raphanus sativus* was purchased from local market and thoroughly washed with distilled water. Further, *Raphanus sativus* was peeled to remove the outermost dry layer and the peeled *Raphanus sativus* was finely cut into small pieces and ground in laboratory grinder to produce a homogenous paste. After filtration through cheese cloth, the filtrate, for example, *Raphanus sativus* bio-extract was collected in an Erlenmeyer flask. The obtained bio-extract was cold centrifuged at 8000 rpm and 0°C for 15 min in order to avoid the degradation of active reducing compounds and remove the remained *Raphanus sativus* biomass. The supernatant obtained was stored at 4°C and then it is used for the synthesis of silver nanoparticles in microreactor.

12.2.3 SYNTHESIS AND CHARACTERIZATION OF SILVER NANOPARTICLES

Silver nitrate ($AgNO_3$) was used as a precursor for silver nanoparticles synthesis. Aqueous solution (1 mM) of silver nitrate was prepared. 5 mL of *Raphanus sativus* bio-extract was added to 50 mL of 1 mM $AgNO_3$ solution with 0.001 g SDS as surfactant at 40°C. The reaction was agitated on a magnetic stirrer. The progress of reaction was monitored using UV-Vis spectroscopy. The reaction mixture turns from colorless to yellow and ultimately to brown color indicating formation of silver nanoparticles. The solution turned from colorless to faint yellow within 30 min of agitation. Figure 12.1 represents the visual observation of colloidal silver nanoparticles as a function of reaction time. As reaction time is increases,

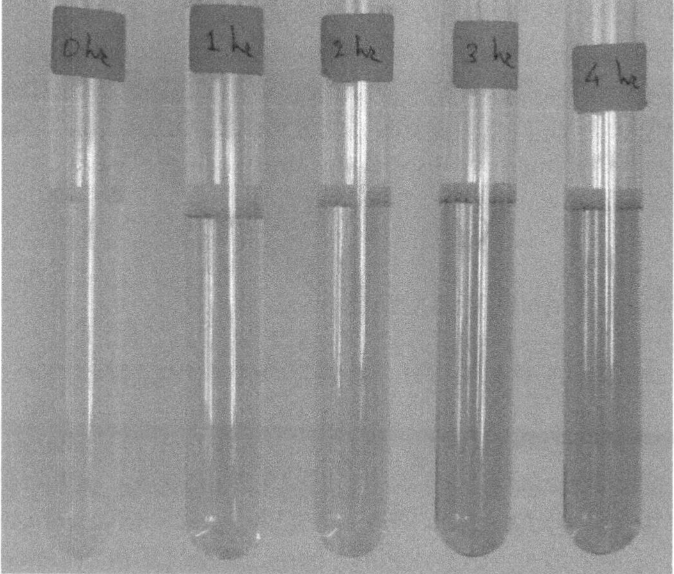

FIGURE 12.1 Visual observations of Ag Nanoparticle Samples indicating progress of reaction in batch reactor.

the yellowish-brown color in the aqueous solution of silver nanoparticles is observed to be darker, an indication of formation of more silver nanoparticles with an increase in the reaction time. The supernatant obtained from cold centrifugation of reaction mixture at 8000rpm for 15 min. to pellet out *Raphanus sativus* biomass and aggregate silver nanoparticles was further used for silver nanoparticles analysis by preliminary test, UV-Vis Spectroscopy, Dynamic light-scattering (DLS Malvern Instruments Ltd, USA) for measurements of particle size distribution. Preliminary test indicating silver nanoparticles formation was done by adding NaCl solution to the reaction mixture, which turned its color from yellow brown to gray with visible precipitation at the bottom of test tube [25]. UV-Vis Spectroscopy is an important technique of silver nanoparticles analysis giving a characteristic peak in the range of 400–450 nm due surface plasmon resonance of AgNPs [12]. The bioreduction $Ag+$ ions in the solution were observed on UV-Vis Spectrophotometer in 200–800 nm range [SHIMADZU 160A model].

Preparation of SEM samples of silver nanoparticles was accomplished by placing a drop of silver hydrosol on 0.5×0.5 mm^2 cover slips allowing water to completely evaporate. High-resolution SEM images were obtained. SEM observations were carried out on LEO-1530 electron microscope (LEO, Germany) and energy dispersive X-ray (EDX) analysis was performed on Tecnai F30 microscope (FEI, The Netherlands). The size distribution and average size of the nanoparticles were estimated on the basis of SEM micrographs with the assistance of Sigma Scan Pro software (SPSS Inc., version 4.01.003).

12.3 RESULTS AND DISCUSSIONS

12.3.1 XRD ANALYSIS OF SILVER NANOPARTICLES

Figure 12.2 depicts the XRD pattern of colloidal silver nanoparticles prepared in batch reactor system. In Figure 12.2, the X-ray diffraction peaks of silver nanoparticles has at 37.5 and 43.8° which are corresponding to the [1–1 1] and [2–0 0] crystalline planes of Ag (JCPDS cards 4-0783) [26]. The broader nature of the XRD peaks is obtained due to the nanosize of the particles. The unassigned peaks might be due to the presence

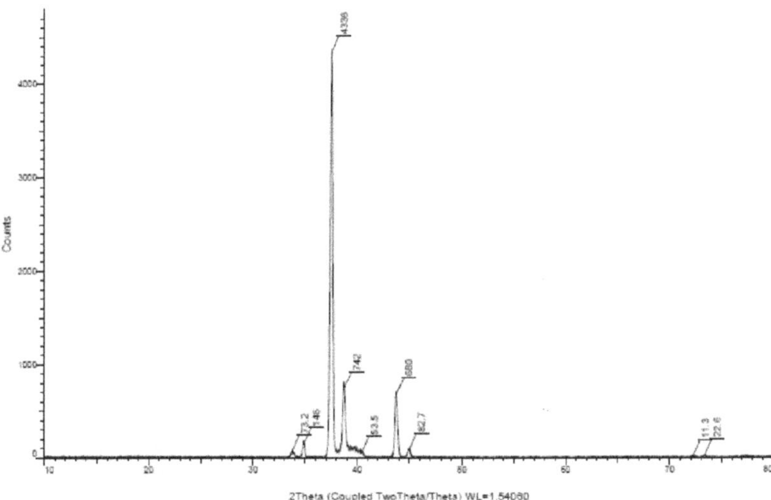

FIGURE 12.2 XRD spectrums for powdered Ag Nanoparticle Samples in batch reactor.

of bioorganic phase on the surface of the Ag nanoparticles. The crystallite size of the colloidal silver nanoparticles prepared by the biosynthesis process in batch reactor was found to be 35 nm. It was determined using Debye Scherrer's equation given by:

$$X_d = \frac{k\lambda}{\beta \cos\theta} \tag{1}$$

where $k = 0.9$, β = full-width at half-maximum height (FWHM) and θ is glancing angle of X rays with the sample holder. CuK_{α} angle $\lambda = 1.5406$ nm radiation was used to obtain XRD patterns.

12.3.2 MORPHOLOGICAL ANALYSIS OF SILVER NANOPARTICLES

Figure 12.3 depicts the SEM images of silver nanoparticles synthesized by batch reactor. The images confirm the formation of nanoparticles capped with its biomoites. The high-resolution study of the nanoparticles using SEM revealed that the AgNPs are polydispersed and spherical in shape. The size of the particles obtained distinct spherical morphology of silver

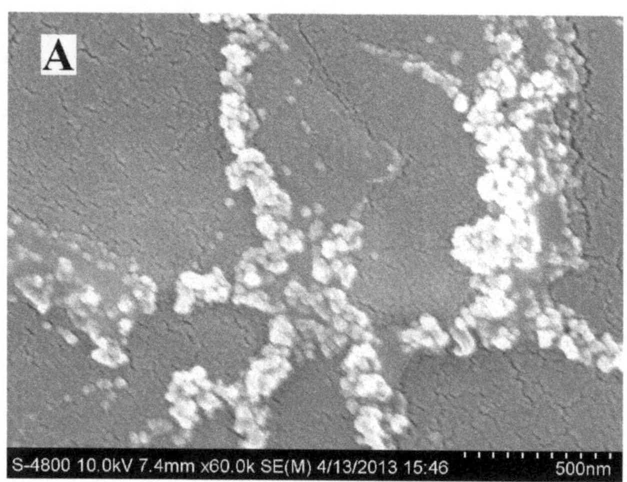

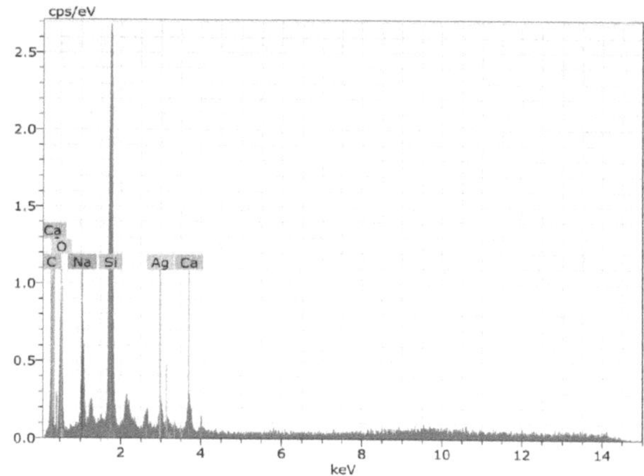

FIGURE 12.3 (a) SEM images; (b) EDAX spectrum of silver nanoparticles of samples in batch reactor.

nanoparticles of particle size in the range of 50–100 nm is observed in the case of Figure 12.3. The particles of silver nanoparticles with significantly smaller size and uniform shape were observed. This is because of the significant enhancement in solute transfer rate, nucleation, and number of nuclei due to physical effects of mixing.

However, Figure 12.3(b) shows the EDAX spectrum of AgNPs synthesized at 40°C. Metallic silver nanoparticles generally show typical absorption peak approximately at 3 keV due to surface plasmon resonance [26]. The presence of the elemental silver due to reduction of silver ions can be observed in the graph obtained from EDAX analysis, which also supports the XRD results. The synthesized nanoparticles were stable in solution over a period of nine months at room temperature.

12.3.3 EFFECT OF REACTION TIME IN BATCH REACTOR

It is known that the presence of silver nanoparticles exhibits the yellowish-brown color in the aqueous solution [27]. Formation of silver nanoparticles was accomplished by the reduction of silver ions with the aid of *Raphanus sativus* extract in batch reactor leading to change in the absorbance, particle size and color change. UV-Vis spectra of the reaction mixture were recorded as a function of time and are depicted in Figure 12.4(a). The absorbance at 470 nm increased to 0.62 *A.U.* for reaction time 120 min from 0.51 *A.U.* at 30 min reaction time. This increase is attributed to increased conversion of silver ions to silver nanoparticles with an increase in the reaction time by using *Raphanus sativus* extract as a reducing agent. The effect of reaction time on particle size distribution is reported in Figure 12.4(b). The larger particle size of silver nanoparticles at 60 min reaction time is in the range from 149 to 266 nm, attributed to limited nuclei formation at the end of 30 min whereas the smaller particle size of silver nanoparticles in the range of 103 to 230 nm range at 60 min attributed to sufficient number of nuclei formation. Further increase in the reaction time (larger residence time) leads to increase in the particles size because of significant crystal growth of silver nanoparticles.

12.3.4 EFFECT OF *Raphanus sativus* BIO-EXTRACT ON SILVER NANOPARTICLES FORMATION

Figure 12.5(a) shows UV-Vis spectra of the colloidal silver nanoparticles obtained for different *Raphanus sativus* bio-extract. The volume of *Raphanus sativus* bio-extract was varied from 5 to 9 mL in order to investigate the effect of *Raphanus sativus* bio-extract concentration on silver nanoparticle formation. It was found that the peaks were shifted towards higher wavelength with an increase in the volume of *Raphanus sativus* bio-extract from 5 to 9 mL (with significant increase in the absorbance 0.46–0.98 *A.U.*). This is possible at lower volume of 6 mL. An increase

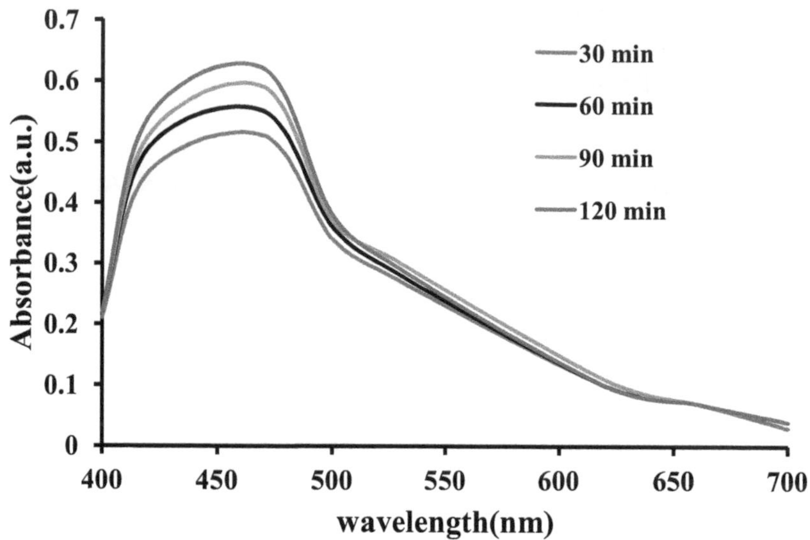

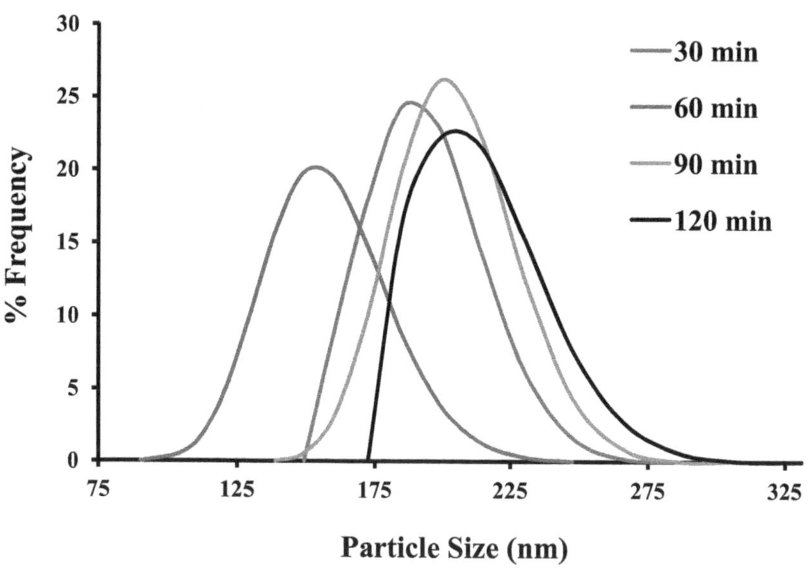

FIGURE 12.4 Effect of reaction time on (a) silver nanoparticle formation and (b) particle size distribution in batch reactor.

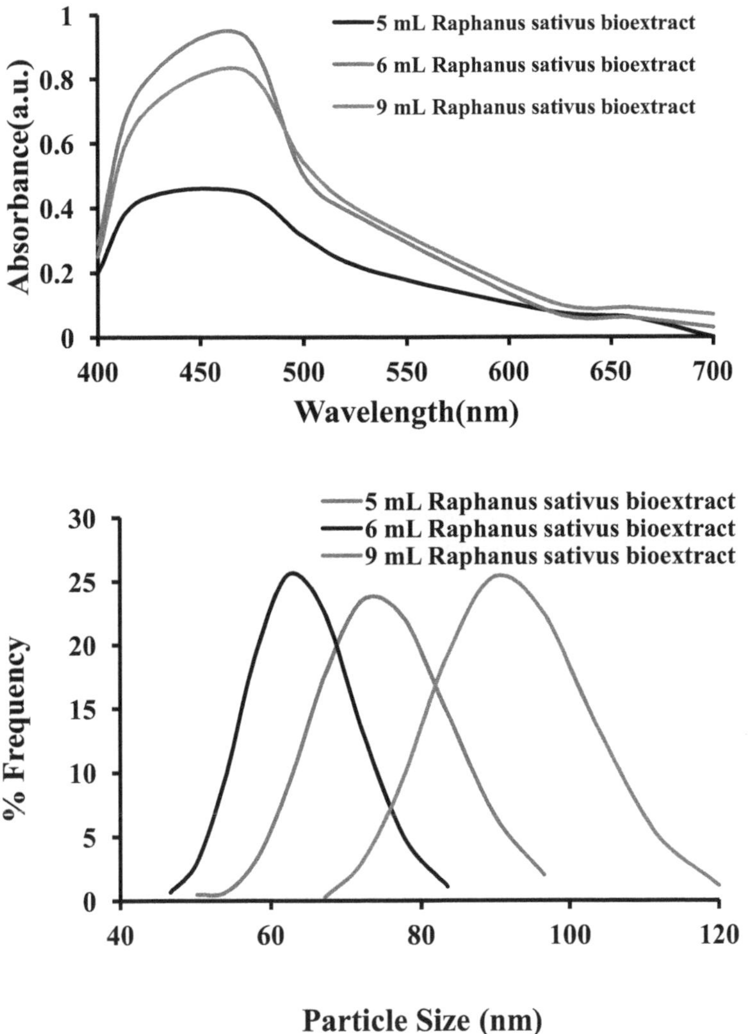

FIGURE 12.5 Effect of volume of *Raphanus sativus* bio-extract on (a) silver nanoparticle formation and (b) particle size distribution in batch reactor.

in the volume of *Raphanus sativus* bio-extract from 5 to 9 results into decrease in the absorbance of the samples from 1.11 to 0.81 *A.U.* It is attributed to the reduction in the number of silver nanoparticles for fixed quantity of silver nitrate in larger volume of bio-extract.

Figure 12.5(b) shows the effect of volume of *Raphanus sativus* bio-extract on particle size distribution in batch reactor. It is observed that the particle size range for 5 mL volume ratio of silver nitrate to bio-extract is 50–98 nm. The particle size range is found to be increased from 48–83 to 68–120 nm with an increase in the volume ratio of bio-extract from 6 to 9 mL. It is attributed to nucleation at limited places at higher volume of bio-extract, which leads to significant growth of particles. However, in the case of lower volume of silver nitrate to bio-extract the more nucleation events take place in the reaction medium leading to limited growth of silver nanoparticles and particle size is found to be decreased.

12.3.5 EFFECT OF $AGNO_3$ CONCENTRATION

The effect of $AgNO_3$ concentration on silver nanoparticles formation was studied by varying the concentration of $AgNO_3$ in microreactor and depicted in Figure 12.6. As discussed in earlier section the absorbance of product sample was observed to be higher for 6 mL *Raphanus sativus bio-extract* volume. The $AgNO_3$ concentration was varied from 0.01 to 0.04 M. All the experiments to study the effect of $AgNO_3$ concentration were carried out with an addition of SDS at 0.001 g/mL concentration. It is found that with an increase in the $AgNO_3$ concentration, the absorbance is found to be increased from 0.42 to 0.95 *A.U.* The decrease in the absorbance with decrease in the $AgNO_3$ concentration is attributed to the electron surface scattering. For smaller particles electron reaches the surface faster and scatters quickly which results in broadening in absorption peak width. The PWHM values are also found to be increased with decrease in the concentration of $AgNO_3$ from 0.02 to 0.04 M.

Further with an increase in the concentration of $AgNO_3$ the particles size is found to be increased. At 0.01 M $AgNO_3$ concentration the silver nanoparticles size range is in between 50 to 98 nm. However, with an increase in the concentration of $AgNO_3$ the particle size is found to be in the range 40 to 130 nm. The possible reason for increased particle size is attributed to increase in the number of silver ions with an increase in the $AgNO_3$ concentration, which is responsible for increase in the particle size of silver nanoparticles.

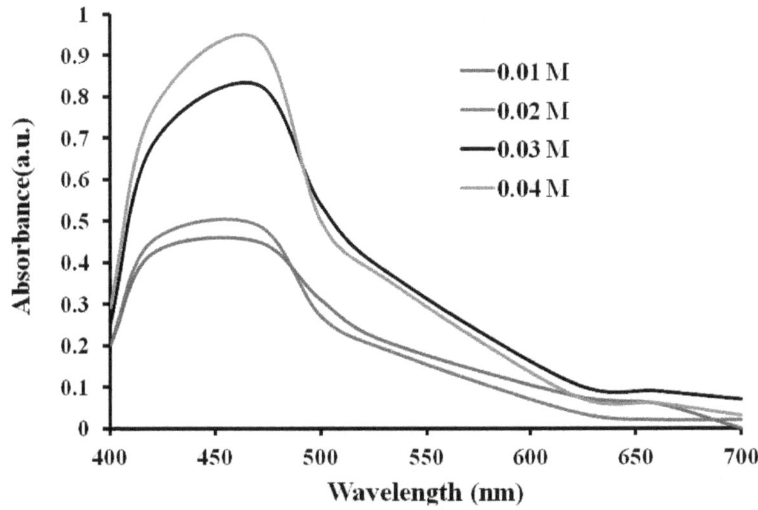

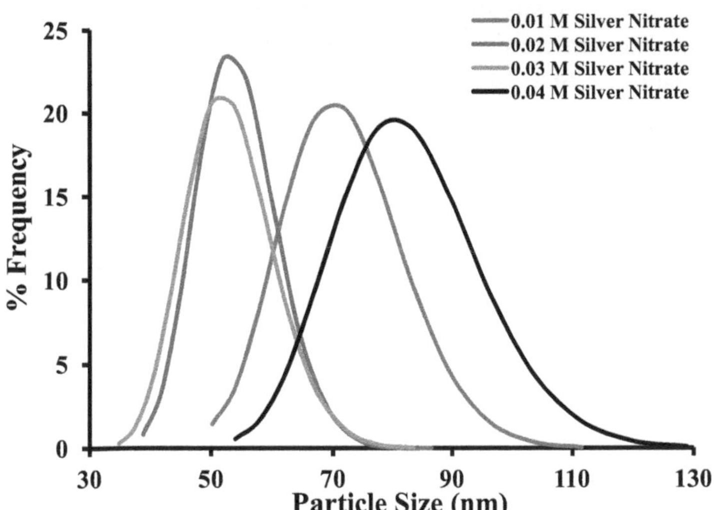

FIGURE 12.6 Effect of $AgNO_3$ concentration on (a) silver nanoparticle formation and (b) particle size distribution in batch reactor.

12.3.6 EFFECT OF SURFACTANT (SDS) LOADING

The effect of surfactant loading was studied in Figure 12.7. During preparation of silver nanoparticles, there is a possibility of agglomeration of silver nanoparticles without use of surfactant. SDS, an anionic surfactant was used to prevent the agglomeration of silver nanoparticles. The repulsive forces due to presence of negative charge on micelles avoid formation of cluster of silver nanoparticles and results into small particle size. Hence, SDS provides a capping effect that can control nucleation and growth of nanoparticles [22]. For the *Raphanus sativus* bio-extract 6 mL, the absorbance was found to be increased with an increase in the SDS quantity from 0 to 0.075 g and further it is found decreased for 0.0125 g SDS loading. It is observed that the absorbance value is increased from 0.23 to 0.39 *A.U.* with an increase in the SDS loading from 0.075 g to 0.0125 g, whereas it is found to be decreased to 0.23 *A.U.* for 0.075 g SDS loading. Also, the particle size of silver nanoparticles was found to be decreased with an increase in the SDS loading. The particle size for 0.1 g and 0.0125 g SDS loading is found to be in the range of 79 to 149 nm. However, it is observed to be decreased for 0.0125 g SDS loading and found in the range of 111 to 185 nm. It is attributed to more stabilization of formed nuclei and silver nanoparticles with significantly reduced aggregation with an increase in the SDS loading resulting into decreased particle size of formed silver nanoparticles [22].

Finally, we can infer that with an increase in the SDS loading, there is an increase in the number of nuclei formation which consecutively leads to decrease in the particle size of the formed Ag nanoparticles and the presence of negative charge because of anionic surfactant, SDS, also leads to repulsion of micelles with negative charges, which significantly reduces an agglomeration effect of silver nanoparticles resulting into smaller sized silver nanoparticles (Figure 12.7).

12.4 CONCLUSION

Synthesis of silver nanoparticles with the use of $AgNO_3$ and *Raphanus sativus* bio-extract was successfully carried out in batch reactor system. The use of natural, renewable and low-cost bioreducing agent has been demonstrated in this work. The observations from different characterization

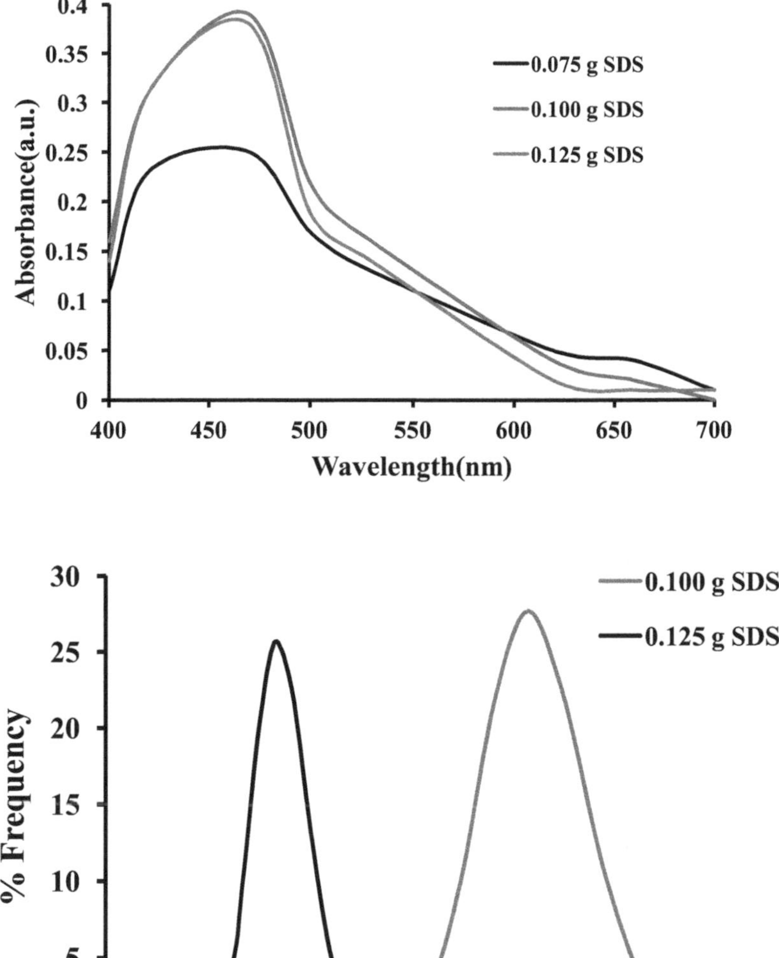

FIGURE 12.7 Effect of SDS concentration on (a) silver nanoparticle formation and (b) particle size distribution in batch reactor.

techniques checked that if sufficient time is given to complete the reaction. The minimum size of the silver particles is found to be 70 nm for 0.0125 g SDS loading with 0.01 M $AgNO_3$ and *Raphanus sativus* bio-extract. Further, the observations are consistent with the expected effect of time, volume ratio, surfactant concentration, $AgNO_3$ concentration in the batch reactor. Also the process devised in this work largely depends on the characteristics of the raw materials used, for example, aqueous silver nitrate and the *Raphanus sativus* bio-extract, regardless of the batch reactors.

KEYWORDS

- **$AgNO_3$**
- **Batch reactor**
- **Green synthesis**
- ***Raphanus Sativus* extract**
- **SDS**
- **Silver nanoparticles**

REFERENCES

1. Sastry, M., Mukherjee, P., Patra, C. R., Ghosh, A., & Kumar, R., Characterization and Catalytic Activity of Gold Nanoparticles Synthesized by Autoreduction of Aqueous Chloroaurate Ions with Fumed Silica. *Chem. Mater.* 2002, *14*, 1678–1684.
2. Albrecht, M. A., Evan, C. W., & Raston, C. L., Green chemistry and the health implications of nanoparticles. *Green Chem.* 2006, *8*, 417–432.
3. Kalu, K., Palanisamy, K., Paramasivam, H., & Loganathan, T., Mechanistic aspects: Biosynthesis of silver nanoparticles from Proteus mirabilis and its antimicrobial study, *Asian Pacific Journal of Tropical Biomedicine* 2012, 1–4
4. Abdul Hameed, M., & Al-Samarrai, Nanoparticles as alternative to pesticides in management plant diseases—A review, *International Journal of Scientific and Research Publications* 2012, *2*, 4.
5. Kumar, R., Singh, S., & Singh, O. V., Bioconversion of lignocellulosic biomass: biochemical and molecular perspectives, *J. Ind. Microbial. Biotechnol.* 2008, *35*, 377–391.
6. Sharma, V. K., Yngard, R. A., & Lin, Y., Silver nanoparticles: Green synthesis and their antimicrobial Activities. *Adv. Colloid Interface Sci.* 2009, *145*, 83–96.

7. Yu, D. G., Formation of colloidal silver nanoparticles stabilized by Na+–poly(γ-glutamic acid)–silver nitrate complex via chemical reduction process. *Colloids Surf. B* 2007, *59*, 171–178.
8. Kaviyaa, S., Santhanalakshmia, J., Viswanathanb, B., Muthumaryc, J., & Srinivasanc, K., Biosynthesis of silver nanoparticles using citrus sinensis peel extract and its antibacterial activity, *Spectrochimica Acta Part A* 2011,*79*, 594–598.
9. Ahmad, N., & Sharma, S., Green Synthesis of Silver Nanoparticles Using Extracts of Ananas comosus, *Green and Sustainable Chemistry* 2012, *2*, 141–147.
10. Mallikarjunaa, K., Narasimhab, G., Dillipa, G. R., Praveenb, B., Shreedharc, B., SREE Lakshmi, C., Reddy, B. V. S., & Deva Prasad Raju, B., Green synthesis of silver nanoparticles using ocimum leaf extract and their characterization, *Digest Journal of Nanomaterials and Biostructures* 2011, *6*, 181–186.
11. Abduz Zahir, A; Bagavan, A., Kamaraj, C., Elango, G., Abdul Rahuman, A., Efficacy of plant-mediated synthesized silver nanoparticles against Sitophilus oryzae. *J. Biopest. Supplementary* 2012, *5*, 95–102.
12. Ahmad, N., Sharma, S., & Rai, R., Rapid green synthesis of silver and gold nanoparticles using peels of Punica granatum. *Adv. Mat. Lett.* 2012, *3*, 376–380.
13. Velavan, S,; Arivoli, P., & Mahadevan, K. Biological reduction of silver nanoparticles using cassia auriculata flower extract and evaluation of their in vitro antioxidant activities, *Nanoscience and Nanotechnology: An International Journal* 2012; *2*, 30–35.
14. Sivakumar, M., Chandran, N., & Renganathan, S., Synthesis of silver nanoparticles using lantana camara fruit extract and its effect on pathogens, *Asian J Pharm Clin Res* 2012, *5*, 97–101.
15. Phanjom, P., Elizabeth Zoremi, D., Mazumder, J., Saha M., and Buzar Baruah, S., Green Synthesis of Silver Nanoparticles using Leaf Extract of Myrica esculenta, *International Journal of NanoScience and Nanotechnology* 2012, *3*, 73–79.
16. Mason, C., Vivekanandhan, S., Misra, M., & Mohanty, A., Switchgrass (Panicum virgatum) Extract Mediated Green Synthesis of Silver Nanoparticles, *World Journal of Nano Science and Engineering* 2012, *2*, 47–52.
17. Thirumurugan, A., Tomy, N., Kumar, H., & Prakash, P. Biological synthesis of silver nanoparticles by Lantana camara leaf extracts, *International Journal of Nanomaterials and Biostructures* 2011, *1*, 22–24.
18. Mukunthan, K. S., Elumalai, E. K., Trupti N. P., & Ramachandra Murty V., Catharanthus roseus: a natural source for the synthesis of silver nanoparticles, *Asian Pacific Journal of Tropical Biomedicine* 2011, 270–274.
19. Thirumurgan, A., Tomy, N.A., Jai Ganesh, R., & Gobikrishnan, S., Biological reduction of silver nanoparticles using plant leaf extracts and its effect an increased antimicrobial activity against clinically isolated organism. *De. Phar. Chem* 2010, *2*, 279–284.
20. Mirkin, C. A., & Taton, T. A., Semiconductors meet biology. *Nature* 2000, *405*, 626–627.
21. Singh, M., Singh, S., Prasad, S., & Gambhir I. S., Nanotechnology in medicine and antimicrobial effect of silver nanoparticles, *J. Nanomaterials and Biostructures* 2008, *3*, 115–122.
22. Vijayaraghavan, K., & Nalini, S. P. K., Biotemplates in the green synthesis of silver nanoparticles. *Biotechnol. J.* 2010, *5*, 1098–1110.
23. Uygur, F., Oncül, O., Evinç, R., Diktas, H., Acar A., & Ulkür, E., Effects of three different topical antibacterial dressings on Acinetobacter baumannii-contaminated full-thickness burns in rats. *Burns* 2009, *35*, 270–273.

24. Blanco-Andujar, C., Le Duc T., Nguyen T., & Thanh, K., Synthesis of nanoparticles for biomedical applications. *Annu. Rep. Prog. Chem., Sect. A* 2010, *106*, 553–568.
25. Ravi Kumar, D., Kasture, M., Prabhune, A., Ramana, C., Prasad, B., & Kulkarni, A., Continuous flow synthesis of functionalized silver nanoparticles using bifunctional biosurfactants. *Green Chem.* 2010, *12*, 609–615.
26. Al-Thabaiti, S. A., Al-Nawaiser, F. M., Obaid, A. Y., Al-Youbi A. O., & Khan, Z., Silver nanoparticles: preparation, characterization, and kinetics. *Colloids Surfs. B: Biointerfaces* 2008, *67*, 230.
27. Ahmad, N., Sharma S., & Rai, R., Rapid green synthesis of silver and gold nanoparticles using peels of Punica granatum, *Adv. Mat. Lett.* 2012, *3*, 376.
28. Shankar S. S., Rai, A., Ahmad, A., & Sastry, M., Rapid synthesis of Au, Ag, and bimetallic Au core Ag shell nanoparticles using Neem (*Azadirachta indica*) leaf broth. *J. Colloid Interface Sci.* 2004, *275*, 496–502.

CHAPTER 13

ACTIVATED CARBON FROM KARANJA (*PONGAMIA PINNATA*) SEED SHELL BY CHEMICAL ACTIVATION WITH PHOSPHORIC ACID

M. L. MESHRAM[1] and D. H. LATAYE[2]

[1]*Department of Civil Engineering, Laxminarayan Institute of Technology, Nagpur, India.*
E-mail: manojlmesh@gmail.com

[2]*Department of Civil Engineering, V.N.I.T., Nagpur, India*

CONTENTS

13.1 Introduction ... 218
13.2 Materials and Methods ... 218
13.2.1 Preparation of Activated Carbon ... 218
13.2.2 Characterization of Activated Carbon ... 219
13.2.2.1 Particle Size Analysis ... 219
13.2.2.2 Proximate Analysis ... 219
13.2.2.2.1 Moisture Content ... 219
13.2.2.2.2 Ash Content ... 220
13.2.2.2.3 Volatiles Content ... 220
13.2.2.2.4 Fixed Carbon ... 221
13.2.2.3 Point of Zero Charge (pHpzc) ... 221

13.2.2.4 Surface Area and Pore Volume Analysis 222
13.2.2.5 X-Ray Diffraction Analysis 222
13.2.2.6 Fourier Transmission Infrared Spectra 222
13.2.2.7 Scanning Electron Microscopy 222
13.3 Results and Discussion 222
13.3.1 Characteristics of the Carbon 222
13.3.2 Point of Zero Charge (pHPZC) 223
13.3.3 X-Ray Diffraction Analysis 223
13.3.4 Fourier Transmission Infrared Spectra 224
13.3.5 Scanning Electron Microscopy 225
13.4 Conclusions 225
Keywords 226
References 226

13.1 INTRODUCTION

Preparation of low cost activated carbons has been always on priority by researchers. Karanja (*Pongamia pinnata*) seed shell is found abundantly as an agricultural waste especially in the rural parts of India. Hence, preparation of activated carbon from the seed shells of *Pongamia pinnata* can be categorized to be one of the most efficient low cost activated carbon. Chemical activation is preferred over physical activation because it requires lower temperature and shorter time for activating the material. Phosphoric acid produces better modification than other acids to the botanic structure by penetrating, swelling and breaking the bonds of lignocellulose materials.

13.2 MATERIALS AND METHODS

13.2.1 PREPARATION OF ACTIVATED CARBON

The Karanja (*Pongamia pinnata*) seed shell was collected from Pauni, Bhandara District, Maharashtra, India. The seed shell waste was sun dried, crushed to desirable size to obtain fine powder and then kept in

1:1 (w/v in g/mL) Orthophosphoric acid (H_3PO_4) for 24 h [1]. This mixture was then charred at 450 °C for 1 h to complete the carbonization and activation process [2]. The resulting carbon was given multiple washing with Double Distilled Water (DDW) till a neutral pH of the slurry was reached. The carbon was then dried at 105 °C in a hot air oven for 12 h [3]. The dried material was ground well to fine powder and sieved in the size range of 150–300 μm (average particle size 225 μm) and stored in airtight container [4].

13.2.2 CHARACTERIZATION OF ACTIVATED CARBON

Characterization of activated carbon can be explored by its physicochemical characteristics. In the present study some of the important characteristics viz. particle size analysis, proximate analysis, surface area and pore volume, XRD analysis and FTIR spectra are determined by using various respective instruments.

13.2.2.1 Particle Size Analysis

The particle size analysis was done by using Indian Standard Sieve IS 460. The material was sieved in the range of 150–300 μm to maintain the uniformity of particle size.

13.2.2.2 Proximate Analysis

Standard procedure as per IS: 1350 (Part 1)–1984 was adopted for conducting proximate analysis. All weight measurements were to the nearest of 0.1 gm.

13.2.2.2.1 Moisture Content

An empty weight, W_1 of crucible was taken. One gram of sample was added in the crucible and weight W_2, was taken. The crucible was tapped gently to spread the sample evenly over the bottom of the crucible. The crucible with sample was kept in oven at a temperature of 105±5°C for one hour. After one hour the crucible was kept in a desiccator and weighed again (W_3). The percentage moisture was calculated as:

$$\%Moisture = \frac{Mass\ of\ water\ removed}{Mass\ of\ orginal\ sample} = \frac{W_2 - W_3}{W_2 - W_1} \times 100$$

where, W_1 = mass of empty crucible; W_2 = mass of crucible plus sample, before heating; W_3 = mass of crucible plus dried sample.

13.2.2.2.2 Ash Content

An empty weight, W_1 of crucible was taken. About 1 g of sample was gradually added and the weight of the crucible and contents was recorded as W_2. The crucible was gently tapped to spread the sample evenly over the bottom of the crucible. Crucible was then placed in a high temperature furnace. The sample was heated to 750°C and left at that temperature for one hour so that all the combustible material was completely burned. The crucible was then removed from the furnace and cooled for about one minute in the laboratory, then placed in the desiccator until it has cooled to room temperature. The sample was reweighed as W_3. The percentage ash is calculated as:

$$\%Ash = \frac{Mass\ of\ residue\ after\ combustion}{Mass\ of\ orginal\ sample} = \frac{W_3 - W_1}{W_2 - W_1} \times 100$$

where, W_1 = mass of empty crucible; W_2 = mass of crucible + sample, before heating; W_3 = mass of crucible + residue, after heating.

13.2.2.2.3 Volatiles Content

An empty crucible plus lid was weighed as W_1. About 1 g of sample was gradually added and the weight of the crucible (plus lid) and contents was recorded as W_2. The crucible was gently tapped to spread the sample evenly over the bottom of the crucible. The covered crucible was then placed into a high temperature furnace, which has already been preheated to 925°C.

The sample was heated for exactly seven minutes. The crucible was then removed from the furnace and cooled for about one minute in the laboratory, then placed in the dessicator until it has cooled to room temperature. The sample was reweighed as W_3. The percentage volatiles quoted in a proximate analysis excludes moisture. However, both water and volatiles are driven off during heating. Thus the volatiles content is given by:

$$\%Volatiles = \frac{W_2 - W_3}{W_2 - W_1} \times 100 - M$$

where, W_1 = mass of empty crucible + lid; W_2 = mass of crucible + lid + sample, before heating; W_3 = mass of crucible + lid + sample, after heating; M = % moisture content.

13.2.2.2.4 Fixed Carbon

The Fixed Carbon was calculated as follows:

% Fixed Carbon = 100 – (% moisture + % volatiles + % ash)

13.2.2.3 Point of Zero Charge (pHpzc)

The point of zero charge (pH_{pzc}) of the adsorbent was determined by solid addition method. For the determination of $pH_{pzc,}$ 0.1 M KNO_3 solution was prepared and the pH of solutions was adjusted in the range of 2 to 12 using either 0.1N HCl or 0.1N NaOH solution. Then the 50 mL solution of each pH was transferred to 250 mL stoppered conical flasks. The pHs of the solutions were then accurately noted as pH_0 values. 0.2 g of adsorbent dose was added in each flask containing 0.1 M KNO_3 solution of different pH values. This adsorbent-KNO_3 solution assembly was then kept for 48 h and was shaken intermittently during this period. After 48 h the pH_f of each solution was measured using digital pH meter. The difference in the initial and final pH of the solution ($\Delta pH = pH_0 - pH_f$) was found out and the graph of pH_0 versus ΔpH was plotted. The point of intersection of the resultant curve represents the pH_{pzc} of the adsorbent.

13.2.2.4 Surface Area and Pore Volume Analysis

The Brunauer Emmet Teller (BET) surface area and Pore volume was found by Nitrogen adsorption and desorption method by static volumetric measurement (volume of Nitrogen adsorbed under low pressure and analyzed volumetrically after desorption).

13.2.2.5 X-Ray Diffraction Analysis

X-ray diffractographs were obtained using MiniFlex II Desktop X-ray diffractometer (Rigaku, Japan) in the 2θ range from 10 to 90° with an interpolated step of 0.02° using Cu-Kα radiation, Continuous scanning.

13.2.2.6 Fourier Transmission Infrared Spectra

IR-Affinity FTIR Spectrophotometer (Shimadzu, Japan), adopting pellet (pressed disk) technique was used for the analysis. Samples were mixed with potassium bromide (KBr) in a mass ratio 1:30 and pellets were prepared in a special mold under pressure [5]. These pellets were then scanned and recorded between 4000 and 400 cm^{-1} wavenumber.

13.2.2.7 Scanning Electron Microscopy

SEM was done by Scanning electron microscope (JEOL 6380A) adopting secondary electron imaging method. Firstly the adsorbent particles were finely gold plated by JFC-1600 (JEOL) Autofine sputter coater to make the sample conductive, thereafter SEM were taken at various magnifications.

13.3 RESULTS AND DISCUSSION

13.3.1 CHARACTERISTICS OF THE CARBON

The characteristics of the Carbon are shown in Table 13.1.

TABLE 13.1 Characteristics of Activated Carbon

Characteristics	Values
Average particle size	225 μm
BET Surface area	922.10 m^2/g
Pore volume	0.211 cm^3/g
Moisture Content	5.90%
Ash Content	11.00%
Volatile Matter	15.90%
Fixed Carbon	67.20%

13.3.2 POINT OF ZERO CHARGE (PHPZC)

The point of zero charge for carbon surface is the pH at which the adsorbent surface has a net neutral charge. The point of zero charge for PPSSC is shown in Figure 13.1 and found to be 6.3.

13.3.3 X-RAY DIFFRACTION ANALYSIS

X-ray diffraction technique is a powerful tool to analyze the crystalline nature of the materials. If the material under investigation is crystalline, well-defined peaks are observed while non-crystalline or amorphous systems show a hallow instead of well-defined peak [6]. The carbon in the present study shows amorphous nature (Figure 13.2).

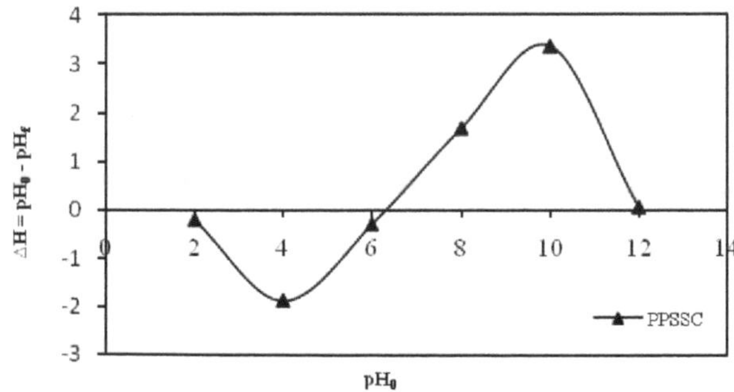

FIGURE 13.1 Point of zero charge (pH_{pzc}) of PPSSC.

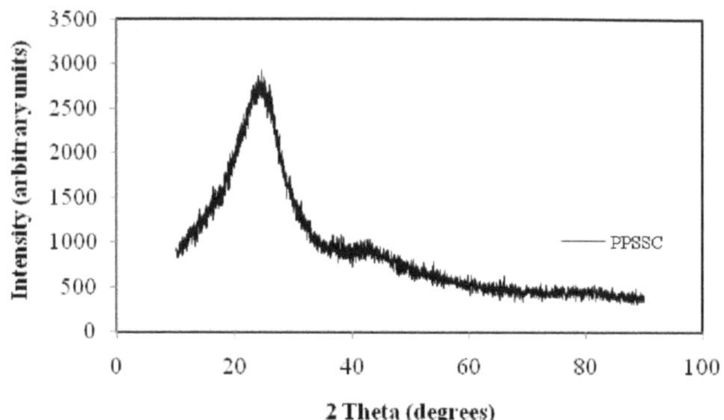

FIGURE 13.2 XRD of *Pongamia pinnata* seed shell carbon.

13.3.4 FOURIER TRANSMISSION INFRARED SPECTRA

Infrared analysis permits spectrophotometric observation of the adsorbent surface in the range 4000–400 cm^{-1}, and serves as a direct means for the identification of the organic functional groups on the surface. IR spectrum of the carbon is shown in Figure 13.3. The peaks around 3749 cm^{-1} to 3527 cm^{-1} indicates the stretching vibrations of aromatic phenolic groups (O−H), and the bands at 3444 cm^{-1} shows the presence of stretching of amines (N−H) groups [7]. Again the peak at 3282 cm^{-1} may be due to hydrogen bonded alcohols and phenols (O−H) [8]. The bands at 1753 cm^{-1} and 1712 cm^{-1} indicates the strong

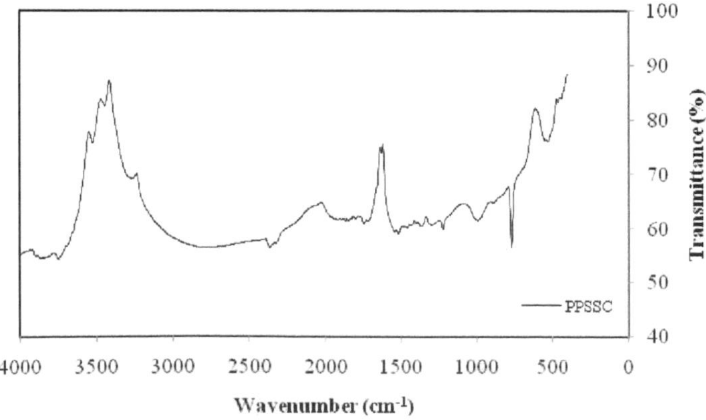

FIGURE 13.3 FTIR spectra of *Pongamia pinnata* seed shell carbon.

carbonyl stretching vibrations of C=O bond probably due to aldehydes and ketones [3]. The peaks observed at 1516 cm^{-1} and 1220 cm^{-1} shows the presence of alkanes (C–H) group [3].

13.3.5 SCANNING ELECTRON MICROSCOPY

SEM is widely used to study the morphological features and surface characteristics of the adsorbent materials [9, 10]. Figure 13.4 shows SEM of the carbon in which rough texture with fine pores can be seen.

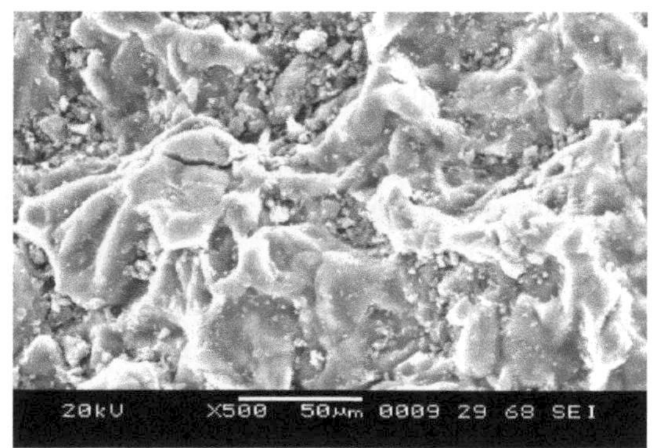

FIGURE 13.4 SEM of *Pongamia pinnata* seed shell carbon.

13.4 CONCLUSIONS

The results of the investigation revealed that the activated carbon prepared from pongamia pinnata seed shell by chemical activation with H_3PO_4 at 450°C carbonization temperature gives high surface area and pore volume which is found to be 922.10 m^2/g and 0.211 cm^3/g, respectively. The point of zero charge is found to be 6.3. XRD study shows amorphous nature of the carbon, while FTIR Spectra shows the presence functional groups, which are essential for carbon. SEM morphology shows rough texture and fine pores. It may be concluded that further studies on PPSSC is encouraged to know its adsorptive characteristics.

KEYWORDS

- **Chemical activation**
- **FTIR spectra**
- **Physico-chemical characteristics**
- ***Pongamia pinnata* seed shell carbon**
- **SEM**

REFERENCES

1. Arivoli, S., Hema, M., Parthasarathy, S., & Manju, N., Adsorption dynamics of methylene blue by acid activated carbon. *J. Chem. Pharm. Research* 2010, *2*, 626–641.
2. Reffas, A., Bernardet, V., David, B., Reinert, L., Bencheikh Lehocine, M., Dubois, M., Batisse, N., & Duclaux, L., Carbons prepared from coffee grounds by H_3PO_4 activation: Characterization and adsorption of methylene blue and Nylon Red N-2RBL. *J. Haz. Mat.* 2010, *175*, 779–788.
3. Hesas, R. H., Niya, A. A., Daud, W. M. A. W., & Sahu, J. N., Preparation and characterization of activated carbon from apple waste by microwave assisted phosphoric acid activation: Application in methylene blue adsorption. *Bioresources* 2013, *8*, 2950–2966.
4. Meshram, M. L., & Lataye, D. H., Adsorption of Methylene Blue Dye onto Activated Carbon Prepared from Pongamia Pinnata Seed Shell. *Int. J. Engg. Res. Tech.,* 2014, *3,* 1216–1220.
5. Vicente, G. S., Fernando, P. A., Carlos, J. D. V., & Jose, P. V., *Carbon* 1999, *37*, 1517.
6. Cullity, B. D., Elements of X-ray Diffraction, Addison-Wesley, Reading, MA, 1978.
7. Namasivayam, C., & Kavitha, D., IR, XRD and SEM studies on the mechanism of adsorption of dyes and phenols by coir pith carbon from aqueous phase. *Microchemical Journal* 2006, *82*, 43–48.
8. Gottipati, R., & Mishra, S., Application of biowaste (waste generated in biodiesel plant) as an adsorbent for the removal of hazardous dye – methylene blue – from aqueous phase. *Brazilian J. Chemical Eng.* 2010, *27*, 357–367.
9. Nelly, J. W., & Isacoff, E. G., Carbonaceous Adsorbents for the Treatment of Ground and Surface water, Marcel Dekker, New York, 1982.
10. Gupta, S., Pal, A., Ghosh, P. K., & Bandyopadhyay, M., Performance of waste activated carbon as a low-cost adsorbent for the removal of anionic surfactant from aquatic environment. *J. Environ. Sci. Health* A 2003, *38*, 381–397.

CHAPTER 14

RICE HUSK BASED CO-FIRING PLANTS IN INDIA: A GREEN PERSPECTIVE

S. U. MESHRAM, A. MOHAN, and P. S. DRONKAR

Laxminarayan Institute of Technology, RTM Nagpur University, Nagpur, Maharashtra, India

CONTENTS

14.1 Introduction 228
14.1.1 Wind Energy 229
14.1.2 Hydrothermal Energy 230
14.1.3 Nuclear Power 230
14.1.4 Biomass 230
14.2 Materials and Methodology 232
14.2.1 Materials 232
14.2.2 Energy From Composite Fuel 232
14.2.3 Production of Zeolite-A and Silica Using Composite Ash 233
14.3 Calculation 234
14.3.1 Energy Balance and Related Analysis for a 15 MW Power Plant 234
14.3.1.1 For Composite Fuel (Rice Husk 80 Parts, Coal 20 Parts) 235
14.3.1.2 For Composite Fuel (Rice Husk 60 Parts, Coal 40 Parts) 236

14.3.2 Cost–Benefit Analysis ... 237
14.3.3 Net Profit Analysis ... 238
14.4 Ash Production ... 239
14.5 Manufacture of Value Added Products From Composite Ash ... 239
14.6 Instrumental Investigations ... 240
14.6.1 XRD Crystallinity ... 240
14.6.2 Scanning Electron Microscopy ... 240
14.6.3 Particle Size Analysis ... 241
14.7 Results and Discussion ... 242
14.7.1 Advantages Allied with the Mini Thermal Power Plants ... 242
14.7.2 Benefits to the Rural Inhabitants ... 242
14.7.3 Limitations of the Process ... 242
14.7.4 Analysis of Composite Ash Based Zeolite-A ... 242
14.8 Conclusion ... 243
Keywords ... 243
References ... 243

14.1 INTRODUCTION

As India is the third largest coal producer in the world after China and USA, coal is identified as the predominant source for energy. The cumulative sum of the Geological Resources of coal has been estimated to be about 301 billion tons in India as of 1st April, 2014 [1]. Two third of the entire energy demand in India is met by coal combustion in Thermal Power Plants, having a capacity of about 250–1100 MW. More than 120 Thermal Power Plants are currently functioning in India, purely based on lignite coal, catering to about 60–70% of the total energy requisite, aggregating to 97,768 MW [2]. It is estimated that close to 454 million metric tons of coal are consumed to produce the aforesaid amount of energy [3]. This fact reveals the severe coal consumption from natural reservoirs.

With the growth in the economic sector, the energy consumption is consistently escalating. It has been reported that nearly 30% of the rural population of India are in critical need of electrical energy and the ever rising costs of coal and its transportation, restrict the setup of major Thermal Power Plants in these sectors. To meet the energy requirement, with the prevailing technologies, the capacity of thermal power plants needs to be increased, leading to increased coal consumption, more discharge of CO_2 into the atmosphere and enormous coal fly ash (CFA) production with restricted disposal techniques. Reportedly, about 110 Million Tons Per Annum (Mt y^{-1}) of coal-ash was generated in India till 2005, from more than 70 thermal power plants [4]. It was predicted to increase to 170 Mt y^{-1} come the year 2012 [5]. At present, the efficient disposal of CFA is a worldwide issue accrediting its massive production, which leads to harmful effects on the environment [6]. As the major chemical compositions contained in the CFA are SiO_2 and Al_2O_3 (55–65 wt. % and 15–27 wt. %, respectively), resource recovery from CFA may be one of the approaches to speed up recycle of CFA [7]. Although CFA has been reused in highway construction, land rehabilitation and restoration of eroded soil; the demand for such applications is still limited [8].

With the present consumption rate, there hardly remains enough coal inventory to satisfy energy requirements for impending four to five decades. The world's energy requirement, to a large extent, depends on liquid petroleum and the earth's oil resources are estimated to range from 1.75 ton to 2.3 ton [9]. Though just about one-third of the reserves has been extracted and consumed, it is predicted that 50% of the reserves would be consumed shortly after the turn of the century [10]. The alarming decline in quantity and quality of indigenous fossil fuels and recognition of their hostile environmental effect has compelled us to stress alternative sources of energy like solar, wind, nuclear, biomass and so forth. Renewable energy sources are perceived as an imperative solution to the problem because they are believed to be environmentally benevolent. Conversely, renewable energy sources may also cause adverse environmental impacts, which must be taken into consideration [11].

14.1.1 WIND ENERGY

With 2980 MW of installed wind power capacity, India ranks fifth in the world in terms of wind power capacity. India's technical potential is estimated at about 13,000 MW assuming 20% grid penetration. However, the gross wind power potential has been assessed at 45,000 MW. As wind power density is not uniform, only certain states have this resource while others do not have sufficient [12]. It has been reported that the occupants in the vicinity of industrial wind turbines are prone to Wind Turbine Syndrome, exhibiting symptoms like dizziness, Vertigo, Tachycardia nausea, visual blurring, etc. [13]

14.1.2 HYDROTHERMAL ENERGY

The development of hydroelectricity in the twentieth century was synchronized with construction of large dams. While creating a major and reliable power supply every bit well as irrigation and flood control benefits, the dams unavoidably flooded large areas of fertile land and displaced many thousands of local inhabitants [14]. Recently, new and improved designs of watermills of capacity 3–5 kW have been developed for electricity generation [15].

14.1.3 NUCLEAR POWER

Nuclear power contributes to about 7% of the global energy demands. As of December 2003, there were 440 nuclear power plants operating in 31 countries around the world [16]. Nevertheless, the nuclear power is twice as expensive as generating electricity from gas or wind and energy conservation measures are around seven times more economical. Moreover, it has serious environmental implications. The mining and handling of primary source Uranium is quite risky and is often associated with radiation leaks. Moreover, the nuclear waste remains hazardous for 240,000 years, which is 20 times longer than the entire history of civilization [17].

14.1.4 BIOMASS

According to late accounts, most of the developing nations have adopted biomass combustion, as a feasible model to get energy. Among thermal applications, an estimated 4119 MW of biomass based grid system, 815 MW off-grid captive systems and about 47 MW of private biogas plants are functioning in India [18]. The primarily emphasized biomass includes coffee husk, bagasse, oil palm waste, rice husk, coconut fibers, wood, short-rotation woody crops, waste paper, grass, waste from food processing, aquatic plants, algae animal wastes [19].

Amongst all the biomass, rice husk has emerged as one of the most economic source concurrently having abundant availability worldwide. It is reported that about 737 Million tons of rice paddies are produced worldwide per annum, producing 147.4 Million tons of rice husk. Interestingly, out of 225 Mt y^{-1} rice paddy from South Asia, India solely produces about 120–130 Mt y^{-1}, out of which 24–26 million tons being the rice husk [21]. In majority of the rice producing countries, husk produced from the processing of rice is either burnt or discarded as waste. India is the second largest producer of rice paddy in the world [22], rice husk has gained substantial consideration as a potential origin of vitality. As the calorific value of rice husk is almost two-third of that of coal [23], it is preferred to blend it with coal for cost effective energy generation. Nevertheless, the capacity of such power plants is limited to 10–15 MW and hence identified as Mini Thermal Power Stations (MTPS). Moreover, rice husk being poor in carbon content, CO_2 emissions during combustion is remarkably curtailed. This fact makes rice husk the most desirable choice for blending with coal, subsequently mitigating the overall output of CO_2 responsible for global warming. Combustion of rice husk in ambient atmosphere leaves a residue called Rice Husk Ash (RHA). Post combustion, 25% of rice husk is converted into RHA comprising about 87% to 97% amorphous silica [24].

The impact assessment results distinctly illustrate the impact of global warming potential of rice husk energy (17.16 kg CO_{2-}eq/MW-h) being far less than the combined value of fossil fuel plants (1,269.91 kg CO_{2-}eq/MW-h) as biomass combustion is considered as greenhouse gas neutral [25]. It has been reported that MTPS with a generation capacity of 2–10 MW, can be set up with confirming solutions. Moreover, the ash produced has widespread industrial applications [26]. Still, the production of a single such plant is limited to 2–10 MW. Hence, it can entirely satisfy the energy needs for a little small town.

There are few reports suggesting the co-firing process of coal with biomass such as wood, bagasse, and agricultural waste and so on for energy generation [27–30]. Nevertheless, the concept of blending of coal with rice husk is relatively new. As the calorific value of rice husk is close to that of coal, it is preferred to mix with coal for cost effective energy generation. Moreover, the combination of husk with coal diminishes CO_2 emissions up to a certain extent [31]. The preference of composition of blend strictly depends upon the seasonal availability of rice husk along with the accessibility of dry coal and their comparative costs. The present article not only deals with the blending of rice husk with coal in a stoichiometry of 80:20 or 60:40 to produce energy economically but also resolves the disposal problem of residual ash. The post combustion residual ash obtained during the process is termed as Composite Ash which can be further explored into the manufacture of commercial value added products like Silica and Zeolite-A, having industrial significance as ion-exchanger/detergent builder, molecular sieve, adsorbent, membrane, etc.

14.2 MATERIALS AND METHODOLOGY

14.2.1 MATERIALS

The Mini Thermal Power Plants procure rice husk and coal from the appropriate sources for the combustion process. The composite ash samples having particular compositions are collected from the Mini Thermal Power Plants easily accessible from Nagpur city. The other chemicals, which are used in the preparation of Zeolite-A and Silica from composite ash, are obtained from Merck India Ltd.

14.2.2 ENERGY FROM COMPOSITE FUEL

Initially, the coal is fed to the pulverizer, where it is crushed to fine particle size and further blended with the rice husk for co-firing using blowers tuned in a specific ratio. This blended material is then combusted to produce energy required for generation of steam, which is further charged onto the turbines to produce electricity. The residue after combustion is removed to the stack, where the composite ash is classified using an Electro Static Precipitator (ESP). The output of such a plant may be merely 15 to 20 MW but can answer the demands of rural sectors in India. If up scaling is required, the concept of the power grid is suggested.

14.2.3 PRODUCTION OF ZEOLITE-A AND SILICA USING COMPOSITE ASH

Initially, the raw ash samples were sieved through suitable mesh size to achieve the uniform particle size. The other volatile matter present in the ash was separated by calcination at 500 °C for 2 hrs. The calcined ash is then refluxed with sodium hydroxide solution in a pre-determined ratio for about 4 h to extract the silica in the form of sodium silicate. By and by the refluxed mixture was filtered and the residual ash was separated and recycled. To this filtrate, calculated amount of sodium aluminate was gradually added to obtain a viscous slurry stirred for 4–6 h. The aged sample was then subjected for hydrothermal crystallization for a period of 2.5 h using stainless steel autoclave having 500 mL capacity. The solid product was separated using a vacuum pump, washed till the desired pH is achieved and finally oven dried. Particular stress has been applied to recycle the residual ash and washed liquor generated in the process to create the process clean and effective. The process layout is displayed in Figure 14.1. It is notable

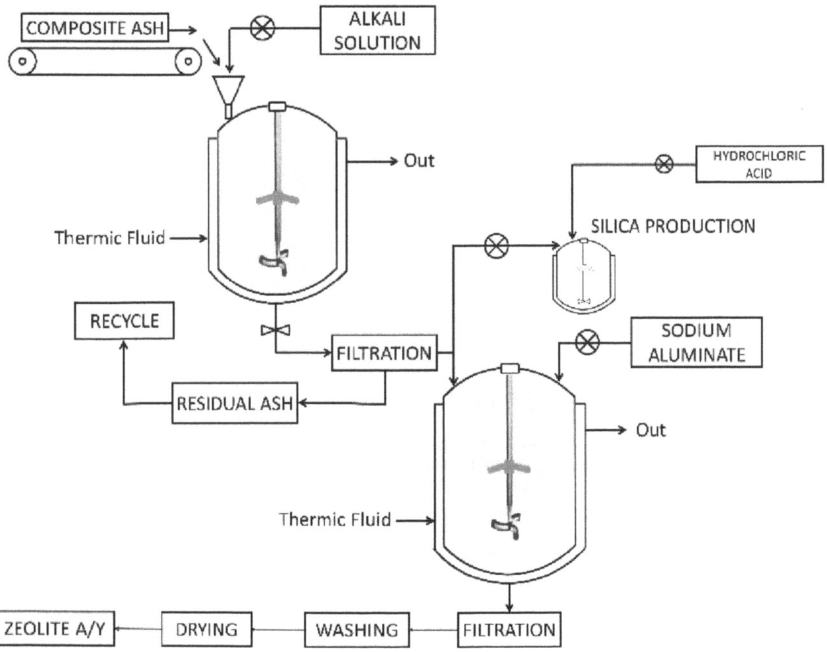

FIGURE 14.1 Process layout for the production of Zeolite A.

that the sodium silicate as such extracted here could be likewise applied to prepare colloidal as well as fine particles of silica with small variation in pH using dilute hydrochloric acid.

14.3 CALCULATION

14.3.1 ENERGY BALANCE AND RELATED ANALYSIS FOR A 15 MW POWER PLANT

The reported calculations are based upon the assumption that a 15 MW plant operated for 24 h. Consequently, the minimal quantity of electrical energy generated will be:

$$15 \text{ MW} \times 24 \text{ h} = 360 \text{ MW-h/day}$$

Further conversion of W-h into Joules is carried out as:

$$360 \text{ MW} \times 3600 \text{ s} = 1.296 \times 10^6 \text{ MJ/day}$$

Considering that about 40% of heat generated will be essentially taken in as electrical output, the boilers need to provide 4.32 × 10^6 MJ/day which can be calculated as:

$$E = 1.296 \times 10^6/0.4 = 3.24 \times 10^6 \text{ MJ/day}$$

Here, E is the minimum expected amount of energy required to produce desired power.

The higher heating value of rice husk (Gross Calorific Value) was detected to be 15,944±55 kJ/kg using Oxygen bomb calorimeter [32]. For calculation purpose, we are considering the net calorific values of rice husk and coal as recommended by one of the Mini Thermal Power Plant is approximately 12,486 kJ/kg or 12,486 MJ/ton and 16,743 kJ/kg or 16,743 MJ/ton, respectively. Therefore, the amount of coal required to meet the aforementioned demand of energy per day can be expressed as:

$$M = 3.24 \times 10^6 \text{ MJ}/16{,}743 \text{ MJ} = 193.5137 \text{ tons}$$

Here, M represents the requisite mass of coal used in traditional combustion.

However, the present route comprises of variable composition of coal and biomass; the actual energy produced ECF using Composite Fuel can be deduced as:

$$\boldsymbol{M}\times F_c\times C_c+\boldsymbol{M}\times F_{Rh}\times C_{Rh}=\boldsymbol{E}_{CF}$$

Here, F_c and C_c represent the mass fraction and calorific value of coal, respectively, whereas F_{Rh} and C_{Rh} are indicating the mass fraction and calorific value of rice husk.

14.3.1.1 For Composite Fuel (Rice Husk 80 Parts, Coal 20 Parts)

As amount of coal is taken to be 20% of fuel (***M*** from Eq. (4)) and amount of rice husk is 80% of the same, then the stoichiometric ratio can be assigned as $F_c= 0.2$ and $F_{Rh}=0.8$.

Accordingly, the amount of coal required (M_c) can be calculated as:

$$M_c = F_c\times \boldsymbol{M} = 0.20 \times 193.5137 \text{ tons} = 39.70274 \text{ tons}.$$

Likewise, the amount of rice husk required (M_{Rh}) can be expressed as:

$$M_{Rh} = F_{Rh}\times \boldsymbol{M} = 0.80 \times 196.685 \text{ tons} = 154.811 \text{ tons}.$$

Therefore, net amount of energy generated $\boldsymbol{E}_1$ using composite fuel with referred stoichiometry per day can be calculated as:

$$\boldsymbol{E}_1 = (M_c\times C_c) + (M_{Rh}\times C_{Rh})$$

$$= (39.70274 \times 16743) + (154.811 \times 12486) \text{ MJ} = 2.581\times 10^6 \text{ MJ (approx.)}$$

But, according to Eq. (3) the total amount of energy required per day is 3.24×10^6 MJ.

Thus, the shortfall of energy is given as:

$$\boldsymbol{E}-\boldsymbol{E}_1 = (3.24 \times 10^6 - 2.581 \times 10^6 \text{ MJ}) = 6.59 \times 10^5 \text{ MJ}$$

This deficit in energy generated is due to the variance in calorific values of rice husk and coal. This problem can be overcome by introduction of adequate amount of biomass, for example, Rice husk.

Therefore, the excess amount of rice husk required is:

$$(\boldsymbol{E} - \boldsymbol{E_1})/C_{Rh} = (3.24\times10^6\,\text{MJ} - 2.581\times10^6\,\text{MJ})/12486 = 52.7815\ \text{ton}$$

The gross rice husk required is 52.7815 tons + 154.811 tons= 207.5925 tons.

14.3.1.2 For Composite Fuel (Rice Husk 60 Parts, Coal 40 Parts)

As amount of coal is taken to be 40% of fuel (***M*** from Eq. (4)) and amount of rice husk is 60% of the same, then the stoichiometric ratio can be assigned as $F_c = 0.4$ and $F_{Rh} = 0.6$.

Accordingly, the amount of coal required (M_c) can be calculated as:

$$M_c = F_c \times M = 0.40 \times 193.5137\ \text{tons} = 77.40548\ \text{tons}.$$

Similarly, the amount of rice husk required (M_{Rh}) can be expressed as:

$$M_{Rh} = F_{Rh} \times M = 0.60 \times 193.5137\ \text{tons} = 116.1082\ \text{tons}.$$

Hence, net amount of energy generated $\boldsymbol{E_2}$ using composite fuel with referred stoichiometry per day can be calculated as:

$$\boldsymbol{E_2} = (M_c \times C_c) + (M_{Rh} \times C_{Rh})$$

$$= (77.40548 \times 16743) + (116.1082 \times 12486)\ \text{MJ} = 2.745 \times 10^6\ \text{MJ}$$

But, according to Eq. (3) the total amount of energy required per day is 3.24×10^6 MJ.

Thus, the shortage of energy is given as:

$$\boldsymbol{E}\text{–}\boldsymbol{E_2} = (3.24 \times 10^6 - 2.745\times 10^6\,\text{MJ}) = 4.95 \times 10^5\ \text{MJ}$$

This deficit in energy generated is due to the variance in calorific values of rice husk and coal. This problem can be overcome by introduction of adequate quantity of biomass, for example, rice husk.

Therefore, the excess amount of rice husk required is

$$(\boldsymbol{E} - \boldsymbol{E}_2)/C_{rh} = (3.24 \times 10^6 \text{ MJ} - 4.95 \times 10^5 \text{ MJ})/12{,}486 = 39.5862 \text{ tons.}$$

In this case, the net amount of excess husk required is:

$$39.5862 + 116.1082 \text{ tons} = 155.6944 \text{ tons.}$$

Based on the aforesaid calculations, the net quantity of both the feed materials according to their required stoichiometry has been evaluated and plotted in a graphical format, which is delineated in Figure 14.2. Using this graph, it is rather easy to derive the volume fraction of either one of the components when the availability of its counterpart is certain.

14.3.2 COST–BENEFIT ANALYSIS

The cost–benefit analysis is based upon the present cost of coal and rice husk per ton. Subsequently, the present price of coal is seen to be Rs. 5500 and that of rice husk to be Rs. 3500.

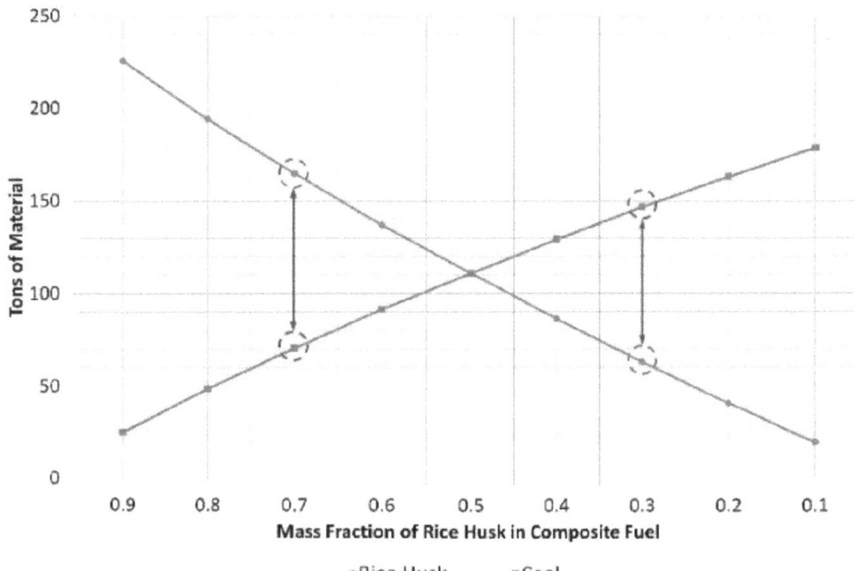

FIGURE 14.2 Graphical presentation of optimization of F_{Rh} against M_{Rh} and M_c.

Cost involved in traditional process where coal is combusted unaided is given as:

$$\text{Rs. } 5500 \times 193.5137 \text{ ton} = \text{Rs. } 10{,}64{,}325$$

Similarly, the cost involved in referring process where composite fuel is consumed in 80:20 ratio = (38.7027 × Rs. 5500) + (207.593 × Rs. 3500) = Rs. 9,39,438

Likewise, the cost involved in the referring process where composite fuel is consumed in 60:40 ratio = (77.4055 × Rs. 5500) + (155.6944 × Rs. 3500) = Rs. 9,70,660

14.3.3 NET PROFIT ANALYSIS

The net profit using composite fuel with 80:20 ratio as compared to pure coal combustion can be evaluated as: Rs. 10,64,325–9,39,438 = Rs. 1,24,887 per day.

Similarly, the net profit using composite fuel with 60:40 ratio as compared to pure coal combustion can be assessed as: Rs. 10,64,325–9,70,660 = Rs. 93,665 per day.

Besides cost savings, the amount of coal saved per day was estimated as well and it was found out that about 154.811 tons and 116.882 tons of coal was conserved per day for 80:20 and 60:40 ratios, respectively.

The quantity of coal saved per day when stoichiometry of 80:20 is used in the composite fuel is 193.5137–38.7027 tons = 154.811 tons.

Also, the quantity of coal saved per day when stoichiometry of 60:40 is used in the composite fuel is=193.5137–77.4055 tons= 116.1082 tons

To determine the cost per unit (per kWh) of electrical energy produced, the cost of fuel is divided by the energy capacity of the plant in kWh, then;

The cost for different compositions of feed is:

i) Pure coal = Rs. 10,64,325/(3,60,000 kW-h) = Rs. 2.96/kW-h
ii) 80:20 Blend = Rs. 9,39,438/(3,60,000 kW-h) = Rs. 2.61/kW-h
iii) 60:40 Blend = Rs. 9,70,660/(3,60,000 kW-h) = Rs. 2.70/kW-h

14.4 ASH PRODUCTION

Surveys have been conducted to evaluate the difference between residual ash generated in these power plants by the combustion of composite fuel and pure coal. It has been noticed that the composite ash production is quite comparable to that of coal ash. Moreover, in certain cases it exceeds the predicted volume of residual ash. Hence, it poses an additional burden over the massive production of coal ash.

14.5 MANUFACTURE OF VALUE ADDED PRODUCTS FROM COMPOSITE ASH

Various experiments have been conducted to synthesize value added products such as Silica and Zeolite-A using composite ash as described earlier. Since, Silica and Zeolite-A have well-established applications as adsorbents, separation materials, catalyst and ion-exchangers, the synthesis of these materials from low-grade resources would be advantageous. Figure 14.3 reveals various routes of utility of composite ash in the manufacture of industrially demanded products.

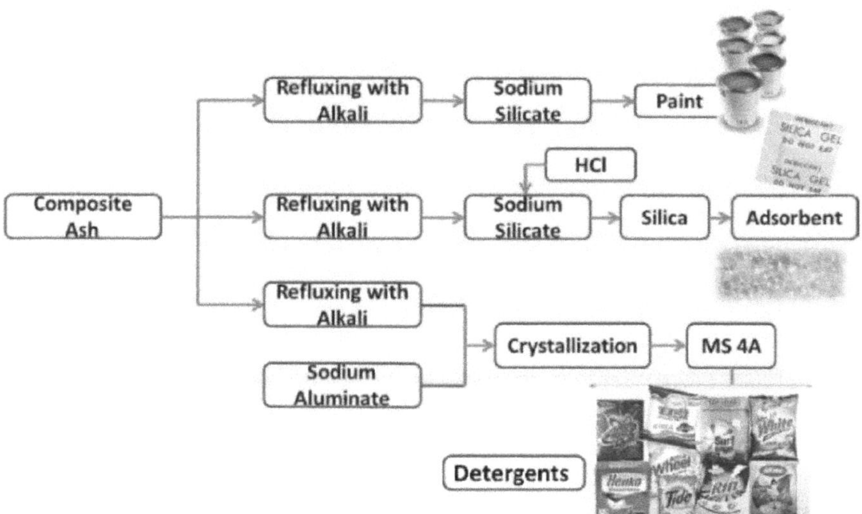

FIGURE 14.3 Commercial applications of composite ash based products.

14.6 INSTRUMENTAL INVESTIGATIONS

14.6.1 *XRD CRYSTALLINITY*

The Zeolite-A sample synthesized using aforementioned route was examined using X-ray diffractometer (Philips PAN Expert-pro) with Cu-K alpha radiation at 40 kV and 30 mA in the scanning range 0–60°. The recorded spectrum is displayed in Figure 14.4.

The quantitative measure of the crystallinity of the synthesized Composite Ash based Zeolite-A was calculated by applying the sum relative intensities of major peaks against their *d*-spacing values as per the reported elsewhere [33]. The sharp peaks obtained at distinct θ values indicate highly crystalline phase of CRHA based Zeolite-A.

14.6.2 *SCANNING ELECTRON MICROSCOPY*

The structural morphology of the synthesized Composite Ash based Zeolite-A was examined using Scanning Electron Microscope (model: FEI-Inspect D 18858). Figure 14.5 exhibits the probed photographic image of Composite Ash based Zeolite-A, revealing the characteristic cubic pattern with partial agglomeration of particles, promising high degree of crystallinity with fineness up to 10 μm.

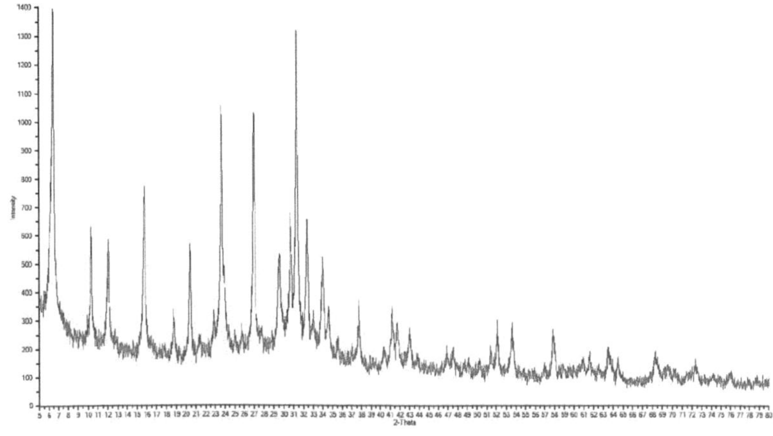

FIGURE 14.4 XRD pattern of Composite Ash based Zeolite-A.

FIGURE 14.5 SEM of CRHA based Zeolite-A.

14.6.3 PARTICLE SIZE ANALYSIS

The average particle size of the samples of Composite Ash based Zeolite A was recorded using a particle size analyzer (Sympatec HELOS (H1004) SUCELL). The result is exhibited in Figure 14.6.

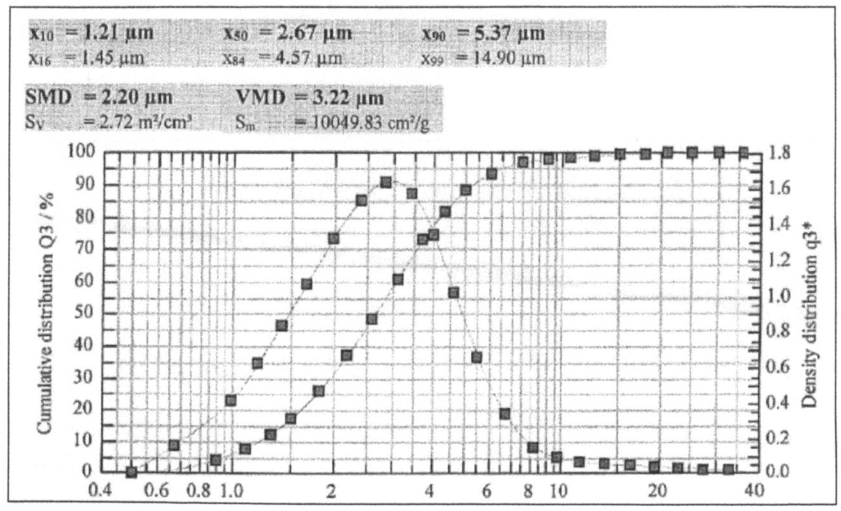

FIGURE 14.6 Particle size analysis of Composite Ash based Zeolite-A.

14.7 RESULTS AND DISCUSSION

14.7.1 ADVANTAGES ALLIED WITH THE MINI THERMAL POWER PLANTS

- There is an abundant availability of biomass against conventional raw material and it is also convenient to procure, store and transport.
- Composite fuel based power generation proves fairly economical against the conventional method. Approximately Rs. 1.25 lakh profit is expected per day by using composite fuel with a complimentary cost–benefit of nearly 11–12% per unit of power generated.
- There is a reasonable decline in CO_2 emissions, causing the power plants eligible for getting government subsidies towards carbon credits.

14.7.2 BENEFITS TO THE RURAL INHABITANTS

- The primary utility of the nominated work is for Mini Thermal Power Plants which generate 15 MW that can satisfy rural power requirements.
- The biomass (rice husk) serves as an additional source of income for the farmers in rural areas.

14.7.3 LIMITATIONS OF THE PROCESS

- The accessibility of rice husk is seasonal and hence, suitable means must be worked out to baggage the biomass feedstock and tune it with coal accordingly.
- The maximum amount of power generation using this route is quite low, for example, 10–15 MW. However, power capacity can be extended by connecting such small-scale plants to the power grid.

14.7.4 ANALYSIS OF COMPOSITE ASH BASED ZEOLITE-A

As per the methodology mentioned earlier, the percent crystallinity has been investigated with reference to commercial standard of Zeolite-A. The sharp peaks obtained in the diffractogram reveal crystallinity up to

88%, which is quite significant. Likewise, the cubical morphology with fine particle size up to 10 μm promotes its applicability in the disciplines of detergent builder, ion exchanger, adsorbent, etc. To boot, the abundant availability of source material (composite ash) essentially with no cost, favors the economic production of Zeolite-A.

14.8 CONCLUSION

The aforesaid technology provides a cost-effective and eco-friendly solution against the conventional methods to produce energy. The virtues of the proposed work lie in the facts that the burning of coal blended rice husk emits relatively low CO_2 and with reduced transportation and warehousing costs, the overall process is economically viable. Optimum ratios of rice husk and coal in the composite fuel are calculated, for varying amounts of either of the components while assuming constant annual availability of the other, to achieve identical energy output. Rice husk being rich in silica, the residual ash, after the combustion process can be turned into value added products viz. Silica and Zeolite-A, hence reducing its impact on the environment. The process not only offers a concomitant solution towards the disposal of residual ash, but also provides a non-tedious route to produce Zeolite-A and Silica having commercial significance.

KEYWORDS

- **Composite ash**
- **Energy**
- **Rice husk**
- **Zeolite A**

REFERENCES

1. Inventory of Coal Resources of India, 2014. Report by GSI, CMPDI, SCCL, MECL, State Govt., www.coal.nic.in/reserve2.htm (7 September, 2014).

2. Report on Review of performance of Thermal Power Stations 2011–12, CEA. http://www.cea.nic.in/reports/yearly/thermal_perfm_review_rep/1112/complete_1112.pdf (7 September, 2014).
3. Executive Summary for the Month of September 2013—Section G 3, All India Yearly Coal Consumption for Power Generation (Utilities), Central Electricity Authority Report, Govt. of India. http://www.cea.nic.in/reports/monthly/executive_rep/sep13/G3.pdf (7 September, 2014).
4. Sarkar, A., Rano, R., Mishra, K. K., & Sinha, I. N., Particle size distribution profile of some Indian fly ash—a comparative study to assess their possible uses. *Fuel Process Tech.* 2005, *86*, 1221–1238.
5. Sushil, S., & Batra, V. S., Analysis of fly ash heavy metal content and disposal in three thermal power plants in India. *Fuel* 2006, *85*, 2676–2679.
6. Twardowska, I., & Szczepanska, J., Solid waste, Terminological and Long-term Environmental Risk Assessment Problems Exemplified in a Power Plant Fly Ash Study. *The Science of the Total Environment* 2002, *285*, 29–51.
7. Hui, K. S., & Chao, C. Y. H., Methane emission abatement by multi-ion-exchanged Zeolite-A prepared from both commercial-grade zeolite and coal fly ash. *Environ. Sci. Tech.* 2008, *42*, 7392–7397.
8. US Environmental Protection Agency, Using Coal Fly Ash in Highway Construction, a Guide to Benefits and Impacts, EPA-530, K-05-002 (2005). http://nepis.epa.gov/Exe/ZyPURL.cgi?Dockey=P100071H.txt (7 September, 2014)
9. Barbiroli, G., & Focacci, A., An appropriate mechanism of fuels pricing for sustainable development. *Energy Policy* 1999, *27*, 625–636.
10. World Energy Council (WEC), 1998. Survey of Energy Resources 1998, London.
11. Georgopoulou, E., Lalas, D., & Papagiannakis, L., A Multi-criteria Decision Aid approach for energy planning problems: The case of renewable energy option, *European Journal of Operational Research* 1996, *103*, 38–54.
12. Bhide, A., & Monroy, C. R., Energy poverty: A special focus on energy poverty in India and renewable energy technologies. *Renew and Sustainable Energy Reviews* 2011, *15*, 1057–1066.
13. Colby, D. W., Dobie, R., Leventhall, G., Lipscomb, D. M., McCunney, R. J., Seilo, M. T. et al., Wind Turbine Sound and Health Effects An Expert Panel Review, 2009.
14. Paish, O., Small hydropower: technology and current status. *Renew and Sustainable Energy Reviews* 2002, *6*, 537–556.
15. Bhattacharya, S., & Jana, C., Renewable energy in India—Historical developments and prospects. *Energy* 2009, *34*, 981–991.
16. Trinnaman, J., & Clarke, A., WEC-World Energy council, 2004. Survey of Energy Resources, Elsevier Publications 2004.
17. Asif, M., & Muneer, T., Energy supply, its demand and security issues for developed and emerging economies. *Renew and Sustainable Energy Reviews* 2007, *11*, 1388–1413.
18. Report by Ministry of New and Renewable Energy (MNRE), Govt. of India, on Renewable energy, http://mnre.gov.in/mission-and-vision-2/achievements/(16 August. 2014).
19. Demirbas, A., Combustion characteristics of different biomass fuels. *Progress in Energy and Combust. Science* 2004, *30*, 219–230.

20. Bhattacharyya, S. C., Viability of off-grid electricity supply using rice husk: A case study from South Asia. *Biomass Bioenerg.* 2014, *68*, 44–54.
21. Gidde, M. R., & Jivani, A. P., Waste to Wealth – Potential of Rice Husk in India a Literature Review. Proceedings of the International Conference on Cleaner Technologies and Environmental Management, PEC, Pondicherry, India. January 4–6, 2007: 586–590.
22. Wang, C., Chandra, S., The use of rice husk ash in Concrete. Tech, Report published by Narayan Sighaniaon, 21 December, 2010, 186.
23. Yalcin, N., & Sevinc, V., Studies on surface areas and porosity of activated carbons prepared from Rice Husk. *Carbon* 2000, *38*, 1943–1945.
24. Yalcin, N., & Sevinc, V., Studies on silica obtained from rice husk. *Ceramics Intern.* 2001, *27*, 219–224.
25. Chungsangunsit, T., Gheewala, S. H., & Patumsawad, S., Emission Assessment of Rice Husk Combustion for Power Production. *World Academy of Science, Engineering and Technology* 2009, 53.
26. Bhhagiyalakshmi, M., Yun, L. J., Anuradha, R., & Jang, H. T., Utilization of rice husk ash as silica source for the synthesis of mesoporous silica and their application to CO_2 adsorption through TREN/TEPA grafting. *Journal of Hazardous Materials* 2010, *175*, 928–938.
27. Hansson, J., Berndes, G., Johnsson, F., & Kjarstad, J., Co-firing biomass with coal for electricity generation—an assessment of the potential in EU27. *Energy Policy* 2009, *37*, 1444–1455.
28. Demirbas, A., Biomass resource facilities and biomass conversion processing for fuels and chemicals. *Energy Convers. and Management* 2001, *42*, 1357–1378.
29. Bridgwater, A. V., The technical and economic feasibility of biomass gasification for power generation. *Fuel* 1995, *74*, 631–653.
30. Gold, B. A., & Tillman, D. A., Wood co-firing evaluation at TVA power plants. Strategic Benefits of Biomass and Wasteful fuels. *Biomass and Bioenerg.* 1996, *10*, 71–78.
31. Hein, K. R. G., & Bemtgen, J. M., EU clean coal technology, co-combustion of coal and biomass. *Fuel Processing Technology* 1998, *54*, 159–169.
32. Shen, J., Zhu, S., Liu, X., Zhang, H., & Tan, J., Measurement of Heating Value of Rice Husk by Using Oxygen Bomb Calorimeter with Benzoic Acid as Combustion Adjuvant. *Energy Procedia* 2012, *17*, 208–213.
33. Rayalu, S., Udhoji, J., Munshi, K., & Hasan, M., Highly crystalline zeolite—A from flash of bituminous and lignite coal combustion. *Journal of Hazardous Materials* 2001, *B88*, 107–121.

CHAPTER 15

FOAMABILITY OF FOAM GENERATED BY USE OF SURF EXCEL AND SODIUM LAURYL SULFATE

P. CHATTOPADHYAY, R. A. KARTHICK, and P. KISHORE

Department of Chemical Engineering, BITS-Pilani, Pilani, Rajasthan–333031, India

CONTENTS

15.1 Introduction 248
15.2 Materials and Methods 249
15.2.1 Materials 249
15.2.2 Foam Characterization 249
15.3 Results and Discussion 250
15.3.1 Foamability Results and Comparison Between SLS and Surf Excel 250
15.3.2 Effect of Increasing SLS and Surf Excel Concentrations on Foamability 253
15.3.3 Comparison of Foam Volume vs Time Plot for SLS and Surf Excel 253
15.4 Conclusions 254
Acknowledgements 255
Keywords 255
References 255

15.1 INTRODUCTION

Foam comprises of gas dispersed in continuous liquid phase. They consist of agglomerations of gas bubbles separated by liquid films [1]. Because of the general interest in foams and foam behavior among researchers, there have been many great developments around the world in the area of foam characterization. Foams can be produced by condensation [2] or by dispersion. In the dispersion methods, the gas originally exists as bulk phase and small volumes of it are introduced into the liquid forming bubbles. During the production of a dynamic foam [3], surfaces are created and are subjected to extension in one place and contraction in another place. Foams are generally produced by mechanical agitation, by passing air or nitrogen gas through the aqueous solution of foaming agents in the laboratory scale [4]. The most logical approach towards describing foams [5] considers the final state of foam to be under the influence of chemical, physical and mathematical constraints during its evolution, further growth and stabilization. These constraints operate as independent pathways leading to the characteristic final foam system. Most foams that are persistent [6] contain a foaming agent which may comprise of a surfactant. The utility of surfactant Sodium Lauryl Sulfate to qualify as a foaming agent was reported in literature [7]. Pure liquids [8] do not foam and for foaming to take place, the presence of surfactants are necessary. The work reported [8] also provides a clear technique for the determination of foaminess or foam production capability of a foaming system- an important parameter that can be used to characterize foam. Surfactants are common chemical agents used to promote foam formation in various industrial processes. They contain hydrophobic and hydrophilic head group structurally. The hydrophilic group can be anionic, cationic or zwitterionic [9]. Foamability of such surfactant generated foams have to be studied and understood so as to produce an optimum foam formulation containing desired foaming agents that suits the application intended [10].

Foams are widely characterized by Bikerman's method and Ross-Miles method [11]. The foaming ability of the foam constituents determines its application in various industrial processes ranging from oil extraction, food processing as well as in pharmaceutical products [12]. Aqueous foams [13] generated by the use of surfactants can be used in many industrial processes and applications including detergents, cosmetics formulation.

Detergents are used in household applications for its better cleansing properties whereas surfactants are widely used as surface-active agents and find applications in various fields [14]. Surfactant generated foams [15] can be used for environmental remediation applications to remove petroleum hydrocarbons or other contaminants from the soil. Foams can play a big role in remediation of contaminated soils [16] by increasing the oxygen content of soil. This also helps in improving the living conditions for plants and reducing the operational costs of cleaning such soils.

The current importance of foamability of aqueous foams in formulations of final product is highlighted in various industries worldwide, starting from soap making to shampoo manufacture. Thus, the evaluation of foamability parameters is an important factor for judging industrial products from customer point of view. The goal of this chapter is to provide more insight in the foaming capability of anionic surfactant generated aqueous foams and how it varies in the case of commercial detergents. As reported here, the aqueous foams have been generated by the use of anionic surfactant Sodium Lauryl Sulfate (SLS) and detergent Surf Excel, commercially available in India. The foamability of such foams generated in either case, have been compared. The mechanisms for the different foam behavior have been explained and the chief conclusions are reported.

15.2 MATERIALS AND METHODS

15.2.1 MATERIALS

Powdered surfactant Sodium Lauryl Sulfate (SLS) of purity 99% and commercially available detergent Surf Excel were used in the study. The surfactant was initially weighed, using a digital weighing balance and then mixed with 100 mL of distilled water. The similar procedure was carried out for detergent Surf Excel.

15.2.2 FOAM CHARACTERIZATION

Aqueous foams of SLS and Surf Excel were characterized by using a Dynamic Foam Analyzer DFA 100 (Kruss GmbH, Germany). All the experiments reported here, were conducted in laboratory at room temperature of

303 ± 2 K. Foams were generated in a glass column of 250 mm length, inside diameter of 40 mm, by a stream of air that was introduced into the aqueous solutions through a porous glass filter (pore size: 16–40 μm) with a constant flow rate of 5 mL/sec. Both the glass column and the glass filter were supplied by Kruss GmbH, Germany. The air was passed for 12 secs from the start of each run to produce the foam. This value of 12 secs was selected, as it was the default setting of the Foam Analysis Software version 1.4.2.3 (Kruss GmbH, Germany) that was used for determination of foam properties. Each foaming experiment was run for a total of 15 mins (900 secs). Each foaming run was repeated three times for better accuracy. The different results for foamability of the solutions like maximum foam volume, foam capacity and foam density, were obtained for each run using the Foam Analysis Software version 1.4.2.3 (Kruss GmbH, Germany). The foam capacity (at the end of air injection) was considered as the ratio of the foam volume to the air volume entered and the foam density [17] was the ratio of the liquid bound in the foam to the foam volume at the end of air injection.

15.3 RESULTS AND DISCUSSION

15.3.1 FOAMABILITY RESULTS AND COMPARISON BETWEEN SLS AND SURF EXCEL

The foamability results obtained for SLS and Surf Excel are represented in Tables 15.1 and 15.2, respectively. Comparison between the foamability

TABLE 15.1 Foamability Parameters of Various Concentrations of SLS in Water

Water (mL)	SLS (g)	Foam capacity	Foam density	Maximum foam volume (mL)
100	0.002	1.1	0.14	63.7
100	0.004	1.2	0.18	69.3
100	0.006	1.3	0.21	75.3
100	0.008	1.4	0.26	85.9
100	0.01	1.5	0.26	89.9
100	0.02	1.5	0.28	91.2
100	0.03	1.6	0.28	92.3

TABLE 15.2 Foamability Parameters of Various Concentrations of Surf Excel in Water

Water (mL)	Surf Excel (g)	Foam capacity	Foam density	Maximum foam volume (mL)
100	0.002	0.2	0.34	7.5
100	0.004	1	0.09	62.9
100	0.006	1.2	0.15	69.4
100	0.008	1.1	0.13	66.6
100	0.01	1.2	0.19	69.6
100	0.02	1.3	0.2	77.9
100	0.03	1.5	0.24	84.8

of SLS and Surf in terms of foam capacity and maximum foam volume (mL) are clearly depicted in Figures 15.1–15.2. The comparison of foam volume vs time plots for one sample run involving 0.008 gms of SLS and 0.008 gms of Surf Excel in 100 mL of water are shown in Figure 15.3. Total run time considered for both cases were 15 mins (900 secs).

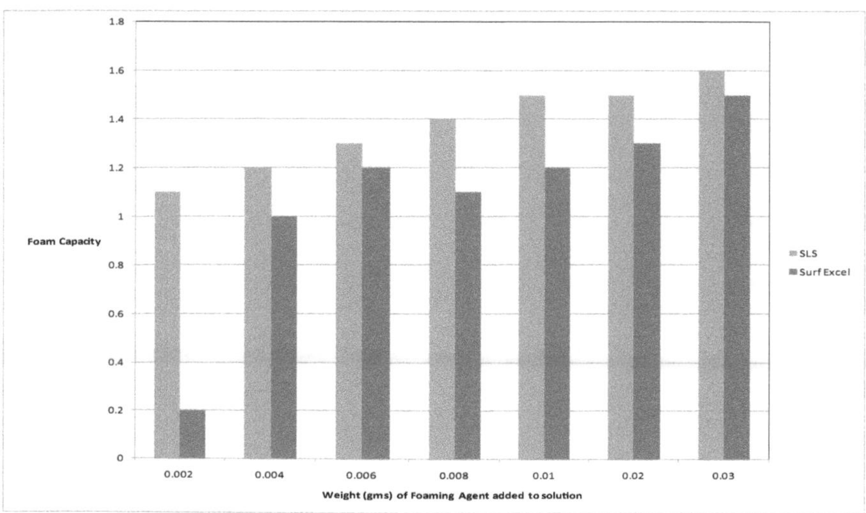

FIGURE 15.1 Comparison of foam capacity for different weights (gms) of SLS and Surf Excel added to solution.

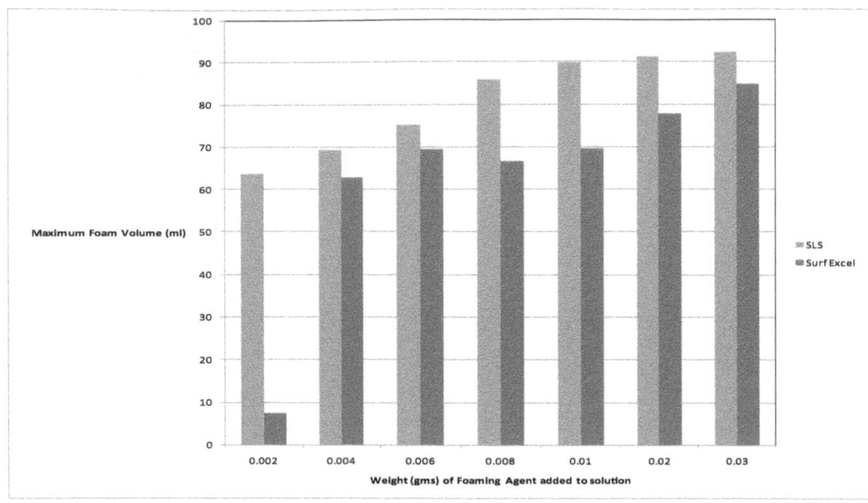

FIGURE 15.2 Comparison of maximum foam volume (mL) for different weights (gms) of SLS and Surf Excel added to solution.

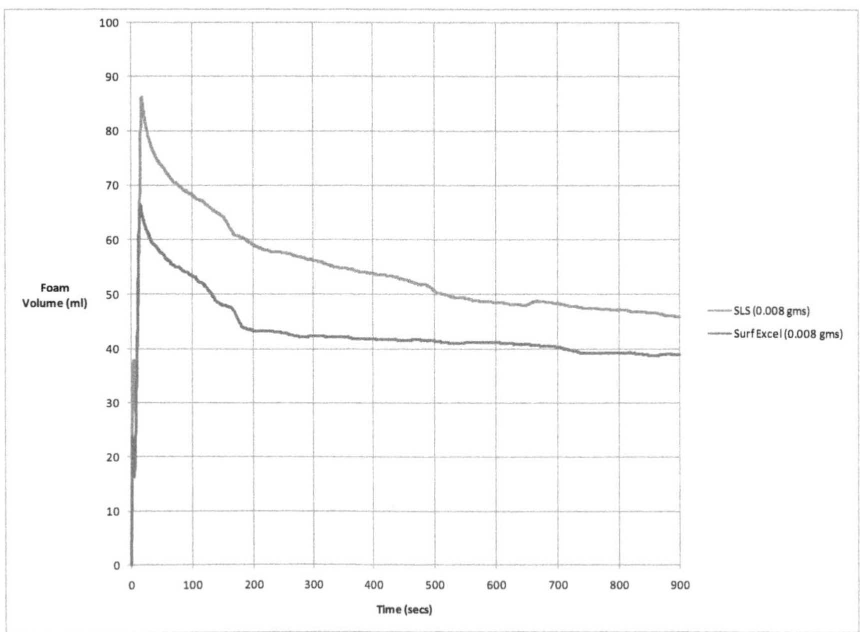

FIGURE 15.3 Comparison of Foam Volume (mL) vs time (secs) for (a) 0.008 gms of SLS added to 100 mL of water (b) 0.008 gms of Surf Excel added to 100 mL of water.

15.3.2 EFFECT OF INCREASING SLS AND SURF EXCEL CONCENTRATIONS ON FOAMABILITY

As can be seen from Table 15.1 and Figure 15.1, the foam capacity for foaming agent SLS gradually increased from 1.1 to 1.6 with increase in SLS concentration (from 0.002 to 0.03 gms added to 100 mL of solution). The concentrations for the foaming agents SLS, Surf Excel were chosen arbitrarily below the CMC (critical micelle concentration). The stabilizing behavior of surfactant SLS below CMC (critical micelle concentration) was responsible for the increasing trend in foamability parameters. With an increase in SLS concentration (from 0.002 to 0.03 gms added), the foam density was found be in the range of 0.14 to 0.28 and the maximum foam volume was found to rise steadily from 63.7 mL to 92.3 mL. The higher maximum foam volumes obtained at higher concentrations of SLS indicated the greater foam performance of the surfactant. The foamability parameters of foaming agent Surf Excel were estimated by varying the concentration of Surf (from 0.002 to 0.03 gms added). The foam capacity increased from 0.2 to 1.5 and foam density was in the range of 0.09 to 0.34. The maximum foam volume obtained by using Surf was in the range of 7.5 mL to 84.8 mL. For the foamability comparison studies as presented in Figures 15.1–15.2, the same range of foaming agent concentrations (from 0.002 to 0.03 gms added to 100 mL of solution) and identical total run time (15 mins) was considered.

15.3.3 COMPARISON OF FOAM VOLUME VS TIME PLOT FOR SLS AND SURF EXCEL

Figure 15.3 (foam volume vs time plot) displayed two distinct stages for the foam:

(i) a foaming phase during the initial air injection period; and
(ii) a final decay phase, where the foam volume decreased.

During the foaming phase, there was an increase in foam volumes, because of foam growth and during the decay phase, the foam volume decreased because of the foam collapse. As per literature [18], foamability depends on the surface tension of the solution. Here, the development of foam took place with the introduction of air bubbles into the solutions tested. This might have resulted in an increase in surface area. Hence, the generation of foam at this stage of air injection would require expenditure of energy against the surface tension forces. The surface tension value relevant for the foam generation process at this stage was thus believed to be dependent on the degree of surface expansion, the rate of surface tension reduction and corroborated by literature [19]. Also it has been reported that greater the rate of surface tension reduction, higher the foamability [20]. Hence, the contributory factors for the greater foaming tendencies for the SLS case, compared to the Surf Excel case, might be the lower surface tension and higher rate of surface tension decrease. This might also explain the initial trend of increasing foam volumes during the air injection for both SLS and Surf Excel case. Also clearly, higher foam volume peaks were observed for the SLS case compared to the Surf Excel case. A gradual decay in the foam volumes after the completion of the air injection, in both the cases shown in Figure 15.3, was attributed to the increase in foam bubble sizes due to coalescence.

15.4 CONCLUSIONS

On comparing the foamability, it was found that the surfactant SLS had higher and better foamability than the commercial detergent Surf Excel at the concentrations tested. As observed from Figure 15.2, the highest of maximum foam volume was attained for SLS (92.3 mL) whereas for Surf Excel, it was found to be 84.8 mL for the same concentration of foaming agent added to solution (0.03 gms). The study revealed that similar foamability parameters can be estimated for number of other detergents and different types of surfactants at concentrations below and above CMC. With further experiments performed on study of complex and higher surfactant systems, more knowledge on foaming characteristics can be obtained. Such studies, if conducted in a systematic fashion, are important in developing novel healthcare and personal care products.

ACKNOWLEDGEMENTS

The authors would like to thank the Department of Science and Technology (DST), India for the financial support of this experimental work (Ref. no. SB/FTP/ETA-208/2012).

KEYWORDS

- **Aqueous foam**
- **Detergents**
- **Foamability**
- **Surfactants**

REFERENCES

1. Azira, H., Tazerouti, A., & Canselier, J., Study of Foaming Properties and Effect of the Isomeric Distribution of Some Anionic Surfactants. *Journal of Surfactants and Detergents*. 2008, *11*, 279–286.
2. Bikerman, J., Foams and Emulsions. *Industrial and Engineering Chemistry*. 1965, *57*, 56–62.
3. Ross, S., & Townsend, D., Dynamic Surface Tensions and Foaminess of Aqueous Solutions of 1-Butanol. *Langmuir*. 1986, *2*, 288–292.
4. Xu, L., Xu, G., Gong, H., Dong, M., Li, Y., & Zhou, Y., Foam properties and stabilizing mechanism of sodium fatty alcohol polyoxyethylene ether sulfate-welan gum composite systems. *Colloids and Surfaces A: Physicochem. Eng. Aspects*. 2014, *456*, 176–183.
5. Rhodes, M., & Khaykin, B., Foam Characterization and Quantitative Stereology. *Langmuir*. 1986, *2*, 643–649.
6. Schramm, L., & Wassmuth, F., Foams: Fundamentals and Applications In The Petroleum Industry. *American Chemical Society*. 1994, *242*, 3–45.
7. Ranjani, G., & Ramamurthy, K., Analysis of the Foam Generated Using Surfactant Sodium Lauryl Sulfate. *International Journal of Concrete Structures and Materials*. 2010, *4*, 55–62.
8. Guitian, J., & Joseph, D., Foaminess Measurements Using A Shaker Bottle. 1996, University of Minnesota, Minneapolis, USA.

9. Martin, V. I., De la Haba, R. R., Ventosa, A., Congiu, E., Ortega-Calvo, J. J., & Moya, M. L., Colloidal and biological properties of cationic single-chain and dimeric surfactants. *Colloids and Surfaces B: Biointerfaces*. 2014, *114*, 247–254.
10. Solorzano, E., Pardo-Alonso, S., De Saja, J. A., & Rodriguez-Perez, M. A., Study of aqueous foams evolution by means of X-ray radioscopy. *Colloids and Surfaces A: Physicochem. Eng. Aspects*. 2013, *438*, 159–166.
11. Tyrode, E., Pizzino, A., & Rojasa, O. J., Foamability and foam stability at high pressures and temperatures. I. Instrument validation. *Review of Scientific Instruments*. 2003, *74*, 2925–2932.
12. Saxena, A., Pathak, A. K., & Ojha, K., Synergistic Effects of Ionic Characteristics of Surfactants on Aqueous Foam Stability, Gel Strength, and Rheology in the Presence of Neutral Polymer. *Ind. Eng. Chem. Res.* 2014, *53*, 19184−19191.
13. Regismond, S., Winnik, F., & Goddard, E., Stabilization of aqueous foams by polymer/surfactant systems: effect of surfactant chain length. *Colloids and Surfaces A: Physicochemical and Engineering Aspects*. 1998, *141*, 165–171.
14. Rosen, M. J., Surfactants and Interfacial phenomena. *Wiley-Interscience*. 2004, Third edition.
15. Sethumadhavan, G., Nikolov, A., Wasan, D., Srivastava, V., Kilbane, J., & Hayes, T., Ethanol-Based Foam Stability As Probed by Foam Lamella Thinning. *Industrial & Engineering Chemistry Research*. 2003, *42*, 2634–2638.
16. Parnian, M., & Ayatollahi, S., Surfactant Remediation of LNAPL Contaminated Soil: Effects of adding alkaline and foam producing substances. *Iranian Journal Of Chemical Engineering*. 2008, *5*, 34–44.
17. Kruss Software Manual for Dynamic Foam Analyzer DFA 100, Kruss GmbH, Hamburg, Germany.
18. Bikerman, J. J., Foams: Theory and Industrial Applications. *Reinhold Publishing Corporation*, 1953, New York.
19. Powale, R. S., & Bhagwat, S. S., Influence of electrolytes on foaming of sodium lauryl sulfate. *Journal of Dispersion Science and Technology*. 2006, *27*, 1181–1186.
20. Garrett, P. R., & Moore, P. R., Foam and Dynamic Surface Properties of Micellar Alkyl Benzene Sulfonates. *Journal of Colloid and Interface Science*. 1993, *159*, 214–225.

CHAPTER 16

PRODUCTION OF ZINC SULPHIDE NANOPARTICLES USING CONTINUOUS FLOW MICROREACTOR

K. ANSARI,[1] S. H. SONAWANE,[1] B. A. BHANVASE,[2] M. L. BARI,[3] K. RAMISETTY,[4] L. SHAIKH,[5] Y. PYDI SETTY,[1] and M. ASHOKUMAR[6]

[1]*Chemical Engineering Department, National Institute of Technology, Warangal, Telangana, India*

[2]*Chemical Engineering Department, Laxminarayan Institute of Technology, Nagpur, Maharashtra, India*

[3]*Institute of Chemical Technology, North Maharashtra University Jalgaon, Maharashtra–425001, India*

[4]*Chemical Engineering Department, Institute of Chemical Technology, Mumbai, Maharashtra, India*

[5]*Chemical Engineering Process Division, National Chemical Laboratory, Pune, Maharashtra, India*

[6]*School of Chemistry, University of Melbourne, Parkville, VIC 3010, Australia*

CONTENTS

16.1 Introduction 258
16.2 Materials and Methods 260
16.2.1 Material 260

16.2.2 Experiments for Residence Time Distribution (RTD) Behavior of Microreactor.................................. 260
16.2.3 Synthesis of ZnS Nanoparticles Using Batch and Continuous Flow Microreactor 261
16.2.4 Predictive Simulation of the Microreactor.................. 262
16.2.4.1 Geometry Definition 263
16.2.4.2 Methods Used in Solver............................. 263
16.2.4.3 Equations Involved in the Formation of ZnS Nanoparticles....................................... 264
16.3 Results and Discussions.. 265
16.3.1 Hydrodynamic Characteristics and Effect of Flow Rate on the Particle Size in Microreactor 266
16.3.2 Effect of Precursor Flow Rate on ZnS Nanoparticles Formation in Microreactor.................... 267
16.3.3 Effect of Precursor Concentration on ZnS Nanoparticles Formation in Microreactor.................... 268
16.3.4 Effect of Temperature on ZnS Nanoparticles Formation in Microreactor 269
16.3.5 Effect of Types of Stabilizers on ZnS Nanoparticles Formation in Microreactor.................... 270
16.3.6 Comparison of Formation of ZnS Nanoparticles in Batch and Microreactor.. 271
16.3.7 Simulations of Microreactor: Understanding Hydrodynamics, Flow and Solute Conversion During ZnS Particle Production 273
16.4 Conclusions.. 276
Keywords 276
References... 276

16.1 INTRODUCTION

Microreactor is one of the most unique modernistic approach for the synthesis of nanoparticles and it has potential advantages over conventional batch reactors in terms of continuous operation, minimal waste, fast reaction rates and better control over particle size as well as the possibility of scaling up for large scale production [1–4]. Microreactors have large surface to volume ratio due to the small dimensions of microreactors and smaller internal volumes, which results into an increase in the rate of reactions as compared to macroscopic devices [5]. Higher aspect ratio in microreactors gives better heat and mass transfer rates and hence better control over the particle size and its distribution compared to conventional reactors [6–8]. Zhao et al. [7] have reported nanoparticle synthesis in microreactors and it have been suggested that the integration of nanoparticle and microreactor technologies has immense opportunity and potential for the development of novel materials and reactors. Further a variety of approaches for synthesizing nanoparticles such as continuous flow, gas-liquid segmented flow and droplet-based microreactors have been discussed.

Continuous production of nanoparticles with narrow particle size distribution and higher conversion rate can be achieved using microreactor. Zinc Sulfide (ZnS) has been attracted an attention of researchers considerably in the field of short wave-length photoelectronic, optical devices, especially for ultra violet laser diodes, photodetectors and fast optical switches because of its wide band gap [9–11]. John and Florence [12] have studied optical, structural and morphological properties of bean like ZnS nanostructures prepared using a single step chemical reaction of $ZnCl_2$ and Na_2S in aqueous solutions in a batch process. The reported particle size of ZnS nanoparticles using batch process by John and Florence [12] is around 12 nm. Hung and Lee [13] have studied microfluidic devices for the synthesis of nanoparticles and biomaterials and it has been indicated that miniaturized reactor provides controlled fluid transport, rapid chemical reactions, and cost-saving advantages over conventional methods for chemical, biological, and medical applications. Patil et al. [14] have prepared mono dispersed colloidal silver nanoparticles in continuous flow microreactor. The effect of type and loading of surfactant, precursor flow rate, and its concentration on size of silver nanoparticles was also studied. Segmented spiral microreactor has been also used by Ravikumar et al. [15] for the preparation of silver nanoparticles. The effect of gas liquid flow rates on particle size distribution of barium sulfate nanoparticles in capillary microreactor was studied by Jeevarathinam et al. [16].

The present work deals with synthesis of ZnS nanoparticles in a continuous flow microreactor system. The effects of various parameters like reaction temperature, precursor concentration, precursor flow rate and surfactant type on ZnS nanoparticles synthesis have also been studied. Simulation was performed in order to study the velocity profile and concentration profile in a microreactor.

16.2 MATERIALS AND METHODS

16.2.1 MATERIAL

Analytical reagent grade zinc chloride ($ZnCl_2$, 99%), sodium sulfide (Na_2S, 99%), sodium dodecyl sulfate (SDS, $NaC_{12}H_{25}SO_{4-}$, 99%), oleic acid ($CH_3(CH_2)_7CH{=}CH(CH_2)_7COOH$) and polyethylene glycol (PEG, Mw. 570–630 g/mol) were procured from SD fine chemicals Ltd Mumbai, India and were used as received. Poly vinyl pyrrolidone (PVP, $(C_6H_9NO)_n$, Mw. 40000 g/mol) was procured from Merck specialties Pvt. Ltd, Mumbai, India. Deionized water (conductivity less than 1 μS, Millipore) was used in all experimental runs. All the experiments were repeated thrice and the deviation in the results was found to be ± 3%.

16.2.2 EXPERIMENTS FOR RESIDENCE TIME DISTRIBUTION (RTD) BEHAVIOR OF MICROREACTOR

The residence time distribution behavior of the microreactor was investigated by injecting tracer material, for example, NaOH as a pulse input. The flow rate of water was maintained in the range of 90 to 180 mL/h for the investigation of residence time distribution. In a typical experiment of RTD analysis, after the system reached steady state, 2 mL pulse input of 0.1 N NaOH was injected using a syringe and response of the tracer was continuously monitored at the outlet. Sample was collected in a span of 15 secs and acid base titration was used to carry out the tracer analysis using 0.1 N HCl.

16.2.3 SYNTHESIS OF ZnS NANOPARTICLES USING BATCH AND CONTINUOUS FLOW MICROREACTOR

Synthesis of ZnS nanoparticles was carried out in batch process using 0.1 M concentration of both the precursors ($ZnCl_2$ and Na_2S) in an aqueous solution. $ZnCl_2$ and Na_2S solutions (0.1 M) 100 mL each were prepared in deionized water. Aqueous solution of sodium dodecyl sulfate (SDS) was separately prepared by adding 0.23 g SDS in 30 mL deionized water and added into Na_2S solution. As prepared $ZnCl_2$ solution was heated (up to 60 °C) under continuous stirring using magnetic stirrer and Na_2S solution was added slowly to $ZnCl_2$ solution in 30 min duration. The conversion of the reactant was monitored using UV spectrophotometer (SHIMADU 160 A). Figure 16.1(a) shows a microreactor set up made up of low-density polyethylene (LDPE) tube (length 150 cm and 0.8 mm inner diameter) and infuser pumps were procured from M/S Universal medical instruments Indore, India. The preparation of ZnS nanoparticles was carried out in microreactor by chemical precipitation method using zinc chloride and sodium sulfide as precursor materials. The concentrations of both the precursor were varied from 0.001 M to 1 M in different experiments. Two dispovan syringes were used in order to feed the two precursors in the microreactor through a 'Y' shape junction and a hot water bath was used to maintain the required reaction temperature. In order to study the effect of reaction temperature, it was varied from 40 to 70 °C. Further various experiments were carried out at different flow rate of both the precursor solutions which was varied from 90 to 180 mL/h. Different types stabilizers like sodium dodecyl sulfate (SDS), polyvinyl pyrrolidone (PVP), oleic acid and poly ethylene glycol (PEG) were used in order to study the effect of it. Stabilizer was added to both the precursor solutions and resultant solution was filled in separate syringes (50 mL each). Na_2S was used in excess (10 wt % excess than the required stoichiometric amount) to ensure the complete conversion so that the reaction will follow pseudo first order kinetics. The final product, for example, colloidal ZnS nanoparticles, was collected at the outlet of the microreactor. The obtained samples of ZnS nanoparticles were characterized by UV-visible spectroscopy, XRD and TEM analysis. The progress of reaction was also monitored using conversion data, which was obtained with the help of UV spectrophotometer.

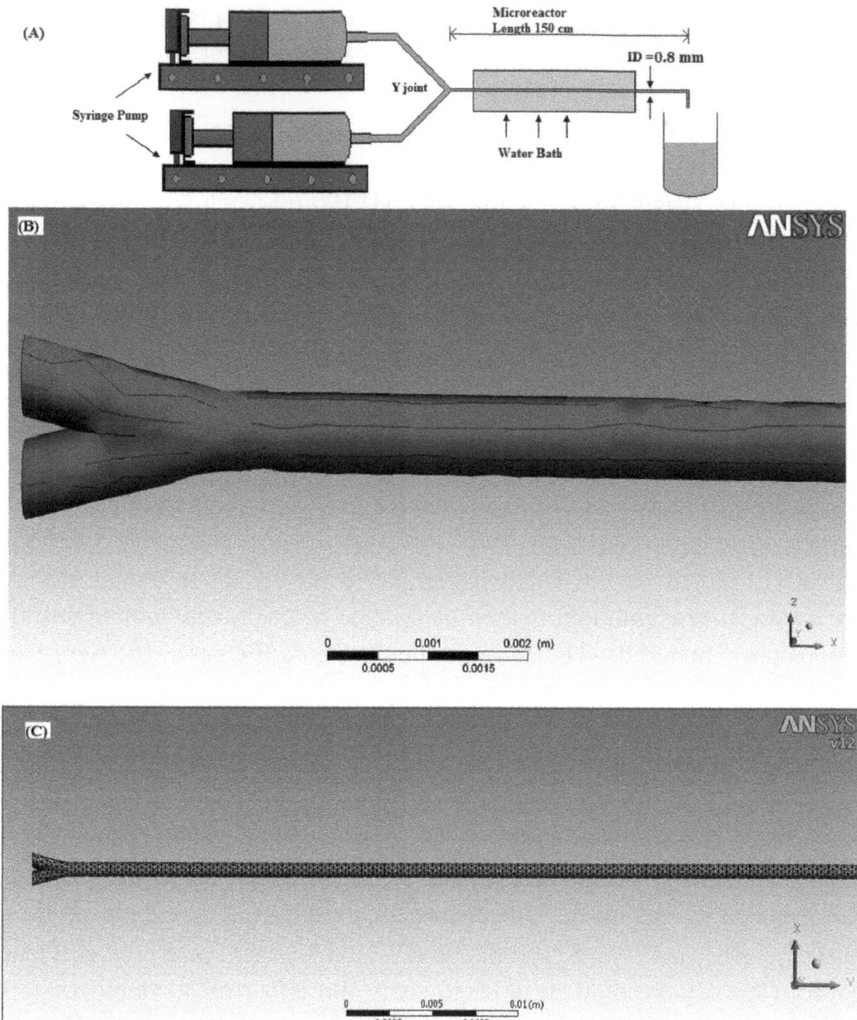

FIGURE 16.1 (a) Schematic diagram of the experimental set up, geometry of microreactor, (b) created in the workbench of ANSYS, and (c) showing generated mesh for the production of zinc sulfide nanoparticles in microreactor.

16.2.4 PREDICTIVE SIMULATION OF THE MICROREACTOR

ANSYS CFX is a commercial Computational Fluid Dynamics (CFD) tool used to simulate CFD to study the hydrodynamics of the microreactor. Based on the predictive conversion of the reactants in this particular case the length of the microreactor was optimized. Methodology of predictive simulation of microreactor is reported in following stages: (i) geometry (physical bounds) definition, (ii) mesh generation (volume occupied by the fluid is divided into discrete cells), (iii) solver (problem definition), and (iv) simulation and post CFD (Results).

16.2.4.1 Geometry Definition

Geometry of microreactor was defined in the workbench of the ANSYS. Microreactor setup as explained earlier consists of a 150 cm long linear low-density polyethylene tube with 0.8 mm inner diameter. A Y-junction (angle of 30°) is made at the entrance of the microreactor to feed the precursor materials. The reaction takes place throughout the microreactor and the product was collected at the outlet of the microreactor. The tube wall was defined to be solid and fluid (precursor) flows inside the tube. Geometry of microreactor created in the workbench is defined as shown in the Figure 16.1(b). Y-joint in the given geometry is clearly seen through at the entrance of the microreactor and after Y-joint microreactor was defined for the simulation. Geometry of microreactor was discretized into 60 thousand nodes using volume mesh generation technique. Figure 16.1(c) depicts the microreactor geometry with generated mesh.

16.2.4.2 Methods Used in Solver

Finite Volume Method: The finite volume method (FVM) is a common approach used in CFD codes, as it has an advantage in memory usage and solution speed. In the finite volume method, the governing partial differential equations (typically the Navier-Stokes equations, the mass and energy conservation equations, and the turbulence equations) are recast in a conservative form, and then solved over discrete control volumes. The finite volume equation yields governing equations in the form,

$$\frac{\partial}{\partial t}\iiint Q dV + \iint F dA = 0 \tag{1}$$

where Q is the vector of conserved variables, F is the vector of fluxes, V is the volume of the control volume element, and A is the surface area of the control volume element.

Finite Element Method: The finite element method (FEM) is used in structural analysis of solids, but is also applicable to fluids. However, the FEM formulation requires special care to ensure a conservative solution. The FEM formulation has been adapted for use with fluid dynamics governing equations. Although FEM must be carefully formulated to be conservative, it is much more stable than the finite volume approach [17, 18]. However, FEM can require more memory and has slower solution times than the FVM [17, 18]. In this method, a weighted residual equation is formed:

$$Ri = \iiint W_i Q dV_e \tag{2}$$

where *Ri* is the equation residual at an element vertex *i*, *Q* is the conservation equation expressed on an element basis, *Wi* is the weight factor, and *Ve* is the volume of the element.

Finite Difference Method: The finite difference method (FDM) is simple to program. It is currently only used in few specialized codes, which handle complex geometry with high accuracy and efficiency by using embedded boundaries or overlapping grids (with the solution interpolated across each grid).

$$\frac{\partial Q}{\partial t} + \frac{\partial F}{\partial x} + \frac{\partial G}{\partial y} + \frac{\partial H}{\partial z} = 0 \tag{3}$$

where *Q* is the vector of conserved variables; *F*, *G*, and *H* are the fluxes in the *x*, *y*, and *z* directions, respectively.

16.2.4.3 Equations Involved in the Formation of ZnS Nanoparticles

The formation of Zinc sulfide nanoparticle is a precipitation reaction, which proceeds through nucleation and crystal growth [19, 20]. The nucleation rate and growth Rate generally expressed as power law model as follows

$$\text{Nucleation Rate} = K_n (C - C_{s)}{}^E \tag{4}$$

$$\text{Growth Rate} = K (C - C_{s)}{}^G \tag{5}$$

where K_n = nucleation coefficient, E = nucleation exponent, C_s = saturation concentration, K = growth coefficient, G = growth exponent, C_s = saturation concentration Reaction:

$$ZnCl_2\ (A) + Na_2S\ (B) \longrightarrow ZnS\ (C) + 2NaCl\ (D) \tag{6}$$

The rate of conversion of the reactant can be expressed as:

$$\frac{dX}{dt} = kC_{A0}(1-X)\left(\frac{C_{B0}}{C_{A0}} - X\right) \tag{7}$$

where X = % conversion of the reaction. Mass deposited on all particles per unit time due to growth can be given by [21–24]:

$$\begin{aligned} dM/dt &= 3 \times \text{Shape Factor} \times \text{density} \\ &\times (\text{Characteristics Length})_2 \times (\text{Number of Particles}) \end{aligned} \tag{8}$$

Therefore, the mass balance equation of 'C' or ZnS becomes

$$\frac{dC_{liq}}{dt} = kC_{A0}(1-X)C_{A0}\left(\frac{C_{B0}}{C_{A0}} - X\right) - \frac{dM}{dt} \tag{9}$$

16.3 RESULTS AND DISCUSSIONS

In batch process, ZnS particles were obtained as a white precipitate by the addition of aqueous solutions of $ZnCl_2$ and Na_2S together in the temperature range of 40–70 °C. In order to prepare stabilized ZnS nanoparticles without any agglomeration and to restrict the growth of nanoparticles surfactant was added in the reaction mixture. ZnS nanoparticles were prepared as per the reaction reported in Eq. (6).

Initially, ZnS nanoparticles were formed and remained in suspended form in the liquid medium by the reaction between $ZnCl_2$ and Na_2S. However, when further ZnS nanoparticles were produced, the particle size increases due to growth and aggregation [21]. The growth of these nanoparticles can be restricted to nano scale with an addition of a suitable surfactant as depicted in Figure 16.2. Stabilization and aggregation of

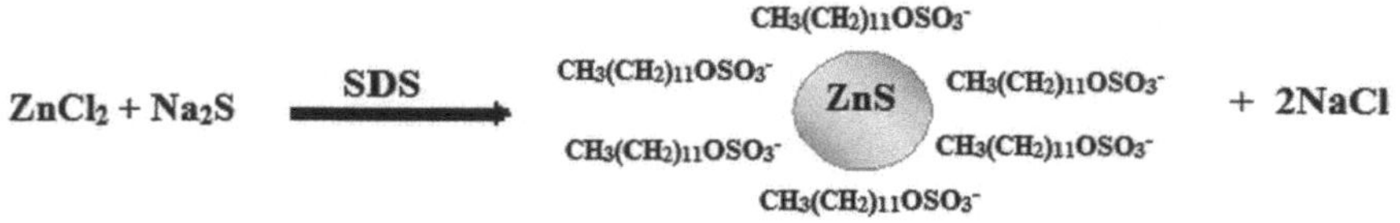

CH3(CH2)11OSO3-
ZnCl2 + Na2S
SDS
CH3(CH2)11OSO3-
ZnS
CH3(CH2)11OSO3-
+ 2NaCl
CH3(CH2)11OSO3-
CH3(CH2)11OSO3-
CH3(CH2)11OSO3-

FIGURE 16.2 Formation of ZnS colloidal nanoparticles in the presence of surfactant.

formed ZnS nanoparticles was accomplished by forming the layer of the surfactant around the ZnS nanoparticles. However, the restricted growth of the ZnS nanoparticles in the microreactor might be due to the uniform mass transport of the precursors.

16.3.1 HYDRODYNAMIC CHARACTERISTICS AND EFFECT OF FLOW RATE ON THE PARTICLE SIZE IN MICROREACTOR

Residence time distribution (RTD) data was used to characterize the mixing and flow pattern within the microreactor, which in turn provides information about the overall performance of the reactor [25]. In present study RTD analysis was carried out for the flow rates of 90, 120, 150 and 180 mL/h. Figure 16.3 shows an external age distribution curve of the tracer

90mL/h
120mL/h
150mL/h
180mL/h
2.5
2
1.5
1
0.5
0
Eθ×10^-4
0
0.5
1
1.5
2
2.5
3
θ

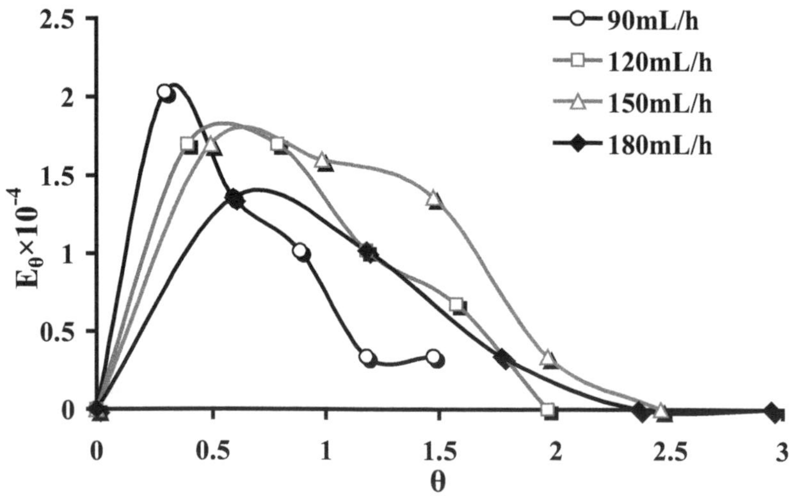

FIGURE 16.3 Residence time distribution curve for the external age distribution curve, E_θ plotted against the dimensionless time θ.

material in the microreactor. RTD obtained (Figure 16.3) in the microreactor shows a broadened symmetry at flow rate of 180 mL/h and channeling for the lower flow rates. Dead zones observed (as shown in Figure 16.3) in microreactor at low flow rates may be due to incomplete filled reactor. RTD results represent the deviation of microreactor from ideal plug flow conditions. This deviation is attributed to the low Reynolds number [25].

As depicted in Figure 16.4, Peclet number is plotted against the Reynolds number. With an increase in the flow rate, the mass diffusion coefficient (or molecular diffusion) is found decreased and the axial mixing is observed to be increased indicating the continuous increase in Pe_{ax} with an increase in *Re* number. Also due to very small lateral dimensions of the microreactor, at higher flow rates a higher flow velocity is observed and hence inertial forces are prevalent over the viscous forces. The Pe_{ax} number was found to be increased from 0.073 at *Re* number 3.44 (90 mL/h) to 0.1638 at *Re* number 6.88 (180 mL/h).

16.3.2 EFFECT OF PRECURSOR FLOW RATE ON ZnS NANOPARTICLES FORMATION IN MICROREACTOR

One of the important hydrodynamic parameter affecting the particle size is the flow rate of precursors in the microreactor [14–16]. The space-time should be less than the reaction time to complete the reaction, which will

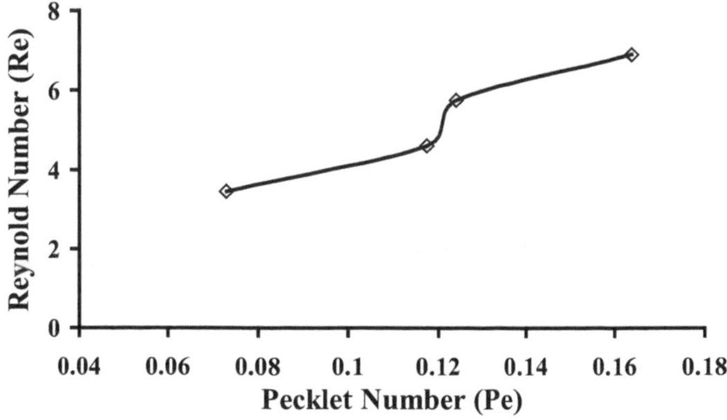

FIGURE 16.4 Variation of Peclet number as a function of Reynold number.

give a narrow particle size distribution. Further the micro mixing time (τ_{mixing}) should be less than the τ for instantaneous precipitation reactions. Hence, the following experiments were carried out to know the effect of space-time on the formation of ZnS nanoparticles which was studied using UV absorption spectra. A study was carried out to understand the effect of flow rate (180, 240, 300 and 360 mL/h) on the formation of ZnS nanoparticles and yield of ZnS nanoparticles. Figure 16.5 shows the absorption spectra of ZnS nanoparticles at different flow rates. The absorbance value was found to be increased with an increase in the flow rate of precursors. It is attributed to more turbulence with an increased flow rate leading to formation of more nuclei resulting into formation of large number of ZnS nanoparticles. These large number of ZnS nanoparticles results into increased concentration of ZnS nanoparticles which increases the absorbance value. There is a rapid increase in the absorbance for a flow rate of 360 mL/h leading to a higher yield of ZnS precipitate. With an increase in the flow rate there is decrease in the space-time and hence there is a possibility of more nuclei formation resulting in smaller particles.

16.3.3 EFFECT OF PRECURSOR CONCENTRATION ON ZnS NANOPARTICLES FORMATION IN MICROREACTOR

The concentrations of the reactants play a very important role in the formation and growth of ZnS nanoparticles and hence experiments were carried

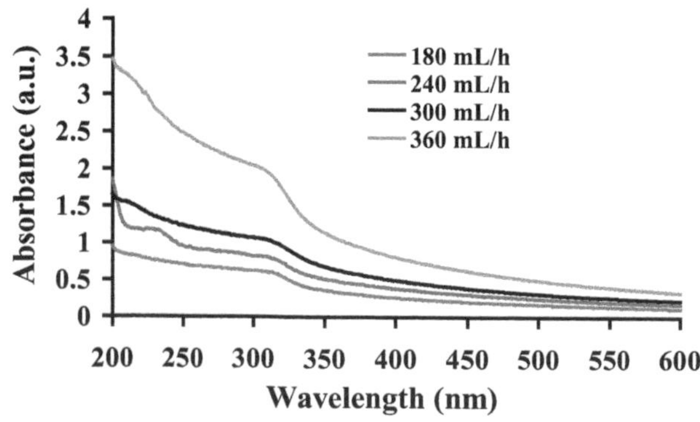

FIGURE 16.5 Absorption spectra of zinc sulfide nanoparticles with varying flow rate at 60–°C temperature.

out using four different concentrations of the precursors ranging from 0.001 to 1 M. Figure 16.6 shows absorption spectra of ZnS nanoparticles formed at different concentrations of the reactants. UV-visible absorption spectra indicate the amount of ZnS produced. In the present study, absorbance was found to be increased with an increase in the concentrations of the reactants. Decrease in the concentration showed decrement in absorbance value corresponds to the decrease in the ZnS particle size. Also with decrease in the precursor concentration broadening of peaks is observed. The possible reason for the decrease in the intensity and broader width of absorption peaks is lesser number of ZnS nanoparticles. Also for smaller ZnS nanoparticles electron reaches the surface faster and scatters quickly which results in broadening in absorption peak width [14].

16.3.4 EFFECT OF TEMPERATURE ON ZnS NANOPARTICLES FORMATION IN MICROREACTOR

The formation of ZnS is driven by the influence of temperature as it is a temperature dependent reaction. A study was carried out to understand the effect of temperature on the particle size and yield of ZnS nanoparticles. Figure 16.7 shows the absorption spectra of the ZnS nanoparticles as a function of different temperatures. It can be seen that the yield is approximately constant at 40 and 50 °C and is substantially increases at

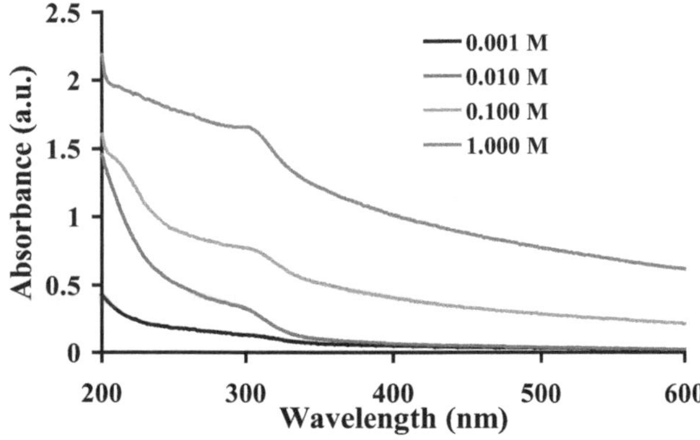

FIGURE 16.6 Absorption spectra of Zinc sulfide nanoparticles with different precursor concentration at 360 mL/h precursor flow rate.

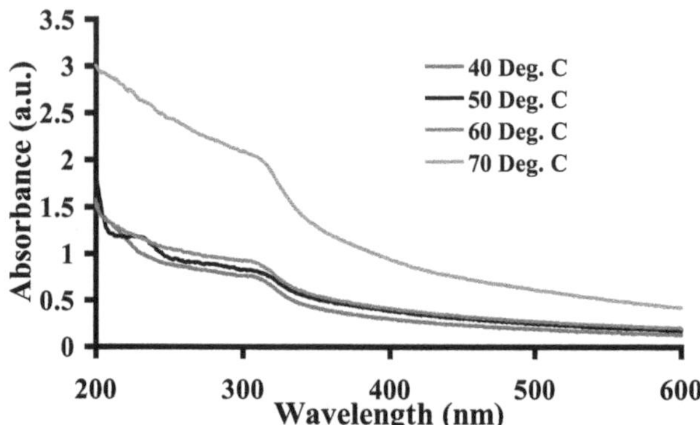

FIGURE 16.7 Absorption spectra of Zinc sulfide nanoparticles with varying temperature at 240 mL/h precursor flow rate.

60 and 70 °C. This may be attributed to large number of nuclei formation at higher temperature because of higher conversion rate. As the temperature increases the absorbance value increases as shown in the Figure 16.7, which is due to large nuclei formation at higher temperature the particle size is also increasing at the cost of higher conversion.

16.3.5 EFFECT OF TYPES OF STABILIZERS ON ZnS NANOPARTICLES FORMATION IN MICROREACTOR

Different stabilizers were used to study their effect on the yield of zinc sulfide nanoparticles. ZnS nanoparticles were prepared in the microreactor and thermal decomposition route in the presence of PVP, SDS, PEG and Oleic acid. Figure 16.8 shows the absorption spectra of the ZnS nanoparticles obtained with different stabilizers. It is observed that PVP stabilized particles are large and SDS stabilized particles are relatively small. The role of surfactant is very important in the formation of ZnS nanoparticles. The surfactant acts as a stabilizing agent. It is reported that the concentration SDS also affects the size and shape of the ZnS [19,26]. In order to achieve the shape of ZnS particles spherical in nature, addition of SDS solution was maintained with the ratio of 1:6 wt % of the $ZnCl_2$. Initially, due to the presence of SDS, precursors $ZnCl_2$ and Na_2S react in a regular manner. SDS also restricts the growth of the nanoparticles and is acting

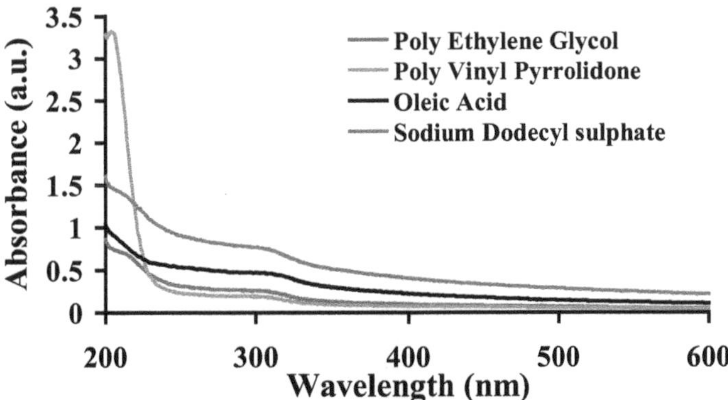

FIGURE 16.8 Absorption spectra of Zinc sulfide nanoparticles using different type of surfactant.

as a capping agent over the formation of the nanoparticles. The absorbance value is found to be lesser for PVP (0.19 a.u.) and is found to be larger for SDS (0.75 a.u.) surfactant. The absorbance is observed in the trend given as PVP (0.19 a.u.) < PEG (0.26 a.u.) < OA (0.46 a.u.) < SDS (0.75 a.u.). This indicates the formation of ZnS nanoparticles is more favored by SDS surfactant compared to other surfactants.

16.3.6 COMPARISON OF FORMATION OF ZnS NANOPARTICLES IN BATCH AND MICROREACTOR

In the present study, the particle size obtained in the microreactor was in the range of 5 to 10 nm as shown in the TEM image (Figure 16.9(a)). Moreover, the TEM image in Figure 16.9(a) shows that the zinc sulfide nano particles are spherical in shape. As discussed in the earlier section, the role of surfactant used has significant effect onto the particle size and shape. It is important to note that SDS addition above critical micelles concentration (CMC) gives the spherical shape. Further the TEM image of ZnS nanoparticles produced in batch reactor at 60°C is reported in Figure 16.9(b). The agglomerated morphology of ZnS nanoparticles is observed with particle size of large aggregates around 20 to 50 nm. This clearly indicates that the use of microreactor for the preparation of ZnS nanoparticles gives lesser size particles without any aggregation compared to batch process.

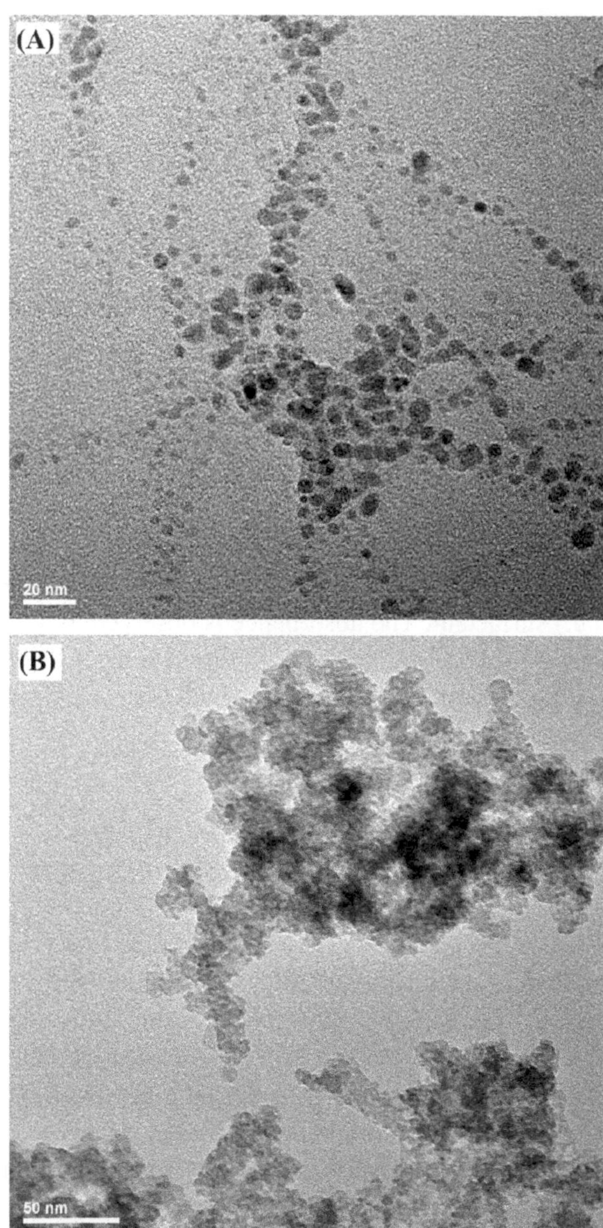

FIGURE 16.9 TEM images of zinc sulfide nanoparticle synthesized in (a) microreactor at precursor flow rate 300 mL/h, and (b) batch reactor at 60 °C temperature using SDS as surfactant.

X-ray diffraction pattern of zinc sulfide nanoparticles show peaks at $2\theta = 43$ and 52° confirms the formation of ZnS nanoparticles. The diffraction peaks of the XRD pattern of as prepared the ZnS samples in the microreactor is shown in Figure 16.10.

The XRD pattern of ZnS nanoparticles exhibit the peaks at scattering angle (2θ) = 29.1, 43.14 and 51.5° corresponds to (111), (220) and (311) planes, respectively. The broadening of XRD peaks indicates crystalline behavior of the ZnS nanoparticles [21, 22]. Further the crystalline behavior is also confirmed by using TEM image reported in the Figure 16.9(a). The obtained ZnS particles are having the wurtzite phase structure and hence it could be used as a semiconductor material. The zinc sulfide nanoparticles produced in batch reactor are large aggregates in the size range of 20 to 50 nm with irregular shapes (Figure 16.9(b)).

16.3.7 SIMULATIONS OF MICROREACTOR: UNDERSTANDING HYDRODYNAMICS, FLOW AND SOLUTE CONVERSION DURING ZNS PARTICLE PRODUCTION

The purpose of the simulation is to understand the behavior of micro channel systems and to provide guidelines for the design of future microreactors. The velocity profile of the fluid flowing through the microreactor is shown in the Figure 16.11(a) and (b) which shows the similarity with the conventional plug

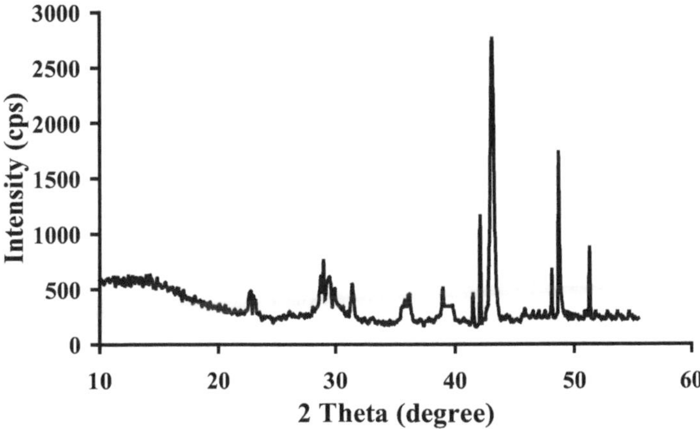

FIGURE 16.10 The XRD pattern of ZnS nanoparticle produced in microreactor at 60 °C with precursor flow rate of 300 mL/h and using sodium dodecyl sulfate as surfactant.

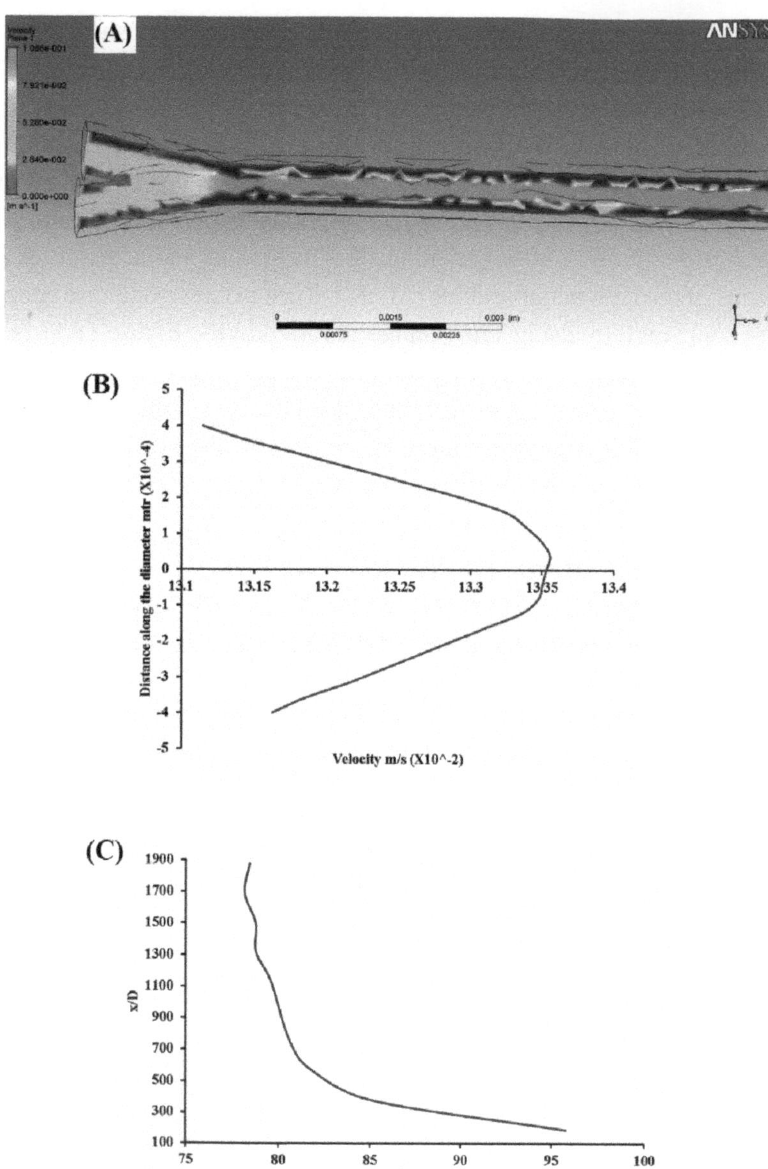

FIGURE 16.11 (a) pictorial view of the velocity profile along the length and diameter, (b) velocity profile along the diameter, and (c) velocity profile U/Umax vs *x*/D along the diameter in the microreactor for fluid flow rate of 180 mL/h.

flow reactor. The velocity at the center is maximum and minimum nearer to the wall. The main hydrodynamic behavior resembles the real physical system. Figure 16.11 shows the pictorial view of the velocity profile along the length and diameter of the microreactor. Figure 16.11 shows the graph plotted for velocity of the fluid flowing in the microreactor along the diameter of the microreactor. Also due to low Reynolds number the velocity at the inlet is maximum and at the outlet is minimum.

Mass transfer is one the major design parameters for any reactor. Microreactor provides a very high surface to volume ratio, which definitely increases the conversion due to high mass transport and high diffusion rate. During the simulation the concentration of the reactant decreases along the length and the concentration of the product increases as shown in the Figure 16.12. It can be clearly seen from the graph that the concentration of $ZnCl_2$ decreases very rapidly and at the same time the concentration of ZnS increases very rapidly along the length. The concentration of $ZnCl_2$ reduces to almost negligible around the 0.7 m length of the microreactor and at the same time this $ZnCl_2$ gets converted to ZnS, which is seen by the increasing graph of ZnS mass fraction. Hence, a microreactor of length 1 m would have been sufficiently large to have 100% conversion.

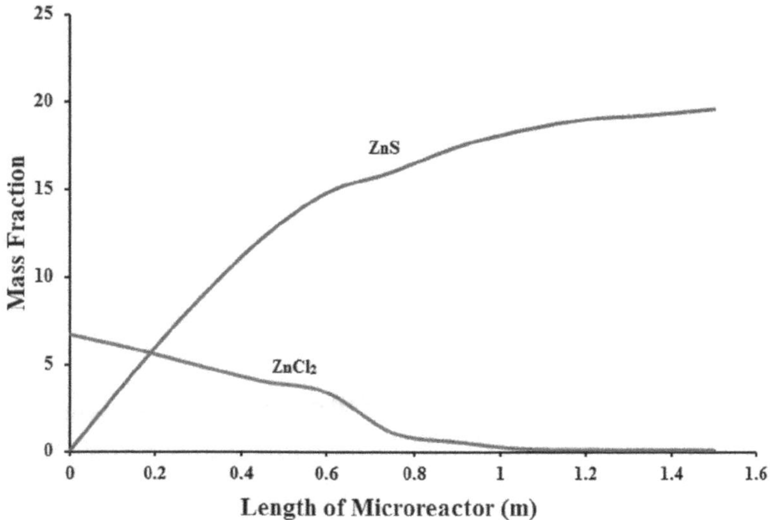

FIGURE 16.12 Concentration profile of ZnS and $ZnCl_2$ showing the mass fraction change along the length of the microreactor for the flow rate of 180 mL/h.

16.4 CONCLUSIONS

The production and intensification parameters of zinc sulfide nanoparticles in a microreactor were investigated. The use of microreactor is successfully demonstrated for the preparation of mono-disperse, nano-sized (5–10 nm) zinc sulfide nanoparticles at optimum process conditions. The effect of stabilizers such as PVP, PEG, Oleic acid and SDS was studied in microreactor. Out of four surfactants, SDS shows the spherical particle shape at concentration 1:6 wt % of the $ZnCl_2$. There is a rapid increase in the absorbance for a flow rate of 360 mL/h leading to a higher yield of product. Also the effect of concentration and temperature were successfully investigated. It was found that with an increase in the temperature and concentration the absorbance value is found increased indicating large number of nuclei formation. Simulation tools have been used to predict the flow pattern inside the microreactor. The concentration profile for the reactants and the products were obtained. The length of the microreactor is optimized and is found to be 0.7 m.

KEYWORDS

- **Batch**
- **Colloidal nanoparticles**
- **Microreactor**
- **Process parameter**
- **Zinc sulfide**

REFERENCES

1. Park, K. Y., Ullmann, M., Suh, Y. J., & Friedlande, S. K., Nanoparticle microreactor: Application to synthesis of titania by thermal decomposition of titanium tetraisopropoxide. *J Nanopart Res.* 2001, *3*, 309–319.
2. Singh, A., Malek, K. C., & Kulkarni, S. K., Development in microreactor technology for nanoparticle synthesis. *Int J Nanosci.* 2010, *9*, 93.
3. Pacławski, A. K., Streszewski, B., Jaworski, W., Luty-Błocho, M., & Fitzner, K., Gold nanoparticles formation via gold(III) chloride complex ions reduction with glucose in the batch and in the flow microreactor systems. *Colloid Surface A.* 2012, *413*, 208–215.

4. Gutierrez, L., Gomeza, L., Irusta, S., Arruebo, M., & Santamaria, J., Comparative study of the synthesis of silica nanoparticles in micromixer–microreactor and batch reactor systems. *Chem Eng J.* 2011, *171*, 674–683.
5. Watts, P., & Wiles, C., Recent advances in synthetic micro reaction technology. *Chem Commun.* 2007, 443–467.
6. Hung, L. H., & Lee, A. P., Microfluidic devices for the synthesis of nanoparticles and biomaterials. *J Med Biol Eng.* 2007, *27*, 1–6.
7. Zhao, C. X., He, L., Qiao, S. Z., & Middelberg, A. P. J., Nanoparticle synthesis in microreactors. *Chem Eng Sci.* 2011, *7*, 1463–1479.
8. Chestnoy, N., Hull, R., & Brus, L. E., Frontiers in electronics: future chips. *J Chem Phys.* 1996, *85*, 2237–2242.
9. Murugadoss, G., & Rajesh Kumar, M., Synthesis and optical properties of monodispersed Ni^{2+} doped ZnS nanoparticles. *Appl Nanosci.* 2012, 1–9.
10. Gilbert, B., Huang, F., Lin, Z., Goodell, C., Zhang, H., & Banfield, J. F., Surface chemistry controls crystallinity of ZnS nanoparticles. *Nano Lett.* 2006, *6*, 605–610.
11. Xu, J. F., Ji, W., Lin, J. Y., Tang, S. H., & Du, Y. W., Preparation of ZnS nanoparticles by ultrasonic radiation method. *Appl Phys A.* 1998, *66*, 639–641.
12. John, R., & Sasi, S., Florance optical structural and morphological studies of bean like ZnS nanostructures by aqueous chemical method. *Chalcogenide Lett.* 2010, *7*, 269–273.
13. Hu, J. S., Ren, L. L., Guo, Y. G., Liang, H. P., Cao, A. M., Wan, L. J., & Bai, C. L., Mass production and high photocatalytic activity of ZnS nanoporous nanoparticles. *Angew Chem Int Ed.* 2005, *44*, 1269–1273.
14. Patil, G. A., Bari, M. L., Bhanvase, B. A., Ganvir, V., Mishra, S., & Sonawane, S. H., Intensification of synthesis of colloidal silver nanoparticles in microreactor: Effect of surfactant and process parameters. *Chem Eng Process: Process Intensification.* 2012, *62*, 69–77.
15. Ravi Kumar, D. V., Prasad, B. L. V., & Kulkarni, A. A., Segmented flow synthesis of Ag nanoparticles in spiral microreactor: Role of continuous and dispersed phase. *Chem Eng J.* 2012, *192*, 357–368.
16. Jeevarathinam, D., Gupta, A. K., Pitchumani, B., & Mohan, R., Effect of gas and liquid flowrates on the size distribution of barium sulfate nanoparticles precipitated in a two phase flow capillary microreactor. *Chem Eng J.* 2011, *173*, 607–611.
17. Surana, K. A., Allu, S., & Tenpas, P. W., k-version of finite element method in gas dynamics: higher order global differentiability numerical solutions, *Int J Numer Meth in Eng.* 2006, *69*, 1109–1157.
18. Huebner, K. H., Thornton, E. A., & Byron, T. D., *The Finite Element Method for Engineers.* 3rd Ed., Wiley Interscience, 1995.
19. Saha, S., Bera, K., & Jana, P. C., Growth time dependence of size of nanoparticles of ZnS. *Int J Soft Comput Eng.* 2011, *1*, 2231–2307.
20. Scott Fogler, H., Elements of Chemical Reaction Engineering. 4th Ed; PHI Publication, 2005.
21. Chen, M., Ma, C. Y., Mahmud, T., Lin, T., & Wang, X. Z., Population balance modeling and experimental validation for synthesis of TiO_2 nanoparticles using continuous hydrothermal process. *Adv Mater Res.* 2012, *508*, 175–179.
22. Myerson, A. S., *Molecular Modeling Applications in Crystallization.* Cambridge University Press, 1999.
23. Mullin, J. W., *Crystallization. 4th Ed.* Butterworth Heinemann, 2001.

24. Ramkrishna, D., *Population Balances: Theory and Applications to Particulate Systems in Engineering. 1st Ed.* Academic Press, 2000.
25. Boskovic, D., Loebbecke, S., Gross, G. A., & Koehler, J. M., Residence time distribution studies in microfluidic mixing structures. *Chem Eng Technol.* 2011, *34*, 361–370.
26. Mehta, S. K., Kumar, S., Chaudhary, S., Bhasin, K. K., & Gradzielski, M., Evolution of ZnS nanoparticles via facile CTAB aqueous micellar solution route: A study on controlling parameters. *Nanoscale Res Lett.* 2009, *4,* 17–28.

PART IV

PROCESSES AND APPLICATIONS

CHAPTER 17

HYDROGENATION WITH RESPECT TO RANCIDITY OF FOODS

D. C. KOTHARI,[1] P. V. THORAT,[1] and R. P. UGWEKAR[2]

[1]*Chemical Engineering and Polymer Technology Department, Shri Shivaji Education Society, Amravati's College of Engineering and Technology, Babulgaon (Jh.), Akola, Maharashtra, India*

[2]*Chemical Engineering Departments, Laxminarayan Institute of Technology, Nagpur, Maharashtra, India*

CONTENTS

17.1 Introduction 282
17.2 Materials and Methods 284
17.2.1 Hydrogenation 285
17.2.2 Characterization and Process Reactions Controls 287
17.2.3 Materials 287
17.2.3 Applications 289
17.3 Results and Discussions 289
17.3.1 Hydrogenator's or Reactor's Working Formulations for the Equipment 289
17.3.2 Hydrogenation Converts Alkenes to Alkanes and the Mass Transfer Rate 291
17.3.3 Applications of the Engineered Products from Vegetable Oils in Daily Life 293

17.4 Conclusion 294
Keywords 295
References 295

17.1 INTRODUCTION

Fats and Oils are considered to be essential nutrients for human being since the primitive ages as they provide most concentrated source of energy of any foodstuff. Unsaturated fatty acids characterized by one (*monounsaturated*) or more (*polyunsaturated*) double bonds in the carbon-chain, as the carbons are double-bonded to each other, there are fewer bonds available for hydrogen, so there are fewer hydrogen atoms, hence it is "*unsaturated.*" In the *cis* arrangement, the chains are on the same side of the double bond and in the *trans* arrangement, the chains are on opposite sides of the double bond, and chain is straight overall as represented in Figure 17.1. It was recognized that the more unsaturated the fatty acid, the more likely it was for the fatty acid to be oxidized, which leds to oxidative rancidity. By removing double bonds from linolenic acid, a monoenoic acid would resist oxidation better [1].

$$\text{Triene} \rightarrow \text{Diene} \rightarrow \text{Monoene} \rightarrow \text{Saturated acid}$$

The development of unpleasant odors, taste, flavors, and color from deterioration in the fat and oil portion of a food which undergoes oxidation reactions producing aldehydes, hydroxyl acids, keta acids, called the rancidity in foods. Apparently this rancidity can be classified in two categories, for example, *Hydrolytic* and *Oxidative* rancidity. Different

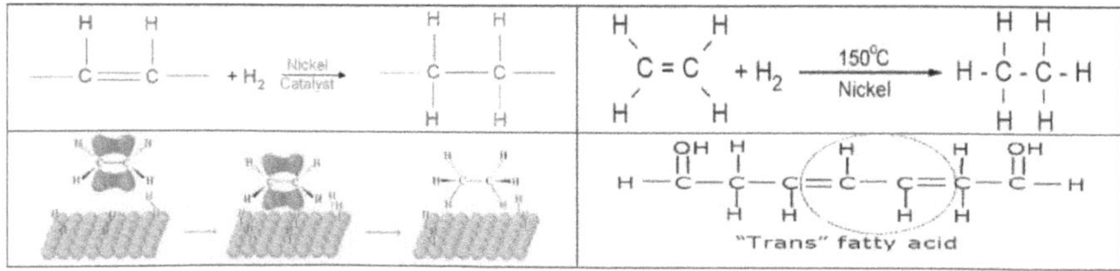

FIGURE 17.1 Representing Oils & Fats with respect to hydrogenation reaction with Nickel Catalyst.

micro-organisms having *enzymes* that could produce ketones and other oxidation products from the fatty acid esters which become responsible for the hydrolytic rancidity. The Oxidative rancidity (deterioration) leads to the destruction of fat soluble vitamins and essential fatty acids as well as concern of toxicological effects of various types.

The lipids and dietary lipids are disposed to oxidative processes in the presence of catalytic systems such as light, heat, enzymes, metals, metalloproteins, and micro-organisms, these giving rise to the development of off-flavors and loss of essential amino acids, fat-soluble vitamins, and other bioactives. Here these lipids could undergo autoxidation, photo-oxidation, thermal oxidation, and enzymatic oxidation under different conditions, most of which could involve some type of free radical or oxygen kind. Rancidity which initiate the foods could occurs when they are exposed to oxygen, are represented in Figure 17.2, which also show that harmful free radicals which are formed in and used all the oxygen present in the food materials [2]. The selection of an antioxidant depends upon its solubility in the products, which need to prevent from the rancidity.

It should not react and forms a new compound within the product; its volatility should be low enough; it could not be extracted by water; it should not be colored; it should be odorless and tasteless; it should not be toxic or irritant to the skin; and it should not cost too much. The filter aids contain pro-oxidants; for example charcoal treatment of oils removes natural antioxidants by absorption, the lipids and the mold's micelle of oils & fats are shown in Figure 17.3.

Some starting materials contain pro-oxidants or those unknown factors that may cause the fat in the finished product to be susceptible to rapid deterioration. As linolenic and linoleic acids are more susceptible to oxidation than are oleic and saturated acids, the selection of fats or

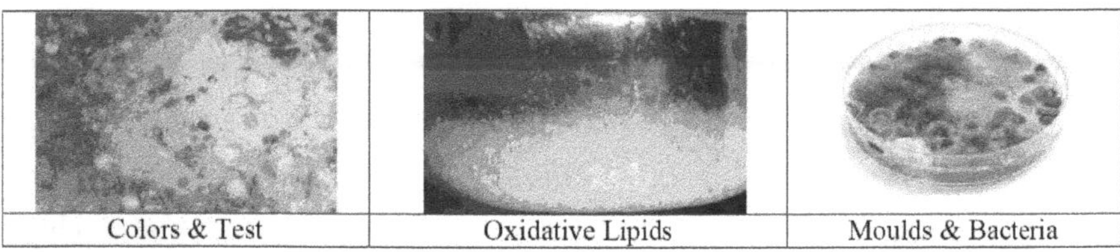

FIGURE 17.2 Rancidity producing molds, lipids, off-color and oxidation of foods, for example, wastages.

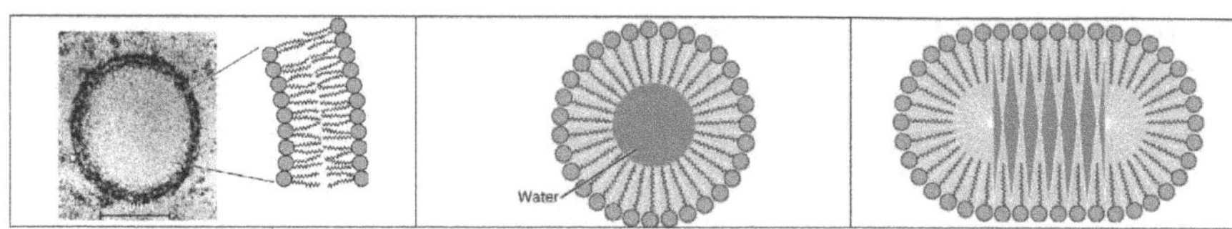

FIGURE 17.3 Lipids & Mould's micelle could develop a water-filled center or flatten out to collapse.

combinations of fats that have lower contents of more highly unsaturated fatty acids may aid in preventing rancidity in some products. Precautions must be taken in the processing of fats and fat products to avoid unnecessary exposure to light, atmosphere, moisture and high temperatures. Contamination with pro-oxidants and particularly metals, from the processing equipment should also be prevented. Some types of filter aids contain pro-oxidants and some of these, for example is charcoal treatment of oils removes natural antioxidants by absorption [3]

17.2 MATERIALS AND METHODS

Fatty acids are characterized as either *saturated* or *unsaturated* based on the presence of double bonds in its structure. If the molecule contains no double bonds, it is said to be saturated; otherwise, it is unsaturated to some degree.

India is a vast country and inhabitants of several of its regions have developed specific preference for certain oils depending upon the oils available in the region. People in the South and West prefer coconut and groundnut oil, while those in East and North use mustard and rapseed oil. Inhabitants of northern plan are basically hard fat consumers and therefore, prefer, Vanaspati. Edible Oil Industry is highly fragmented industry with over 600 oil extraction units and 166 vanaspati manufacturing units; Table 17.1 is showing market shares of edible oil and fats in India. In vanaspati Dalda is the oldest and largest brands, while other brands are *Rath, Gemini, Jindal, Gagan, Dhara, Postman, Sundrop, Suffola, Amul, Ruchi, Adani, Parachute, Vansada, Parmpara, Gowardhan, Swatic*, etc. These Oils are low in saturated fats and the double bonds within unsaturated acids are in the *cis* configuration. To improve their oxidative stability and

TABLE 17.1 Showing Market Share of EDIBLE Oil and Fats in India

Oil	Percentage USES %	Oil	Percentage USES %
Cotton	6%	Palm	38%
Soybean	21%	Peanut	14%
Sunflower	8%	Rapseed	13%

to increase their melting points, vegetable oils are hydrogenated. The process of hydrogenation is intended to add hydrogen atoms to *cis*-unsaturated fats, eliminating a double bond and making them more saturated. Partially hydrogenated oils give foods a longer shelf life, and more stable flavor, but melts upon baking (or consumption). Most applications show saturates increase well below this in Figure 17.4. Frying main fat functionality is to provide oxidative stable structure and solidity [4].

The oil bearing seeds contain naturally substances like tocopherols, which act as antioxidants and prevent the development of rancidity in the seed oil, however, in the process of refining of oil, the natural antioxidants are get removed.

17.2.1 HYDROGENATION

Hydrogenation is the process in which the hydrogen is added directly to points of un-saturation in the fatty acids. Hydrogenation of fats has being developed as a result of the need to convert liquid oils to the semi-solid

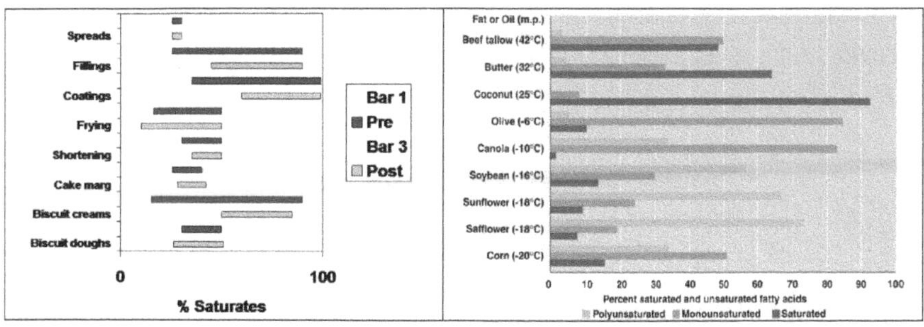

FIGURE 17.4 Representation of oils-partially saturated and saturated fats these are (Tailor made).

form for greater utility in certain food uses and increase the oxidative and thermal stability of fat or oil are shown in Figure 17.5. Hydrogenation is a strongly exothermic reaction. In the hydrogenation of vegetable oils and fatty acids, for example, heat released is about 25 kcal per mole (105 kJ/mol), sufficient to raise temperature of oil by 1.6–1.7 °C per drop [5]. In edible oil hydrogenation processes the feedstock is heated under vacuum a specific temperature at which hydrogen is introduced into the reactor. From time zero onward, bulk hydrogen concentration is determined by mass balance; $V_{oil}\, d\, C_{bulk}\,(t)/dt = k_L\, a\, [C_{max} - C_{bulk}\,(t)]\, V_{oil\text{-}} V_{cat} = k_L\, a\, [C_{max} - C_{bulk}\,(t)]\, V_{oil} - k_{rp}\, [IV\,(t) - 75]\, \eta_p\,(t)\, C_{bulk}\,(t)\, V_{cat}$.

At t = 0 the bulk hydrogen concentration is 0. Upon integration equation could be obtained; $C_{bulk}\,(t)/C_{max} = K/K + L\{1 - \exp\,[-\,(K + L)\, t]\}$ in $K = k_L\, a$ and $L = k_{rp}\, [IV\,(t) - 75]\, \eta_p\,(t)\, €_{cat}$.

Once hydrogen concentration has reached its equilibrium value, will occur after a time $t = 4/k_L\, a$, the derivative becomes almost zero are represented in Figure 17.6 with different parameters. This derivation neglects the fact that in the initial phase the specific gas/liquid interface is also function of time. It may be expected that this time will be much shorter than the time constant $1/k_L\, a$ for the absorption [6].

Hydrogenation Reactor. | Seeds to Oils to Fats. | Different TYPES.

FIGURE 17.5 Hydrogenation reactor in which oils to fats and partial to full hydrogenation of oils.

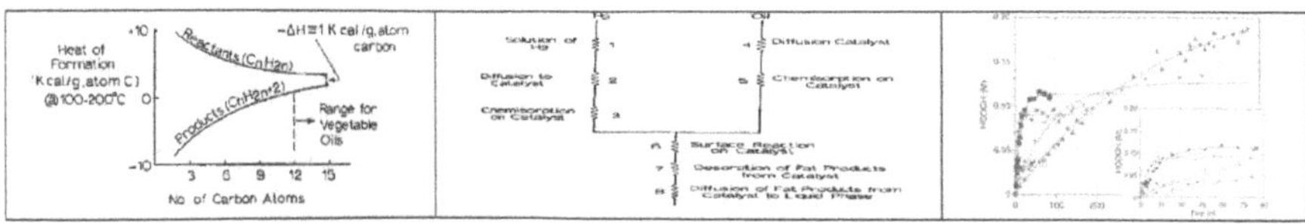

FIGURE 17.6 Different operating and reactive parameters in hydrogenation process of oils and fats.

17.2.2 CHARACTERIZATION AND PROCESS REACTIONS CONTROLS

Hydrogenation could be defined as the conversion of various unsaturated fatty glycerides into completely saturated glycerides by the addition of hydrogen in presence of a catalyst. The objective of the hydrogenation is not only to raise melting point but also to improve the reaping qualities, taste, and odor for many oils. It is faccompanied by isomerization with a increase in melting point, for example, by oleic (**cis**) isomerizing to olaidic (*trans*) acid. Importantly, as the reaction itself is exothermic, the reaction could be generalized as:

$$(C_{17}H_{31}COO)_3C_2H_3 + H_2\ (Ni) -$$
$$((C_{17}H_{33}COO)_3C_2H_5\ \text{Exothermic}\ \Delta H = -420.8\ \text{kJ/kg}$$

The testing for unsaturation carbon-carbon double bonds in unsaturated oils could be detected using the elements *bromine* or *iodine*. These elements react with the double bonds in the oils, the more double bonds in the reactors the more bromine or iodine should be used. During the test Bromine water is a dilute solution of bromine, which is normally orange-brown in color. It becomes colorless when shaken with an alkene, or with unsaturated fats. When shaken with alkanes or saturated fats, its color remains the same.

17.2.3 MATERIALS

Rany Nickel Catalyst is special form of spongy, finely divided Nickel is a powerful Hydrogenation Catalyst. Due to its ability to hydrogenate virtually any type of functional group. Raney Nickel is produced when a block of nickel-aluminum alloy is treated with concentrated sodium hydroxide. his treatment called "activation" dissolves most of the aluminum out of the alloy in which a high surface area sponge catalyst obtained by the chemical reaction. Macroscopically, Raney nickel is a finely divided gray powder using the leaching process represented in Figure 17.7 in which each particle of this powder with pores of irregular size and shapes.

The Figure 17.8 showing Energy released by catalyst and hydrogenation phase diagram. Raney nickel is notable for being thermally and structurally stable, as well has having a large surface area [6].

Raney Nickel is highly pyrophoric when dry, for example, catches fire when allowed to dry in atmosphere and is therefore always kept submerged under water or a suitable solvent such as Ethanol, Cyclohexane, Dioxane, etc. The surface area, which could be determined by a BET measurement, which carried out using a gas, should provide the active metal to react in particle of the catalyst. The Rany nickel catalyst have had an average Ni contains surface area of 100 m^2 per gram are described in Figure 17.9.

FIGURE 17.7 Representing nickel catalytic surface of hydrogenation and filtration of fats and oil.

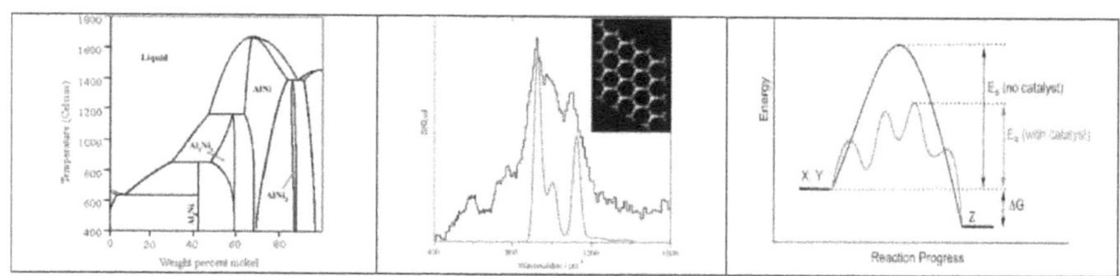

FIGURE 17.8 Representing energy released by catalyst and phase diagram of Ni – Al catalyst.

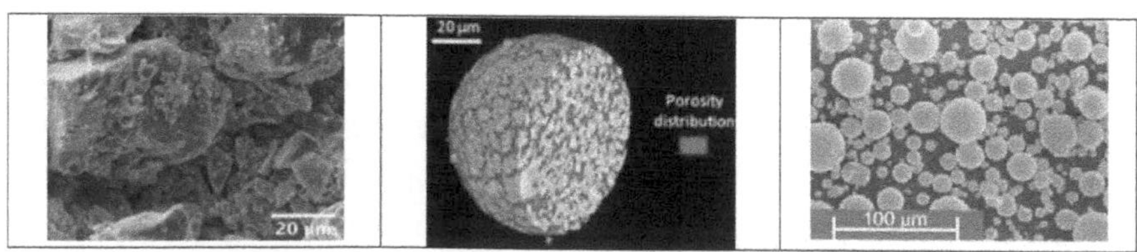

FIGURE 17.9 SEM of Raney nickel catalyst in which small cracks of approximately 1–100 nm.

For catalyst particles that have a mean square diameter of 5 μm, the corresponding value of solid phase mass transfer coefficient is 0.9×10^{-2} m/s. It is, in fact mainly determined by 2 are from [6].

17.2.3 APPLICATIONS

The manufactured form of trans fat, known as partially hydrogenated oil, is found in a variety of food products, including. **Baked goods:** Most cakes, cookies, pie crusts and crackers contain shortening, which is usually made from partially hydrogenated vegetable oil. Ready-made frosting is another source of trans fat. **Snacks:** Potato, corn and tortilla chips often contain trans fat. And while popcorn can be a healthy snack, many types of packaged or microwave popcorn use trans fat to help cook or flavor the popcorn. **Fried food:** Foods that require deep-frying French fries, doughnuts and fried chicken can contain trans fat from the oil used in the cooking process. **Refrigerator dough:** Products such as canned biscuits and cinnamon rolls often contain trans fat, as do frozen pizza crusts. **Creamer and margarine:** Nondairy coffee creamer and stick margarines also may contain partially hydrogenated vegetable oils. Industry can design almost any fat or oil for a specific application by the use of various modification processes, such as hydrogenation, inter esterification, fractionation or blending.

17.3 RESULTS AND DISCUSSIONS

17.3.1 HYDROGENATOR'S OR REACTOR'S WORKING FORMULATIONS FOR THE EQUIPMENT

The purpose of the equipment used for the gas-liquid operations is to provide intimate contact of the gas to oils in order to permit interphase diffusion of the constituents. The rate of mass transfer is directly dependent upon the interfacial surface exposed between the phases, and the nature and degree of dispersion of one oil in other are therefore of prime importance. The working equations could be as follows for transfer of oil:

$$R = k\,A\,(C^* - C_l) = k'\,A'\,(C_l - C_s) \text{ Diffusion to Catalyst}$$
$$= k''\,A''\,C_s\,U \text{ Reaction at Catalyst}$$

where R is the rate transfer of chemical reaction, C^* is the saturation hydrogen concentration in oil, C_l and C_s are the hydrogen concentration at the liquid and catalyst interphases respectively. U (Le, L, O) is the unsaturated fatty acid and concentration at the catalyst Surface k and k' are the mass transfer coefficient, k" is the chemical constant and k is the overall transfer coefficient. A and A' are the specific interface (bubble to reactor volume and particle surface to reactor volume, respectively) – areas of the mass transfer and A" is specific active nickel (surface to volume) area.

$$\text{Rate} = \text{all driving forces/resistances}$$
$$= C^* - C_l + C_l - C_s + C_s/(1/kA) + (1/k'A') + (1/k''A'')$$
$$= C^*/(1/k)$$

The reaction rate is always proportional to saturation concentration or partial pressure [7].

The rising single bubbles of hydrogen gas that could behave like rigid spheres should represented in Figure 17.10, by Sherewood number as $Sh = 2 + 0.6\,Re^{0.5}\,Sc^{0.33}$, apparently, it should consider to be 2 if there is no flow around the bubble. A commercial hydrogen reactor that uses one Rushton type six-blade impeller and has a tank diameter T = 2.5 m and impeller diameter D = 0.9 m, a liquid height H = T, hg = 1.7 and no recirculation of hydrogen, then No = 0.75^{-1}, and if the power $p = 6\,\rho_{oil}\,N^3D^5$ is $2kW/m^3{}_{oil}$ then N = 2.8 s^{-1}. Hence, V_L = 0.68 m/s and $€_{gas}$ ranges between 0.09 and 0.06, if the linear gas velocity vs drops from 0.004 to 0 m/s, which occurs

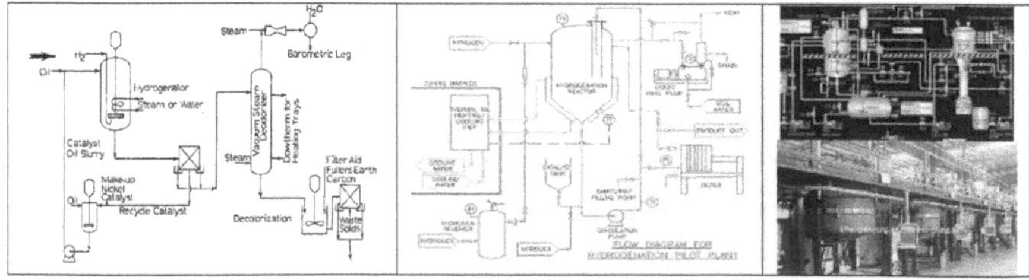

FIGURE 17.10 Hydrogenation of vegetable oil; final products oil, partial and full fats like DALDA.

at a result of the decrease in the hydrogen rate. The value of 0.004 m/s is calculated from an initial hydrogenating rate of 1ΔIV/min. Therefore, for mean bubble diameter is $d_{3.2} = (0.416/\rho_{oil}\ g)^{0.5} \pm 25$ = A gas hold up of 9% ($€_{gas} = 0.09$) and sauter diameter of 1.1 mm are the values calculated from the above equation. The specific interfacial area a will then be 490 m^2/m^3_{oil} as described previously [8].

17.3.2 HYDROGENATION CONVERTS ALKENES TO ALKANES AND THE MASS TRANSFER RATE

Hydrogenation is typically carried out by bubbling H_2 gas through the heated oil, in presence of a nickel catalyst. Alkene + Hydrogen –> Alkane, All in all, we could say that hydrogenation rate per unit volume of catalyst can be approximated as;

$$R\,(t) = R_{poly} = k_{r,\,p}\,[IV\,(t) - 75]\,C_{bulk}\,(t)\,\eta_p\,(t) \text{ for } IV > 80$$

$$R\,(t) = R_{mono} = k_{r,\,m}\,IV\,(t)\,C_{bulk}\,(t)\,\eta_m\,(t) \text{ for } IV < 80$$

And $R\,(t) = J\,(t)/€_{cat}$ (mol H_2/m^3 cat.S), in which C_{bulk} (t) is also the hydrogen concentrating at the liquid/Catalyst interface, because we demonitrated that $k_s\,a_s >> k_L\,a$, η (t) is the effectiveness factor, which is a function of the Thiele modulus, and $€_{cat}$ is the volume fraction of catalyst particles. Above equations reflect the correlation between reaction rate R and the iodine value [9].

If we eliminate unknown concentration C_{bulk} (t) from above equation, we get hydrogenation rate:

$$J\,(t) \text{ for } IV > 80\ J\,(t) = C_{max}\,[1/k_L\,a + 1/X^n \\ \{\,k_{r,}\,p\,[IV\,(t) - 75]\,\eta_p\,(t)\,\}\,€_{cat}]^{-1}$$

These equations express the absorption rate as a function of the driving force, which is the maximum hydrogen concentration C_{max} and the term in brackets, which expresses the total resistance. The interfacial area a_s between the catalyst particles and the oil is given by equation;

$$a_s = 6\,€_{cat}/d_{3.2\ cat} \text{ in which; } €_{cat} = P_{oil}/P_{part}\ (\text{mg Ni in Oil}/(\%\ \text{Ni}\ (10^4)\ \rho_{oil});$$
$$= (W_{ppm}\ \text{Ni})\ (\rho_{oil}/\rho_{part})/(\%\ \text{Ni}\ (10^4)$$

If 100 W_{ppm} Ni is used the $€_{cat}$ is 1.4 $\times 10^{-4}$ m^3 Cat/m^3 oil and a_s is 170 m^2 cat/m^3 oil. Hence, the characteristic rate constant k_s a_s is 1.5 s^{-1}. It is one order of magnitude larger than the characteristic rate constant k_L a for the absorption rate of hydrogen into the oil phase. The maximum obtainable value of characteristic rate constant k_L a = 1.2 s^{-1} depends very much on type and size of the reactor. Four baffles, hydrogen sparger below the bottom impeller, as the sparger is a device for introducing a stream of hydrogen gas in the form of small bubbles into a oil. On the other hand, it required a simply a device for agitation. The size of hydrogen gas bubbles depends upon the rate of flow through the orifices, the orifice diameter, the oil properties, and the extent of turbulence prevailing in oil. The driving force is the extent of difference between actual concentrations and the equilibrium concentrations. The rate J at which the hydrogen is transferred from the bulk oil phase to the catalyst particles is directly proportional to the driving forces;

$$C_{bulk} - C_{cat}; \; J = k_s \, a_{cat} \, (C_{bulk} - C_{cat})$$

In this equation product k_s a_{cat} (s^{-1}) is again characteristic rate constant. It is product of the solid phase mass transfer coefficient k_s (m/s) and specific contact area between catalyst particles and oil a_{cat} ($m^2{}_{cat}/m^3{}_{oil}$). The H_2 concentration C_{cat} is that existing at the outer surface of the catalyst particles. The solid phase mass transfer coefficient k, is a property that in well stirred solid–liquid system. The specific interfacial area of hydrogen bubbles in oil, a = 6 $€_{gas}/d_{3.2}$, where $€_{gas}$ is the gas hold up, $d_{3.2}$ is sauter mean diameter (Volume/Surface ratio) of the gas bubbles.

A typical hydrogenation batch process operates as follows: oil is charged to the reactor, heated and evacuated to drive off air and water. Catalyst-oil slurry is pumped in; the catalyst concentration is 5 to 15 kg per ton of oil. Hydrogen is introduced and the steam is turned off since reaction is mildly exothermic. After the desired hydrogenation has ensued as checked by an iodine number iteration, the batch is cooled and filtered to recover the catalyst for reuse. The chemical, physical and sensory properties of the final product strongly depends on the number of residual double bond and on the contents of cis-trans isomers present in the mixture, which depend on various operating factors, including temperature, hydrogen pressure, catalyst and circulation rate/agitation.

The effect of temperature on the hydrogenation was clearly observed. Once the hydrogenation reaction is completed to the desired degree, the reaction materials are removed from the hydrogenator (autoclave) and passed through a filtration system to remove the inorganic solids from the hydrogenated edible oil. Different inorganic materials could be added at this time to the oil to enhance its filterability all are described in Figure 17.11.

In most cases the hydrogenator is interested in the time required to get the iodine value reduced from the initial value could be IV_o to a particular IV_{end}. The equation express in the hydrogenation time HT can be represented as; $HT = 0.6\ (IV_o\ IV_{end})/J_{mean}$ (min). The proportionality constant of 0.6 is required if HT is in minutes and J_{mean} is in moles per cubic meter of oil per second. The overall rate of hydrogenation process (which is usually expressed as (Δ IV/min) is reflected by the rate at which hydrogen is absorbed by the oil. It can be derived from definition of iodine value that IV = 36 mol H_2/m^3_{oil} and IV/min = 0.6 mol H_2/m^3_{oil} s), which generates = 17 kCal/(m^3_{oil} s) = 71 kJ/(m^3_{oil} s) or kW/m^3_{oil} end. This overall rate is determined by a number of physical steps, such as transfer of hydrogen from the gas phase the oil phase and diffusion of hydrogen through intricate texture of catalyst particles, and the reaction ratio on nickel surfaces.

17.3.3 APPLICATIONS OF THE ENGINEERED PRODUCTS FROM VEGETABLE OILS IN DAILY LIFE

It has been claimed by the different industries that reformulated Crisco has same cooking properties and flavor as original version of product before

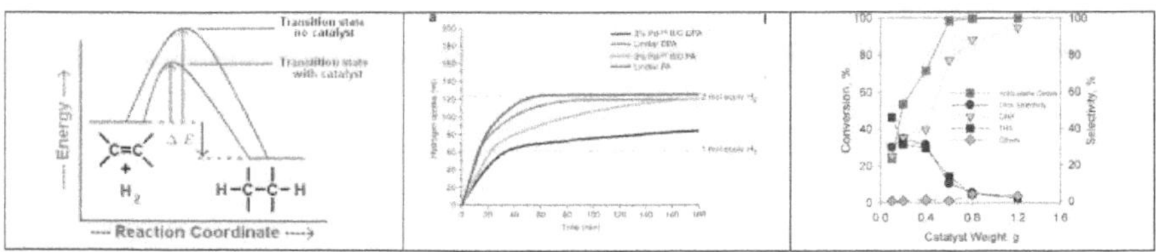

FIGURE 17.11 Representing catalyst weight reaction with exothermic energy evolved in examples.

the processing. Hydrogenation converts liquid Changing the degree of saturation of the fat changes some important physical properties such as the melting range, which is why liquid oils become semi-solid. Solid or semi-solid fats are preferred for baking because the way the fat mixes with flour produces a more desirable texture in the baked product as their main packing shown in above Figure 17.12. Vanaspati ghee is a cheaper substitute for clarified butter that is made from milk.

Vanaspati means any refined edible vegetable oil or oils subjected to a process of hydrogenation in any form. Dalda (5.14 g/100 g) was found with the lowest trans fat followed by Gagan (6.80 g/100 g) and Anchal (8.27 g/100 g) [12].

Shaktibhog (26.83 g/100 g) was found with highest trans fat content followed by Raag (16.64 g/100 g) and Scooter (13.18 g/100 g). Dalda had the lowest and Shakti Bhog the highest amount of Trans fat.

17.4 CONCLUSION

Producing hydrogenated vegetable oil exhibiting superior thermal stability; hydrogenation process incorporating high shear; better appearance, texture, and stability; useful in frying, confectionery baking. The hydrogenation enables a reduction of hydrogenation time, and operation at lower temperatures if we followed the customer's supply guide line for liquid to semi-solid to solid oils and fats. The hydrogenated vegetable oil could be particularly useful in frying, confectionery baking, and other applications where the product with a low trans fat content or higher thermal stability are desirable. The hydrogenated oil produced may comprise less than 10 weight % of trans fatty acids with less than 5 weight % of linolenic acid.

FIGURE 17.12 Representing the Crisco and other hydrogenated products with the applications.

KEYWORDS

- **Edible oils**
- **Fats**
- **Foods**
- **Hydrogenation**
- **Rancidity**

REFERENCES

1. Austin, G. T., Shreve's Chemical Process Industries, Fifth Edition, Tata McGraw-Hill Education Private Limited, Chemical Engineering Series, New Delhi, 2012, 508–528.
2. Rao, M. G., & Sttig, M., Dryden's Outlines of Chemical Technology—the Twenty-First Century, 3rd Ed., East-West Press, Private Limited, New Delhi, 2012, 282–293.
3. Shigeo, N., Handbook of Heterogeneous Catalytic Hydrogenation for Organic Synthesis, 1st ed. New York: Wiley-Inter-Science, 2001, 7–19.
4. Rao, D. G. Fundamentals of Food Engineering, PHI Learning Private Limited, New Delhi – 110001, 2010.
5. Patterson, H. B. W., Hydrogenation of Fats and Oils: Theory and Practice, American Oil Chemists' Society, Champaign, 1994, 39–40.
6. Fernandez, M. B., Tonetto, G. M., Crapiste, G. H., & Damiani, D. E., Revisiting the hydrogenation of sunflower oil over a Ni catalyst. *J. Food Eng.* 2007, *82*, 199–208.
7. Fillion, B., Morsi, B., Heier, K., & Machado, R., The kinetics for soybean oil hydrogenation using a commercial Ni/Al_2O_3. *Ind. Eng. Chem. Res.* 2002, *4*, 697–709.
8. Pocklington, W. D., Determination of the iodine value of oils and fats, Result of a Collaborative Study. *Pure & Appl. Chem.*, 1990, *62*, 2339–2343.
9. Treybal, R. E., Mass-Transfer Operations. McGraw Hill Education, 3rd Indian Edition, 2012.
10. Gavhane, K. A., Mass Transfer – II, Nirali Prakashan, Pune, 2014.
11. Schmidt, A., & Schomaker, R., Partial hydrogenation of sunflower oil in a membrane reactor. *J. Molecular Cat. A: Chem.* 2007, *271*, 192–199.
12. Gunstone, F. D., Vegetable Oils in Food Technology: Composition, Properties and Uses. Blackwell Publishing, 2002.

CHAPTER 18

EXPERIMENTAL STUDIES ON A PLATE TYPE HEAT EXCHANGER FOR VARIOUS APPLICATIONS

V. D. PAKHALE and V. A. ARWARI

Chemical Engineering Department, MIT Academy of Engineering, Alandi (D), Pune–412105, Maharashtra, India

CONTENTS

18.1 Introduction 297
18.2 Experimental Set-Up 298
18.3 Experimental Procedure 299
 18.3.1 For Water–water System 299
 18.3.2 Glycerol-Water System 300
18.4 Result and Discussion 300
18.5 Conclusion 304
Keywords 304
References 305

18.1 INTRODUCTION

Heat exchangers are devices that are designed to allow a heat transfer between two fluids. These types of devices are used in many applications such as chemical process industries, Pharmaceutical industries, Food industries,

Textile industries, waste heat recovery systems, refrigeration, cooling engine systems and air conditioning [1]. In order to create the adequate configuration, various parameters are taken into account, such as the fluid properties, the pipe setup and the direction of the fluid. If the flow is directed in the opposite direction, it is called counter flow. In contrast, if the flow follows the same direction of both cold and heat fluids, then it is called a parallel flow.

Heat exchangers are of two type direct contact and indirect contact. In case of direct contact heat exchangers we can take the example of cooling towers where two fluids are in contact with each other. While in case of indirect contact heat exchangers two fluids are separated by wall medium. Example of indirect contact heat exchanger is a plate type heat exchanger [1–11]. A plate type heat exchanger or compact heat exchanger is one type of heat transfer device commonly used. It consists of various plates stacked one behind the other. These plates are the path through which heat will be transferred from the hot fluid to the cold fluid [1].

A particular advantage for this configuration is the compact ness of the device. The original idea for the plate heat exchangers was patented in the latter half of the nineteenth century, the first commercially successful design being introduced in 1923. The basic design remains unchanged, but continual refinements have boosted operating pressures from 1 to 25 atmospheres in current machines.

An important, exclusive feature of the plate heat exchanger is that by the use of special connector plates it is possible to provide connections for alternative fluids so that a number of duties can be done in the same frame. Plate type heat exchangers have many advantages compared to many other heat exchangers. Plate type heat exchangers can be used for high viscosity applications, because turbulence is induced at low velocities which leads to effective heat transfer [1, 3]. The rate of heat transfer obtained is quite higher and maintenance of these kinds of heat exchangers is also very simple.

18.2 EXPERIMENTAL SET-UP

The apparatus consists of a plate type heat exchanger (Figure 18.1). The hot fluid enters at one side of the fixed end cover and flows through alternate channels between the plates and leaves the exchanger through a connection at other fixed end cover. The cold fluid enters at one side of the fixed cover and flows in co current flow through alternate channels

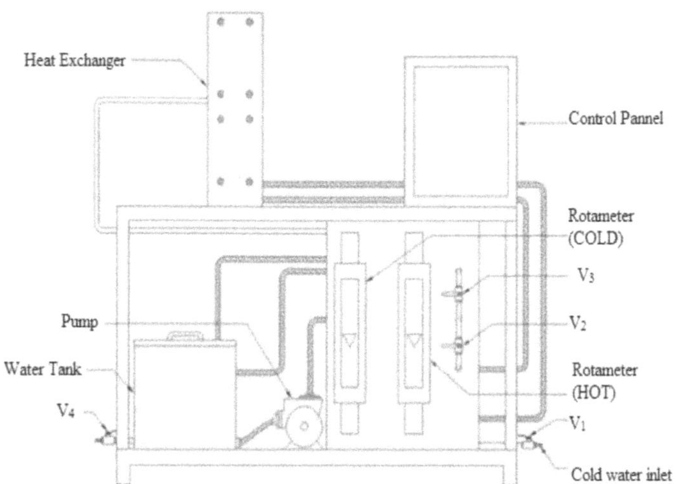

FIGURE 18.1 Front view of plate type heat exchanger experimental set up [12].

TABLE 18.1 Details of Plate Type Heat Exchanger [12]

Specification of PTHE	
Length of Plate, L	0.425 m
Number of Plates, N	6
With of Plate, B	0.125 m
Area	0.32 m^2

between the plates and leaves the exchanger at the other side of the fixed end cover. Temperature sensors are installed at various and appropriate locations to record the temperature of fluids. A magnetic drive pump is used to circulate the hot water from a recycled type water tank. Heater is provided to heat the water in water tank. Valves are provided for flow control and drainage [12].

18.3 EXPERIMENTAL PROCEDURE

18.3.1 FOR WATER–WATER SYSTEM

Experiments were carried out on above-mentioned setup. In the first experiment, water was used on both sides of hot & cold channels. Both hot & cold fluids enter from the bottom of the heat exchanger and co-current fluid flow pattern is achieved. The inlet temperature of cold water was kept at 27 °C and that of hot water was kept at 57 °C. Thus a temperature difference of 30 °C was obtained. The cold-water flow rate was kept at constant value of 1.7 L/min and readings were taken by varying the hot water flow rate from 1.3 L/min to 4.2 L/min. A heater assembly was used to heat the water in a water tank to keep it at constant temperature of 57 °C.

18.3.2 GLYCEROL-WATER SYSTEM

In order to study the viscous liquid effects experiments were carried out on glycerol water system on the above-mentioned setup. In this experiment water was used as a cold side fluid and glycerol was used as a hot side fluid. Both hot & cold fluids enter from the bottom of the heat exchanger & Co-current fluid flow pattern is achieved. The inlet temperature of cold water was kept at 28.7 °C and that of glycerol was kept at 57 °C. Thus, a temperature difference of 28.3 °C was obtained. The cold-water flow rate was kept at constant value of 1.7 L/min and readings were taken by varying the hot glycerol flow rate from 1.3 L/min to 4.2 L/min. A heater assembly was used to heat the glycerol in a tank to keep it at constant temperature of 57 °C. For experimentation purpose we prepared 10% glycerol solution. The viscosity of solution is more than water. Viscosity of mixture obtained at 57 °C is 6.2×10^{-4}Pa. Sec, density of mixture is 1012.3 kg/m^3 and that of water is 4.8×10^{-4}Pa.Sec.

18.4 RESULT AND DISCUSSION

The difference of inlet & outlet temperatures of hot & cold stream is plotted against hot water flow rate, V_{HW} and results are shown in Figures 18.2 and 18.3. For both the systems (water–water and glycerol–water), the ΔT_{HW} decreases with increasing flow rate and is minimum at highest flow rate. The ΔT_{HW} is maximum at the lowest flow rate because the hot water gets more time to exchange heat with cold water. The ΔT_{CW} is maximum at maximum V_{HW} because hot water stream continuously supplies heat energy to the cold-water stream. In both the systems results obtained are quite similar. A similar trend for variation of ΔT against hot water flow rate has been reported in Ref. [1].

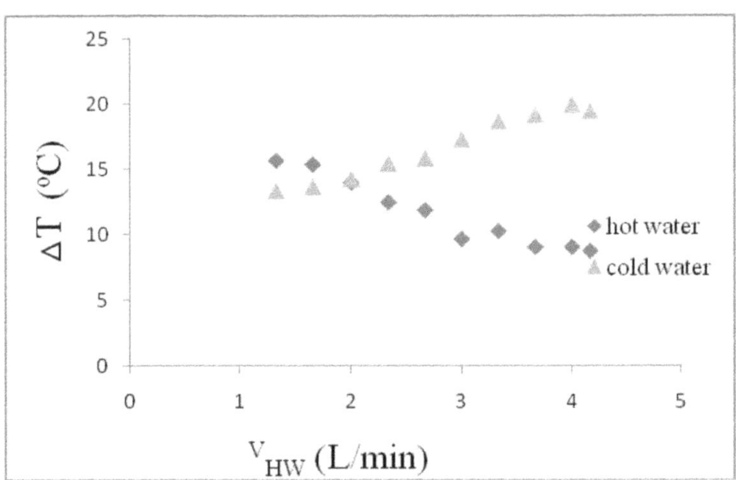

FIGURE 18.2 The plot of difference of inlet & outlet temperatures of the respective streams and V_{HW} for water–water system.

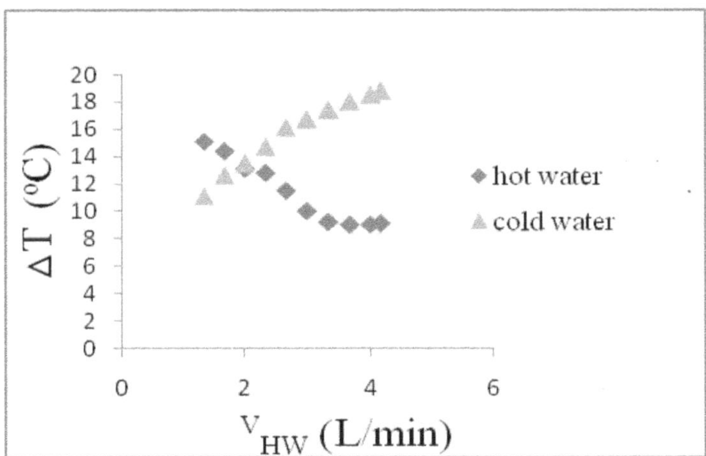

FIGURE 18.3 The plot of difference of inlet & outlet temperatures of the respective streams and V_{HW} for water–glycerol system.

Figure 18.4 shows the temperature difference between the hot & cold streams ΔT_{outlet} measured at the exit of the heat exchanger for both water–water & water–glycerol system. The ΔT_{outlet} value obtained are almost constant. The temperature difference between cold & hot fluid at inlet is 30 °C and 28.3 °C.

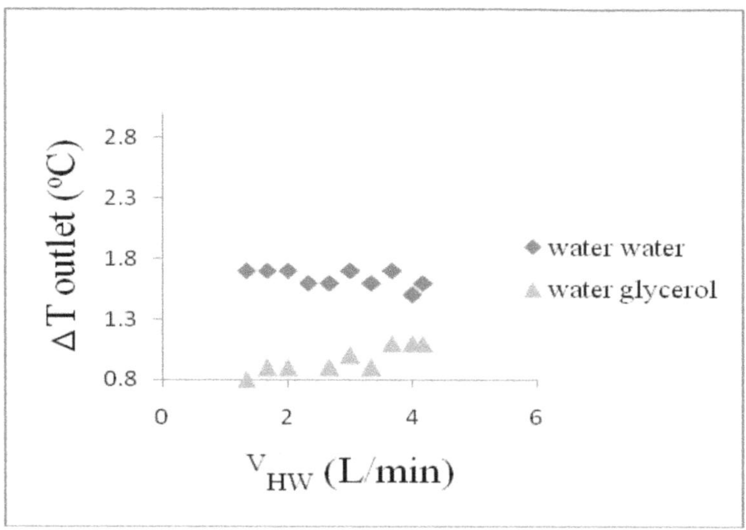

FIGURE 18.4 The plot of difference of hot & cold streams at the exit of heat exchanger and V_{HW}.

From Figure 18.5 the average heat transferred between two streams increases with increasing V_{HW} because of high turbulence at high velocities, causing a much higher heat transfer. Because of the viscosity effects the average heat transferred is quite less in case of water–glycerol system. A similar trend for variation of $Q_{average}$ against hot water flow rate has been reported in Ref. [1].

The overall heat transfer coefficient calculated by taking $Q_{average}$, which is arithmetic mean of the Q_{HW} and Q_{CW}. From figure 18.6 the U value increases with increasing value of V_{HW} because of turbulence. In case of water–glycerol system the overall heat transfer coefficient value obtained is also near about same to the water–water system thus viscosity effect get balanced in plate type heat exchanger. A similar trend for variation of overall heat transfer coefficient against hot water flow rate has been reported in Ref. [1].

The Figure 18.7 shows the variation of average thermal length with respect to hot water flow rate. A higher thermal length means that the heat transfer and the pressure drop are large, where as a lower thermal length means that heat transfer and pressure drop are low [1].

In the following diagram average thermal length value obtained is almost constant therefore pressure losses are also constant and heat transfer obtained is quite good.

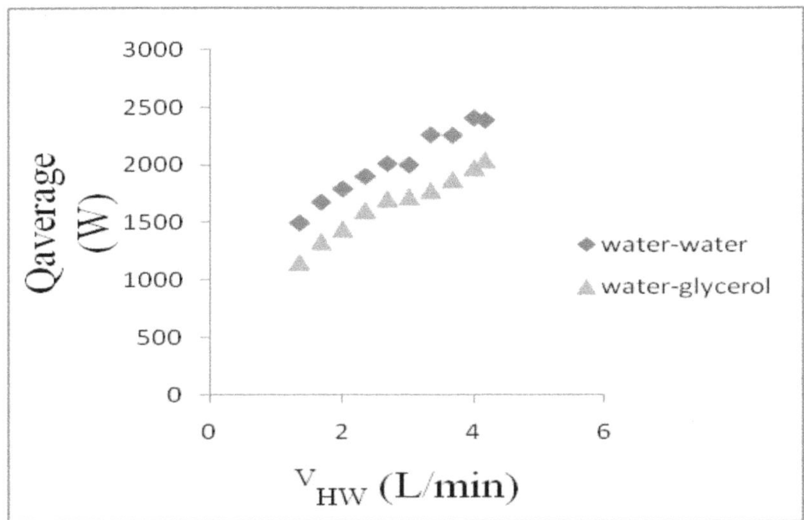

FIGURE 18.5 The plot of average heat transfer between hot & cold streams against varying V_{HW}.

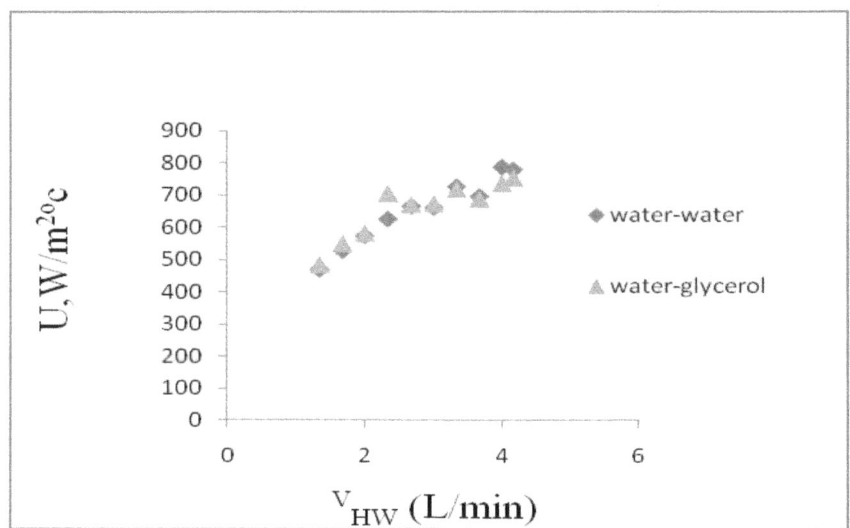

FIGURE 18.6 The plot of overall heat transfer coefficient against varying V_{HW}.

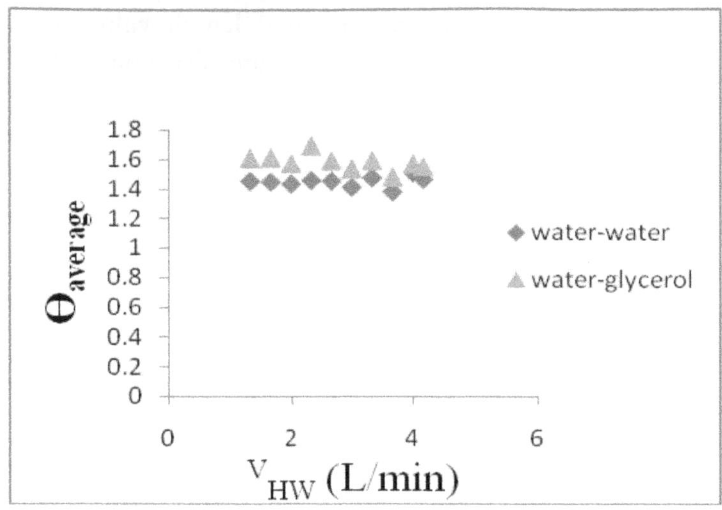

FIGURE 18.7 The plot of average thermal length against varying V_{HW}.

18.5 CONCLUSION

A comparative study is done on a plate type heat exchanger for non-viscous fluid system such water–water and viscous fluid system such as water–glycerol. From the above experimental results, with increasing hot water flow rate V_{HW} the average heat transfer, $Q_{average}$ increases in both the systems due to turbulence. $Q_{average}$ is slightly less in viscous system because of viscosity and therefore less heat transfer. Similar pattern is observed for overall heat transfer coefficient U, it also increases with hot-water flow rate and matches with plain water–water system. Thus plate type heat exchangers are much more efficient for heat transfer applications for non-viscous as well as viscous fluids. The results from the work will be helpful for modeling and simulation of plate type heat exchangers.

KEYWORDS

- **Heat transfer rate**
- **Overall heat transfer coefficient**
- **Plate type heat exchanger**

REFERENCES

1. Faizal, M., & Ahmed, M. R., Experimental studies on a corrugated plate heat exchanger for small temperature difference applications. *Journal of Experimental Thermal and Fluid Science* 2012, *36*, 242–248.
2. The theory behind heat transfer – Plate Heat Exchangers http://www.distribution-chalinox.com/produits/alfa-laval/echangeurs/heat-transfer- brochure.pdf.
3. Kumar, D. S., *Heat and Mass Transfer, 3rd Edition*, New Delhi: SK Kataria & Sons, 1990.
4. Kuppan, T., *Heat Exchanger Design Handbook.* New York: Marcel Dekker Inc. 2000, pp. 347–377.
5. Gut, J. A. W., & Pinto, J. M., Optimal configuration design for plate heat exchangers. *Int. J. Heat Mass Trans.* 2004, *47*, 4833–4848.
6. Tauscher, R., & Mayinger, F., Heat transfer enhancement in a plate heat exchanger with rib-roughened surface http://www.td.mw.tum.de/tum-td/de/forschung/pub/CD_Mayinger/302.pdf.
7. Ciofalo, M., Piazza, D., & Stasiek, J. A., Investigation of flow and heat transfer in corrugated – undulated plate heat exchangers. *Heat and Mass Transfer* 2000, *36*, 449–462.
8. McCabe, W., Smith, J., & Harriott, P., *Unit Operations of Chemical Engineering. 7th Ed.* New York: McGraw Hill, 2005, pp. 455–459.
9. Warnakulasuriya, F. S. K., & Worek, W. M., Heat transfer and pressure drop properties of high viscous solutions in plate heat exchanger. *Int. J. Heat Mass Trans.* 2008, *51*, 52–67.
10. Wang, B. S., & Manglik, R. M., *Plate Heat Exchangers–Design, Applications, and Performance*, WIT Press, 2007.
11. William, H. A., & Chih, W., *Renewable Energy from the Ocean – A Guide to OTEC*, Oxford: Oxford University Press, 1994.
12. Plate Type Heat Exchanger Manual, (H. T. 124), Ambala: K. C. Pvt. Ltd., pp.1–9.

CHAPTER 19

ULTRASOUND ASSISTED EXTRACTION OF BETULINIC ACID FROM LEAVES OF SYZYGIUM CUMINI (JAMUN)

S. V. ADMANE,[1] S. M. CHAVAN,[1] and S. G. GAIKWAD[2]

[1]*Chemical Engineering Department, Sinhgad College of Engineering, Vadagaon, Pune, Maharashtra, India*

[2]*Chemical Engineering Department, National Chemical Laboratory (NCL), Pune, Maharashtra, India*

CONTENTS

19.1 Introduction 308
19.2 Materials and Methods 309
19.2.1 Materials 309
19.2.2 Soxhlet Extraction Technique 309
19.2.3 Ultrasonicated Assisted Extraction Technique 310
19.2.4 Analysis 310
19.3 Results and Discussions 310
19.3.1 Selection of Solvent 310
19.3.2 Effect of Partical Size 310
19.3.3 Effect of Solid Loading 311
19.3.4 Effect of Temperature 312
19.3.5 Mathematical Modeling 313

19.4 Conclusion 314
Keywords 315
References 315

19.1 INTRODUCTION

Syzygium cumini (family Myrtaceae) is also known as Syzygium Jambolanum and Eugenia Cumini. It is an evergreen tropical tree, which can grow above 30 m height [1]. The leaves contain various acids such as β-sitoterol, betulinic acid, mycaminose, n-nanocosane and flavonol glycosides myrecetin. Syzygium cumini (Jamun) having promising therapeutic value with its various phytoconstituents, such as Tannins, Alkaloids, steroids, Flavonoids, Terpenoids, Fatty acids, Phenols, Minerals, Carbohydrates and Vitamins. Other common names are Jambul, Black plum, Java Plum, Indian Blackberry, Jamblang, Jamun, etc. The tree fruits once in a year and the berries are sweetish sour to taste. The various types of Phytochemicals are present in Jamun plant. Betulinic acid is present in Jamun stem bark. The bark contains 8–19% tannin, gallic acid, 1.67% resins small amounts of ellagic acid and myricetin. Human body requires various macro and micronutrients such as protein, carbohydrates, fat or lipid as macronutrients and vitamins, minerals, water and fiber as micronutrients. Human food is mostly derived from plant and animal sources. Jamun fruit is one of those, which contain a variety of important nutritional compositions. Vitamin C is one of the most crucial vitamins in human that plays a large role in hundreds of the body's functions.

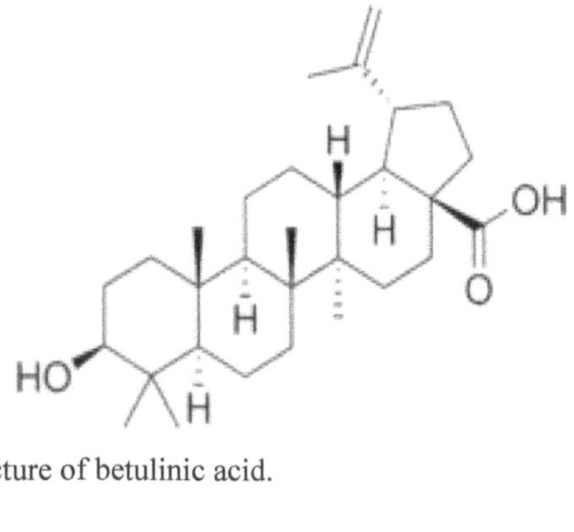

FIGURE 19.1 Structure of betulinic acid.

Betulinic acid is a naturally occurring pentacyclic triterpenoid, which has many medicinal applications such as antiretroviral, antimalarial and anti-inflammatory properties, as well as a more recently discovered potential as an anticancer agent [2–3]. The physical properties of betulinic acid are: molecular formula: $C_{30}H_{48}O_3$, molar mass: 456.70 g mol^{-1}, density: 1.065 g/cm^3, boiling point: 550°C at 760 mmHg, melting point: 295–298°C, flash point: 300.5°C.

Extraction is a technique, which is used for separation of components by using suitable solvent. The use of ultrasound assisted extraction techniques reduces the extraction at least to half of the time needed by conventional extraction technique without any change in composition of extracted cells [4].

19.2 MATERIALS AND METHODS

19.2.1 MATERIALS

Betulinic acid standard is purchased from Aldrich. The leaves of Jamun are collected from nearby area of Dhayri, Pune. These are washed with water and rinse it though rally. The powdered mass is then sorted into various sizes by screening technique such as 75 micron, 105 micron, 1000 micron and 1410 micron. 105-micron size is taken into experiment and carried out all the experiment. The experiment is performed by varying various parameters such as agitation speed, particle size, temperature and solid loading. Ultrasonication set up is used for experimentation. The samples are collected at specified time interval and analyze on UV Spectrophotometer.

19.2.2 SOXHLET EXTRACTION TECHNIQUE

The maximum recoverable betulinic acid is obtained by Soxhlet extraction using methanol. 105-micron particle size leaves were used for Soxhlet apparatus with 200 mL methanol. Extraction was carried out for 72 hr. samples collected were analyzed on UV Spectrophotometer.

19.2.3 ULTRASONICATED ASSISTED EXTRACTION TECHNIQUE

Extraction experiment was carried out at 200 mL glass vessel. 15 KHz ultrasonic waves are generated. Samples were collected at different time interval and analyzed betulinic acid on UV Spectrophotometer. The selection of organic solvent, particle size, solid loading and temperature were investigated to select optimum condition for UAE extraction.

19.2.4 ANALYSIS

The extracted samples were analyzed by spectrophotometrically using (Thermofischer 840–210800) UV–VIS spectrophotometer at wavelength of 210 nm.

19.3 RESULTS AND DISCUSSIONS

19.3.1 SELECTION OF SOLVENT

Figure 19.2 shows the concentration of betulinic acid in solvents respectively after 60 min of extraction at 30°C. The maximum extraction 1.896 mg/mL was achieved than other solvents. It may be due to role of size of solvent molecule and diffusion mechanism by which the solvent penetrated the solid matrix of natural material and increases the polarity of solvents. Ultrasonic waves also penetrated the solvent into solid matrix. Methanol was selected for the further experimentation.

19.3.2 EFFECT OF PARTICAL SIZE

Figure 19.3 shows extraction of betulinic acid with varying particle size. Small particle size has more extraction rate. This is happened because decreasing size of solid matrix through grinding treatment leads to higher surface area, making extraction process more efficient.

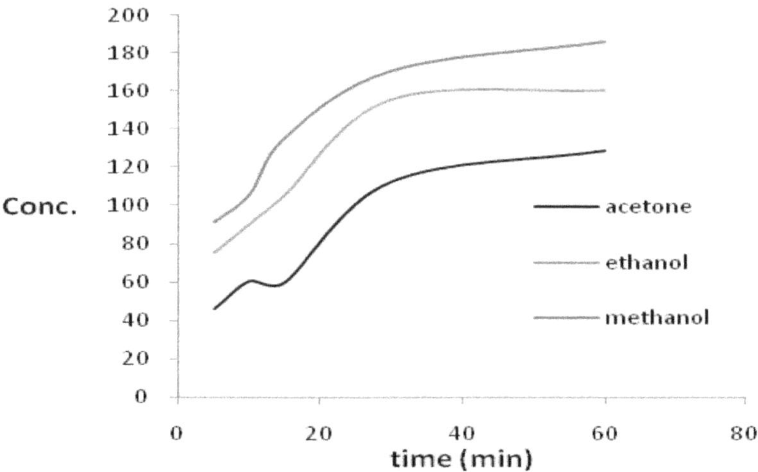

FIGURE 19.2 Selection of solvent for extraction of betulinic acid.

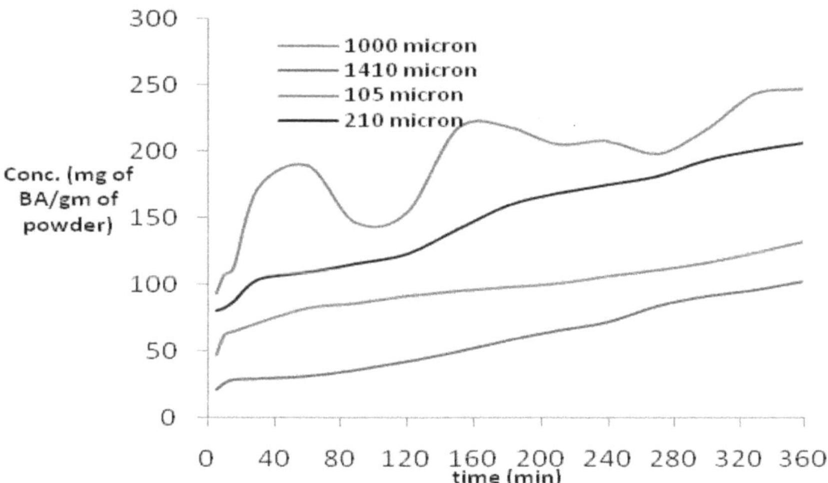

FIGURE 19.3 Effect of particle size on extraction.

19.3.3 EFFECT OF SOLID LOADING

Solid loading is also play an important factor for extraction. Small solid loading gives higher extraction. It is seen that at small solid loading more extraction takes place, because it directly affect the matrix of the leaves.

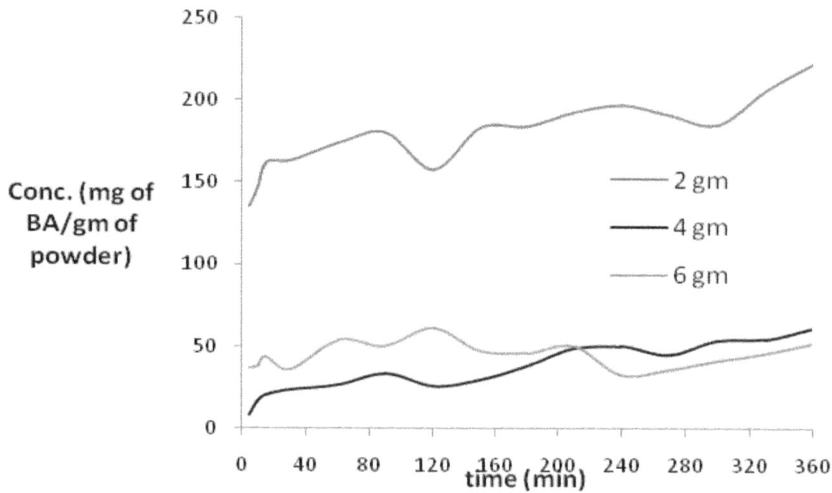

FIGURE 19.4 Effect of solid loading on extraction.

19.3.4 EFFECT OF TEMPERATURE

Temperature plays an important role in extraction process. As temperature increases, extraction also increases. As after increase in the temperature, bulk solution directly penetrates through matrix of leaf and here again extraction rate becomes higher.

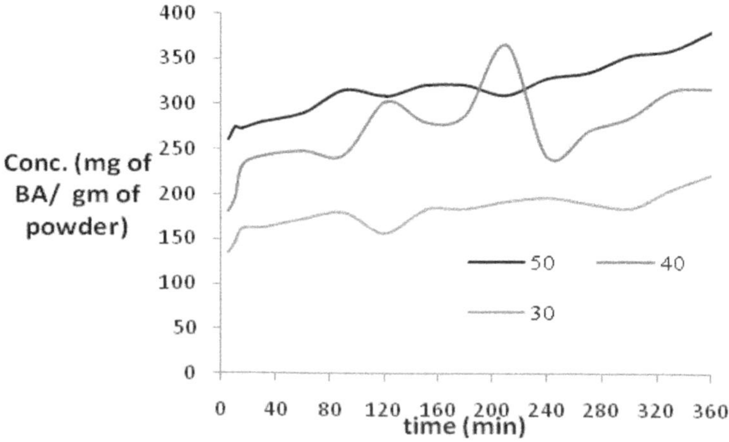

FIGURE 19.5 Effect of temperature on extraction.

19.3.5 MATHEMATICAL MODELING

A Pelegs kinetic model is used for solid-liquid extraction techniques by many researchers. The model describes the sorption isotherms of material as follows:

$$C_t = C_0 + \frac{t}{K_1 + K_2 t} \tag{1}$$

where, C_t is the concentration of betulinic acid at time t (mg of BA/g of powder); K_1 is the Peleg's rate constant (min g/mg); K_2 is the Peleg's capacity constant (g/mg).

C_0 term can be omitted from Peleg's equation as initial concentration of target solution is zero in the extraction solvent as fresh solvent is used at beginning. The extraction process, at the beginning behaves like first order and then process decreases to zero order. The modified Peleg's equation representing concentration of target solute in extraction solvent against time is shown below:

$$C_t = \frac{t}{K_1 + K_2 t} \tag{2}$$

where C_t can be calculated using Eq. (2) by determining K_1 (Peleg's rate constant) and K_2 (Peleg's capacity constant) values by plotting the graph between $1/C_t$ vs 1/t.

Using Peleg model parameters, the temperature dependence on initial extraction rate $1/K_1$ (1/Peleg rate constant) is represented by the linearized Arrhenius equation. Following equation represents the relation between ($1/K_1$) and extraction temperature.

$$\ln\left(\frac{1}{K_1}\right) = \ln A - \frac{E_a}{RT} \tag{3}$$

where K_1 is the Peleg rate constant (min g/mg BA), A is constant as frequency factor (min^{-1}), E_a is the activation energy (J/mol) R the universal gas constant (8.314 (J/mol K)) and T is the absolute temperature in (K). The plot of the $\ln(1/K_1)$ vs (1/T) would result in straight line with the

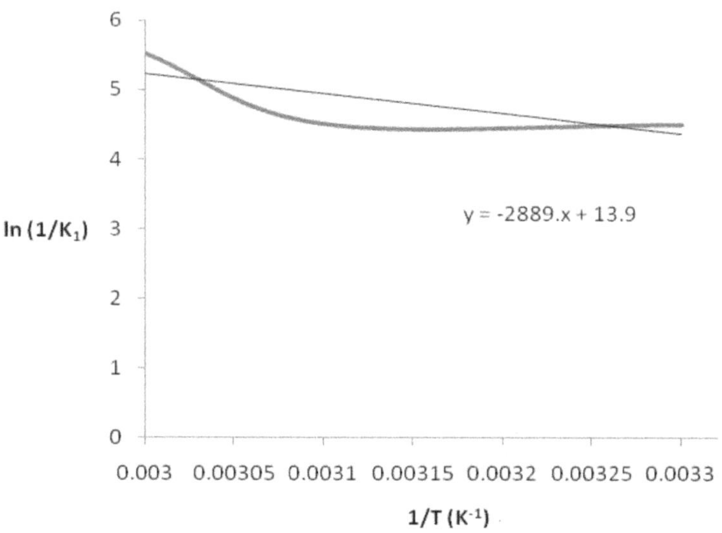

FIGURE 19.6 Plot of ln $(1/K_1)$ vs 1/T.

negative slope equal E_a/R and intercept equal ln$A$. The activation energy of the ultrasonication extraction was found to be 24.019 KJ/mol.

19.4 CONCLUSION

In this study, ultrasonicated assisted extraction technique was carried out. Influence of operating parameters such as solvent, solid loading and temperature were investigated. Methanol has shown greatest release on extraction of betulinic acid. Particle size 105 micron, temperature 60°C and 2 gm solid loading is said to be optimize condition from the experimental work. The Pelegs kinetic model is used for comparing experimental data. The activation energy of the betulinic acid in methanol was found to be 24.019 KJ/mol. The percentage yield was obtained from ultrasonicated extraction process is 64%. From various experiment, it was observed that ultrasonication process is used for reducing extraction time and increasing yield of extraction.

KEYWORDS

- **Cost effective**
- **Extraction**
- **Nutrients**
- **Pentacyclic**
- **UV spectrophotometer**

REFERENCES

1. Vetal, M. D., Lade, V. G., & Rathod, V. K., Extraction of Ursolic acid from Ocimum sanction by ultrasound: process intensification and kinetics. *Chem. Engg. Journal,* 2013, *69*, 24–30.
2. Grakal, D. J., Taralkar, S. V., Kulakrni, P., Jagtap, S., & Nagawade, A., Kinetic model for extraction of Eugenol from leaves of Ocium sanctum Linn (Tulsi). *International Journal of Pharmaceutical Applications* 2012, *3,* 267–270.
3. Pal, S. R., Nimbalkar, M. S., Pawar, N. V., & Dixit, G. B., Optimization of exraction techniques and quantification of Betulinic acid (BA) by RP-HPLC method from *Ancistrocladus heyneanus* wall. Ex. Grah. *Industrial Crops and Products* 2011, *34*, 1458–1464.
4. Yogeswari, P., & Sriram, D., Betulinic acid and its derivatives: A review on their Biological Properties. *Current Medicinal Chemistry,* 2005, *12*, 657–666.

PART V

ANALYTICAL METHODS

CHAPTER 20

SEPARATION, ANALYSIS AND QUANTITATION OF HESPERIDIN IN CITRUS FRUITS PEELS USING RPHPLC-UV

S. KULKARNI[1] and B. A. BHANVASE[2]

[1]*Chemical Engineering Department, Vishwakarma Institute of Technology, Pune, Maharashtra, India*

[2]*Department of Chemical Engineering, Laxminarayan Institute of Technology, Nagpur–440033, Maharashtra, India*

CONTENTS

20.1 Introduction ... 320
20.2 Materials and Methods ... 322
  20.2.1 Materials ... 322
  20.2.2 Preparation of Sample ... 322
  20.2.3 Identification of Hesperidin by HPLC ... 322
  20.2.4 Precision and Accuracy ... 323
  20.2.5 Instrumentation ... 323
20.3 Results and Discussion ... 323
20.4 Conclusion ... 325
Keywords ... 326
References ... 326

20.1 INTRODUCTION

The genus citrus belonging to the family of Rutaceae comprises of about 40 species, which are distributed over a spectrum of countries namely, India, China, Malaysia, Srilanka, and Australia. Citrus is one of the most important worldwide fruit crops and is consumed either as fresh fruit or in the form of juice because of its nutritional value added upon by species characteristic flavor [1].

Citrus fruit extracts are found to exhibit a plethora of activities like anti-oxidant, anti-inflammatory, anti-tumor, anti-fungal and blood clot inhibition, etc. [2–4]. The health benefits of citrus fruit have mainly been attributed to the presence of bio-active compound such as ferulic acid, hydrocinnamic acid, cyanidine, glucoside, hesperidin, Vitamin C, carotenoid and naringin content [4–6]. Further, concern about the safety of the commonly used synthetic antioxidant such as butylated hydroxy-anisole (BHA) and tertiary butylhydroquinone (TBHQ) have led to the increased interest on natural antioxidants which occur in plants as secondary metabolites.

Flavonoids are naturally occurring polyphenolic compounds. Chemically, hesperidin consists of a flavanone component joined to a sugar group bonded at 7-position to oxygen atom to give the rhamnoglucoside. However, after ingestion, the sugar is found to be getting removed to leave the parent flavanone, 3′,5,7-trihydroxy-4′-methoxyflavanone-7-rhamnoglucoside), also known as hesperetin. Flavonoids are found to exist in two enantiomeric forms, R-and S-hesperidin, respectively (Figures 20.1 and 20.2).

There has been a necessity of developing new but more powerful methods for better analysis of hesperidin in biological fluids. Given the chiral nature of hesperidin (HD) which was hitherto not given much

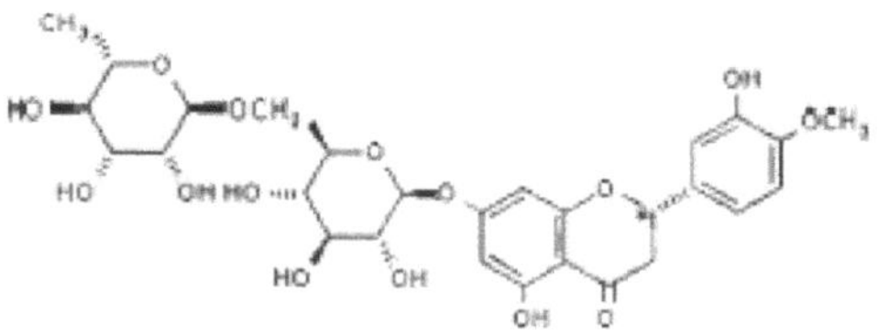

FIGURE 20.1 Structure of Hesperidin, (*) denotes chiral center.

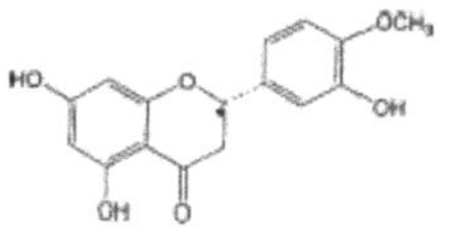

FIGURE 20.2 Structure of Hesperidin.

consideration while evolving at conventional analytical methods, an effort has been made here by using Octyl decyl Silane (ODS, C18) as a non polar stationary phase and further using mobile phase consisting of water: Acetonitrile: Acetic acid 20:78:2, v/v) at room temperature for maintaining p^H 4.5. The detection was done by using on column UV detector at variable wavelengths namely 205 nm and 298 nm. Calibration curves were obtained and were found to be reasonably linear. Sample preparation and processing was done meticulously by using methods like vacuum filtration and sonication etc. before subjecting it to reverse phase high performance liquid chromatographic analysis. It is more of an analytical purification than preparative one.

Previous studies on biochemical activities from citrus were mainly focused on its essential oils, which include antimicrobial properties [7] and anti-aflatoxigenic activity [8]. Resent accumulative evidences suggest that citrus contains several possible anti-cancer agents such as flavonoids [9–12].

Flavonoids are naturally occurring polyphenolic compounds usually present in oranges, other citrus fruit and herbal products. They are conjugated with B-glucosides, for example, in hesperidin the sugar is bounded at position 7 of the flavonone (3,'5,7-trihydroxy-4′methoxyflavonone-7-rhamnoglucoside (Figure 20.1) [13–16]. However, after juice ingestion, the rutinose sugar molecule is rapidly cleaved off during its journey through gastrointestinal tract and liver leading to formation of aglycone molecule, hesperitin (3,5,7-trihydroxcy-4′-methoxyflavonone, HT) (Figure 20.2), a chiral flavonoid, which exists in two enantiomeric forms [17, 18].

HPLC analytical methods are currently used either alone or hyphenated with MS for authentication of foodstuffs, for example, the determination of ingredients a given foodstuff should have. This kind of exercise is of vital importance from the economic and quality point of view.

20.2 MATERIALS AND METHODS

20.2.1 MATERIALS

All chemicals used in this study were of analytical grade. Water used for extraction, washing and dilution was Milli Q while solvents used in the making of mobile phase were HPLC grade Acetonitrile, and Milli Q water.

20.2.2 PREPARATION OF SAMPLE

Fresh fruits belonging to the genus Citrus like Lemon, Sweet lime and Orange at the commercial mature stage were made available from local market in the months of March-April from Pune, India. Healthy fruits were selected randomly for uniformity of shape and color. The fruits were washed thoroughly in potable water followed by Millipore water. The extract was collected by crushing the peel and membranous part of each of the fruit. Slurry (7.2 g) was weighed and centrifuged at 3000 rpm in a centrifuge. The residue was washed with Milli Q water (3 × ~5 mL) and the aqueous solutions were combined in the volumetric flask. The extract was filtered using vacuum filtration technique using 0.2 μm filter and the filtrate was stored at −20°C till further analysis.

20.2.3 IDENTIFICATION OF HESPERIDIN BY HPLC

High performance liquid chromatography coupled with UV-Vis detector was used to analyze hesperidin content of citrus samples using separation module equipped with a C18 column (inertsil, phenomenex, 250 × 4.6 mm, 5 μm particle size). The sample was eluted using isocratic system of water:acetonitrile:2% acetic acid (20:78:2 v/v) as the mobile phase at the flow rate of 1.0 mL/min. The temperature of the column was ambient and the injection volume was 20 μL. The peak of standard hesperidin was accomplished by comparing the retention times of peaks in sample to those of standard. Calculation of hesperidin concentration was carried out by external std. method using calibration curves of standard hesperidin.

20.2.4 PRECISION AND ACCURACY

The precision and accuracy studies were performed with replicate assays (n=6) on the same day (within –run precision) and over 3 consecutive days (between-run precision) at different level of concentration of Hesperidin. The precision was evaluated calculating the relative standard deviation (RSD) of the enantiomer concentrations. While accuracy was estimated based on the mean percentage error of measured and actual concentration.

20.2.5 INSTRUMENTATION

Spectrophotometer conditions: System: UV-1650PC (Shimadzu Make), Software: UV Probe, Source: 50W Deuterium Lamp, Wavelength Range: 190–1100 nm, Detector: Silicon Photodiode, Chromatography Conditions, System: LC-10AT vp (Shimadzu Make), Software: Spinchrome, Column: Inertsil C-18 (ODS) (250 × 4.60 mm, 5 µm particle size), Flow rate: 1 mL/min, λ_{max}: 205 nm and 298 nm, Mobile phase: water:acetonitrile:2% acetic acid (20:78:2, v/v) (isocratic elution), Column temperature: Ambient, Run time: 5 min. Ingestion volume: 20 µL of the prepared standard and sample solution.

20.3 RESULTS AND DISCUSSION

The Figure 20.3 shows a chromatogram for Lemon at 298 nm, which results into separation of one of the enantiomers similar to one witnessed in chromatogram for Hesperidin at same wavelength and with comparable retention time. The Figure 20.4 shows a chromatogram for external standard, Hesperidin at 298 nm which results into separation of one of the enantiomers similar to one witnessed in chromatogram for Hesperidim at same wavelength and with comparable retention time. The Figure 20.5 shows a chromatogram for Lemon at 205 nm, which results into separation of one of the enantiomers similar to one witnessed in chromatogram for Hesperidin at same wavelength and with comparable retention time. The Figure 20.6 shows a chromatogram for external standard, Hesperidin

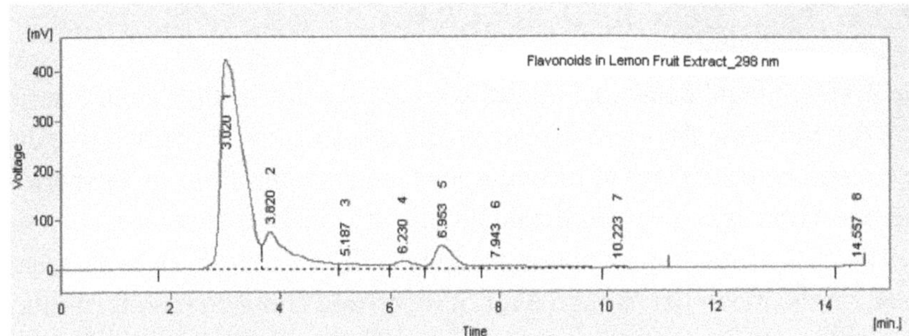

FIGURE 20.3 Chromatogram for Lemon fruit peel and membranous part extract at 298 nm.

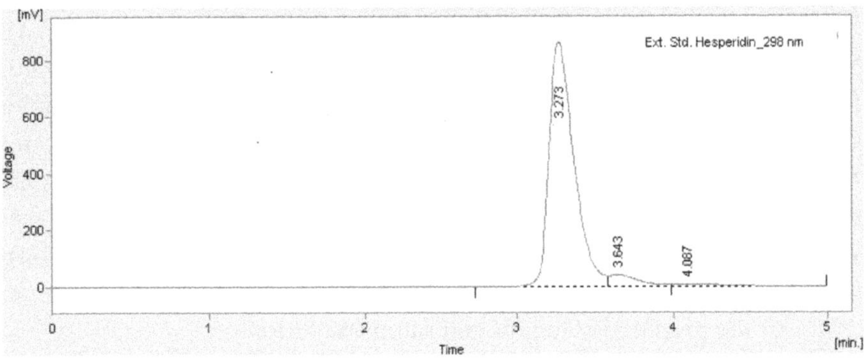

FIGURE 20.4 Chromatogram for Ext. Std. Hesperidin at 298 nm.

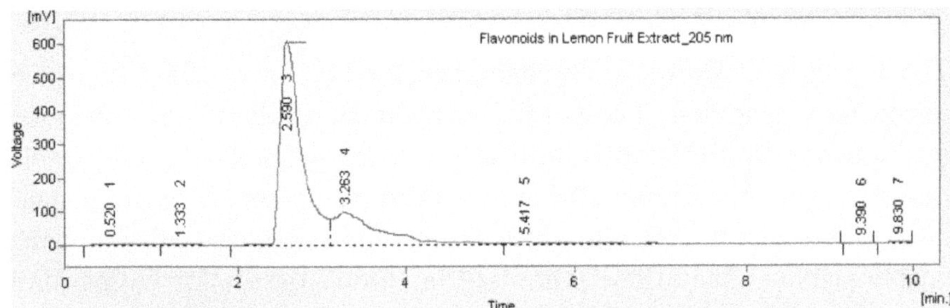

FIGURE 20.5 Chromatogram for Lemon fruit peel and membranous part extract at 205 nm.

at 205 nm which results into separation of one of the enantiomers similar to one witnessed in chromatogram for Hesperidim at same wavelength and with comparable retention time. It is interesting to note that retention

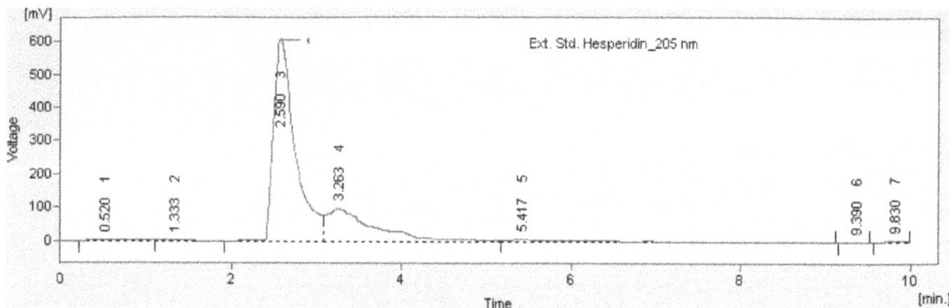

FIGURE 20.6 Chromatogram for Ext. Std. Hesperidin at 205 nm.

characteristics for different citrus fruits vary slightly and this variation may be attributed to differential interaction between a range of compounds in citrus fruits apart from flavonoids.

20.4 CONCLUSION

Liquid chromatography is a precise method for obtaining reliable data about the bioactivity of citrus fruits grown under the same geographical and climatic conditions. A simple stereoselective, isocratic, reversed-phase, liquid chromatography (HPLC) method for the determination of the enannomers of hesperidin (a health tonic) The extraction, analysis and separation of flavonoids from citrus fruits was undertaken at variable wavelengths from the peel and membranous part of the citrus fruit successfully. Two enantiomers R- & S- are found to absorb at different wavelengths and the advantage of these absorption characteristics was taken for their separation purpose. The relative standard deviation did not exceed 1.5%. As low as 45 and 30 ng (per injection) could be detected in less than 6 min with a high degree of accuracy. Lemon contain about 5.34 mg while lime contain about 4.5 mg/100 gm of fruit. Hesperidin if separated recovered in a bulk by employing preparative chromatography, the product can be of immense help from human health point of view with citrus fruits as a source. The improved, simple, selective and sensitive reversed-phase HPLC methods were developed for the determination of hesperidin separately, using external or internal standard techniques, respectively.

KEYWORDS

- **Flavonoid**
- **Hesperidin**
- **RPHPLC**

REFERENCES

1. Dugo, G., & Giacomo, A., Citrus: the genus citrus. Taylor & Francis, New York, 2002.
2. Garg, A., Garg, S., Zaneveld, L. J., & Singla, A. K., Chemistry and pharmacology of the citrus bioflavonoid hesperidin. *Phytother. Res.* 2001, *15,* 655–669.
3. Kaur, C., & Kapoor, H. C., Antioxidants in fruits and vegetables-the millennium's health. *Int J. Food Sci. Technol.* 2001, *36,* 703–725.
4. Abeysinghe, D. C., Li, X., Sun, C. D., Zhang, W. S., Zhou, C. H., & Chen, K. S., Bioactive compounds and antioxidant capacities in different edible tissues of citrus fruit of four species. *Food Chem.* 2007, *104,* 1338–1344.
5. Kelebek, H., Canbas, C., & Selli, S., Determination of phenolic composition and antioxidant capacity of blood orange juices obtained from cvs. Moro and Sanguinello (*Citrus sinensis* (L.) Osbeck) grown in Turkey. *Food Chem.* 2008, *107,* 1710–1716.
6. Xu, G., Liu, D., Chen, J., Ye, X., Ma, Y., & Shi, J., Juice compounds and antioxidant capacity of citrus varities cultivated in China. *Food Chem.* 2008, *106,* 545–551.
7. Chanthaphon, S., Chanthachum, S., Hongpattarakere, T., Antimicrobial activities of essential oils and crude extracts from tropical Citrus species against food- related microorganisms. *Songklanakarin J. Sci. Technol.* 2008, *30,* 125–131.
8. Razzaghi-Abyaneh, M., Shams-Ghahlarokhi, M., Rezaee, M. B., Jaimand, K., Alinezhad, S., Saberi, R., & Yoshinari, T., Chemical composition and anti-aflatoxigenic activity of *Carum carvi* L., Thymus vulgaris and Citrus aurantifolia essential oils. *Food Control* 2009, *20,* 1018–1024.
9. Tanaka, T., Maeda, M., Kohno, H., Murakami, M., Kagami, S., Miyake, M., & Wada, K., Inhibition of azoxymethane-induced colon carcinogenesis in male F344 rats by the citrus limonoids obacunone and limonin. *Carcinogenesis* 2001, *22,* 193–198.
10. Tian, Q., Miller, E. D., Ahmed, H., Tang, L., & Patil B. S., Differential inhibition of human cancer cells proliferation by citrus limonoids. *Nutrition and Cancer* 2001, *40,* 180–184.
11. Poulose, S. M., Harris, E. D., & Patil, B. S., Citrus limonoids induce apoptosis in human neuroblastoma cells and have radical scavenging activity. *Journal of Nutrition* 2005, *135,* 870–877.
12. Londoño-Londoño, J., Lima, V. R., Lara, O., Gil, A., Pasa, T. B. C., Arango, G. J., & Pineda, J. R. R., Clean recovery of antioxidant flavonoids from citrus peel: optimizing an aqueous ultrasound-assisted extraction method. *J. Food Chem.* 2010, *119,* 81–87.

13. Maria, K., Evagelos, G., Sofia, G., Michael, K., & Irene, P., Selective and rapid liquid chromatography/negative-ion electrospray ionization mass spectrometry method for the quantification of valacyclovir and its metabolite in human plasma. *Journal of Chromatography B* 2008, *864*, 78–86.
14. Vanamala, J., Reddivari, L., Yoo, K. S., Pike, L. M., & Patil, B. S., Variation in the content of bioactive flavonoids in different brands of orange and grapefruit juices. *J. Food Compos. Anal.* 2006, *19*, 157–166.
15. Peterson, J. J., Dwyer, J. T., Beecher, G. R., Bhagwat, S. A., Gebhardt, S. E., Haytowitz, D. B., & Holden, J. M. Flavanones in oranges, tangerines (mandarins), tangors, and tangelos: A compilation and review of the data from the analytical literature. *J. Food Compos. Anal.* 2006, *19*, S66–S73.
16. Jing, X., Akira, K., Hideki, H., & Fumiyo, K. Determination of hesperidin in Pericarpium Citri Reticulatae by semi-micro HPLC with electrochemical detection. *J. Pharm. Biomed. Anal.* 2006, *41*, 1401–1405.
17. Majo, D. D., Giammanco, M., Guardia, M. L., Tripoli, E., Giammanco, S., & Finotti, E., Flavanones in Citrus fruit: Structure antioxidant activity relationships. *Food Res. Int.* 2005, *38*, 1161–1166.
18. Benavente-García, O., Castillo, J., Marin, F. R., Ortuño, A., & Delrío, J. A., Uses and properties of citrus flavonoids. *J. Agr. Food Chem.* 1997, *45*, 4505–4515.

CHAPTER 21

QUANTIFICATION OF ALUMINUM METAL IN COSMETIC PRODUCTS BY NOVEL SPECTROPHOTOMETRIC METHOD

S. B. GURUBAXANI and T. B. DESHMUKH

Department of Chemistry, Institute of Science, Civil Lines, Nagpur–440001, India. E-mail: sevakgurubaxani@gmail.com

CONTENTS

21.1 Introduction ... 329
21.2 Material and Method ... 330
21.3 Results and Discussion ... 331
21.4 Conclusion ... 334
Acknowledgment ... 334
Competing Interests ... 334
Keywords ... 334
References ... 335

21.1 INTRODUCTION

Aluminum, an established neurotoxin, is the main cause of Alzheimer's disease. Today it is found in almost everything we eat, drink and absorb. Not only it is used for packaging of eatables but is also added as raising agent or as additive in cakes, biscuits, wines, and fizzy drinks. Most medicated drugs use it as buffering agent. Even vaccines contain aluminum to micro levels. Nowadays cosmetics, sunscreen lotions and deodorants, which are high in demand use it as one of its ingredient. It is a part under investigation as to what levels its ingestion is safe since there is no government legislation in this regard [1–5]. Thus it is important to estimate human exposure to this toxic metal using novel methods facilitating quick estimation. Complexometric methods using 8-hydroxyquinoline and aluminon have limited accuracy; the alternate Spectrophotometric technique proposed by Syed Raashid et al. [6] gives quick and reliable results as confirmed by indirect EDTA titration. We have used their technique to estimate aluminum in few cosmetic products.

21.2 MATERIAL AND METHOD

An experimental set up was established. Commercially available cosmetic products were used for analysis. About 1 g of each sample was taken and then dissolved in 25 mL of 2 N HNO_3 kept overnight at room temperature. The solution was filtered on the next day and a stock solution was prepared by diluting it to 100 mL distilled water [7, 8]. Analysis was carried out on Systonics 2203 UV-Vis Spectrophotometer and reliability of method was further checked by indirect EDTA titration. Samples tested with brand names are listed as under Table 21.1.

When xylenol orange was added with aluminum a pinkish complex appeared. The mole ratio of the complex formed depends upon temperature and pH. A temperature of 35°C and pH 5 was maintained and shift in »$_{max}$ for free xylenol and aluminum xylenol orange complex was observed by addition of aluminum to the solution [9]. Absorbances were taken at

TABLE 21.1 List of Cosmetic Products for Aluminum Quantification

Sr no.	Sample	Sample brand	Make
1	Skin cream	Nivea cream	Nivea India Pvt. Ltd.
2	Active Deo Roll on	Rexona Roll on skin light	Rexona Pvt. Ltd.
3	Face pack	Joy 24 carat gold kit	Joy Beautycare Pvt.ltd.
4	Nourishing cream	Shahnaz chocolate kit	Shahnaz Ayurvedic
5	Mask	Ever youth peel off mask	Zydus Wellness Ltd.

λ_{max} of aluminum xylenol complex. The pH was maintained using acetate buffer. The molar composition of the complex was found by using mole fraction method. The calibration curve gives us molar absorbtivity coefficient (ε) if graph plotted between known aluminum ion concentrations vs. absorbance. The results obtained from spectrophotometric method were further verified by back EDTA titration. For that a fixed volume of sample was taken in conical flasks and 15 mL of 1.0 M EDTA was added. The pH 5 was maintained by using acetate buffer [10, 11]. For this titration xylenol orange was used as an indicator. Finally the mixture was kept for 10 min and back titrated with 2.0 M Zn^{2+} solution to get pinkish red end point.

21.3 RESULTS AND DISCUSSION

The absorbance maximum of Xylenol orange was found at the wavelength of 476 nm, which on complexation with aluminum shifts to 555 nm. The shift in wavelength was observed in free xylenol and complexed form as shown in the Figure 21.1.

Aluminum ion solutions of known concentrations were complexed with xylenol orange in 1:3 molar ratios. Absorbances at the obtained λ_{max}

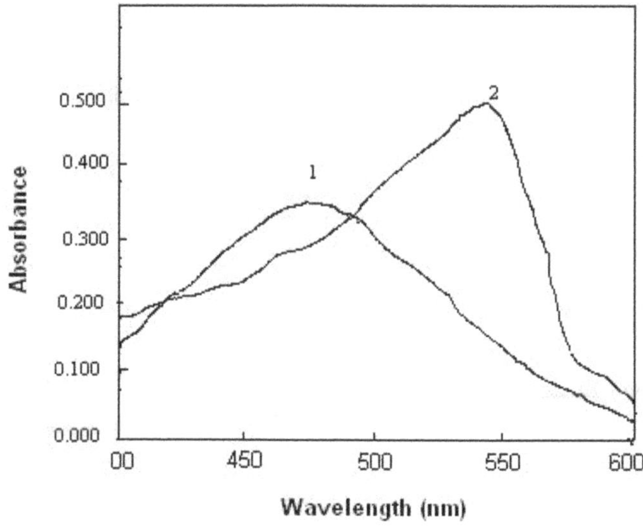

FIGURE 21.1 Shifting of λ_{max} for free xylenol and complexed form of solution. (a) Curve for free xylenol orange solution. (b) Curve for complexed form of xylenol orange solution.

of aluminum xylenol complex of all the solutions were recorded as shown in Table 21.2.

From the above absorbances, molar absorbance coefficient (ε) for the complex was calculated by using Beer-lamberts law. The graph was plotted between absorbances vs. concentrations for known aluminum ion solutions as shown in Figure 21.2. The molar absorbtivity coefficient (ε) of the complex was calculated from the slope of the calibration curve.

The value of molar absorbtivity coefficient was found to be 116.66 lit cm^{-1} mol^{-1}. Absorbances of unknown sample solutions were recorded under similar experimental conditions. The concentration of Al^{3+} ions in each sample was then calculated from the reported absorbance values as shown in Table 21.3.

TABLE 21.2 Absorbance at λ_{max} for the Known Concentration Al^{3+}

S.no.	Sample content (*x* mL of Al^{3+} 10^{-2} moles)	Concentration of solutions on diluting to 25 mL (*y* mL of Al^{3+} 10^{-3} moles)	Absorbance at λ_{max}
1	3	1.2	0.146
2	6	2.4	0.280
3	9	3.6	0.432
4	12	4.8	0.550
5	15	6.0	0.716
6	18	7.2	0.846

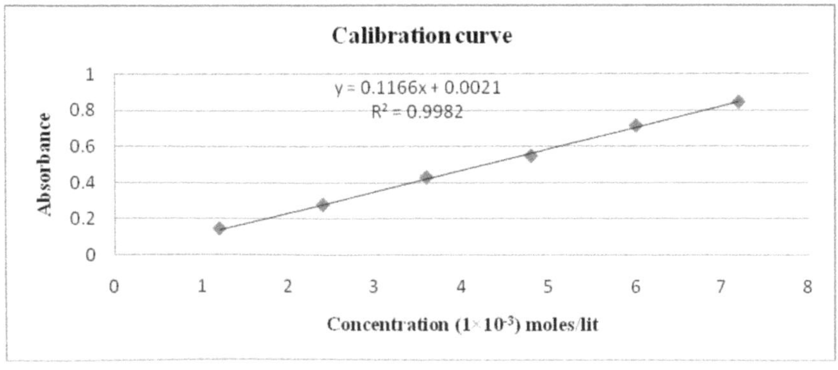

FIGURE 21.2 Absorbance of aluminum xylenol complex vs. concentration of Al^{3+} at λ_{max} (correlation coefficient = 0.9982).

The complexometric back titration results were recorded and the concentration of aluminum ion in each sample was then calculated and reported in Table 21.4.

A reliability check of spectrophotometric technique for aluminum estimation is as under Table 21.5.

TABLE 21.3 Absorbance and Concentration of Al^{3+} Ions in the Samples

S.no.	Sample	Absorbance at λ_{max}	Concentration of Al^{3+} ions moles/liters
1	Skin cream	0.682	5.84×10^{-3}
2	Active Deo Roll on	0.773	6.62×10^{-3}
3	Face pack	0.716	6.13×10^{-3}
4	Nourishing cream	0.991	8.49×10^{-3}
5	Mask	0.798	6.84×10^{-3}

TABLE 21.4 Aluminum Concentration of the Samples Determined by Indirect EDTA Titration

S.no.	Sample	Volume of 2.0 M Zn^{2+} used in titration V_b (mL)	Concentration of Al^{3+} ions moles/lit
1	Skin cream	4.6	5.80×10^{-3}
2	Active Deo Roll on	4.2	6.60×10^{-3}
3	Face pack	4.5	6.00×10^{-3}
4	Nourishing cream	3.3	8.40×10^{-3}
5	Mask	4.1	6.80×10^{-3}

TABLE 21.5 Comparison of Spectrophotometric and Complexometric Methods for Aluminum Content in Cosmetic Products

S.no.	Sample	Concentration of Al^{3+} ion concentration moles/liters	Spectrophotometry Complexometry
1	Skin cream	5.84×10^{-3}	5.80×10^{-3}
2	Active Deo Roll on	6.62×10^{-3}	6.60×10^{-3}
3	Face pack	6.13×10^{-3}	6.00×10^{-3}
4	Nourishing cream	8.49×10^{-3}	8.40×10^{-3}
5	Mask	6.84×10^{-3}	6.80×10^{-3}

From the results it is concluded that all these samples contains milimoles quantity of aluminum. As both spectrophotometric and complexometric methods shows close resemblance in the amount of aluminum estimation, the reliability of the technique is justified.

21.4 CONCLUSION

This novel Spectrophotometric method provides rapid estimation of aluminum with reliability and gives sufficient information about toxicity level of aluminum in various cosmetic products. It has wide applicability and can be extended for estimation of other substances containing aluminum. Cosmetic products containing aluminum is to be avoided and more of the natural products to be used. Cosmetic products of various brands need further research as to what quantity they contain aluminum so as to aware users about the toxic effects of such products.

ACKNOWLEDGMENT

We deeply acknowledge Department of Chemistry and Department of Environmental sciences, Institute of Science, Nagpur for providing necessary facilities.

COMPETING INTERESTS

The authors declare that they have no competing interests.

KEYWORDS

- **Aluminum**
- **Cosmetic**
- **EDTA**
- **Spectrophotometric**

REFERENCES

1. Bestic, L., The telegraph: Is aluminum really a silent killer. 05 Mar 2012.
2. European Aluminum Association: Does Aluminum Play a Role in Alzheimer's Disease? 2001 (http://www.alufoil.org).
3. Massey, R. C., Aluminum in food and Environment, (Special Publication No. 3) Royal Society of Chemistry, London, 1989.
4. Darbre, P., Potential link between aluminum salts in Deodorants and Breast cancer, *J. Applied Toxicol*, 2006.
5. Yokel, R. A., & Golub, M. S. (Eds.). Research Issues in Aluminum Toxicity; Taylor & Francis: Bristol, PA, 1997.
6. Raashid, S., Rizvi, M., & Khan, B., An Alternate Spectrophotometric Method of Aluminum Estimation in Some Daily Use Products, 2011.
7. Robinson, G. (Ed.). Coordination Chemistry of Aluminum; VCH Publishers: New York, 1993.
8. Vogel, A. I., *Quantitative Inorganic Analysis, 4th Ed.* ELBS & Longman, London, 1978, 415–428.
9. Miller, F. I., & Fog, H. M., Reversible formation of aluminum xylenol orange by temperature variation. An experimental demonstration of the entropy effect. *J. Chem. Educ.*1973, *50*, 147.
10. Yang, S. P., & Tsai, R. Y., Complexometric Titration of Aluminum and Magnesium Ions in Commercial Antacids. An Experiment for General and Analytical Chemistry Laboratories. *J. Chem. Educ.* 2006, *83*, 906.
11. Sangale, M. D., Daptare, A. S., & Sonawane, D. V., Determination of Aluminum and Magnesium Ions in Some Commercial Adsorptive Antacids by Complexometric Titrations. 2014.

INDEX

A

Absorbance value, 212, 268–271, 276, 332
Absorption capacity, 59, 176
Accidental spills, 57
Acetone, 20, 74, 89, 144, 187
Acoustic cavitation, 175, 184, 185
Activation, 121, 219, 225, 287, 313, 314
Adeq Precision, 44
Adsorbant dose, 84
Adsorbate, 74, 79–81, 84, 89–93, 133, 173
  adsorption, 79
  concentration, 81, 84, 89, 91
Adsorbent, 32, 36–38, 48, 50, 60, 61, 64, 67, 73, 74, 79–84, 89–94, 99–116, 122, 123, 126, 128, 129, 132–134, 137, 138, 143, 150–152, 158, 159, 163, 173, 174, 221–225, 232, 243
  dose, 107, 109–111, 116
  effect, 103, 104, 107, 108
Adsorption
  capacity, 73, 85, 90, 93, 106, 109–111, 115, 121, 122, 128–134, 138, 143, 150–152, 158, 163, 171
  equilibrium, 80, 130, 172
  experiments, 79, 106, 110, 114, 155
  isotherm, 92, 122, 132, 134, 144, 151, 174
  isotherm models, 122
  kinetics, 78, 82, 90, 136, 137, 155
  parameters, 84
  process, 36, 90–94, 99, 105, 110, 113, 123, 150, 157, 159, 173
  sites, 108, 129, 130, 174
  technique, 73, 89
  time, 91, 122, 123, 130, 131
Aerogel, 59, 61
Agglomeration, 143, 190, 192, 212, 240, 265
Agitation speed, 40, 89, 93, 309
Aglycone molecule, 321
$AgNO_3$, 202, 203, 210–214
Air conditioning, 4, 298
Airborne nickel compounds, 142
Algorithm/computational technique, 25
Aliphatic carbon chain, 124
Alkaloids, 308
Allergic reactions, 36
Alumina-silicate minerals, 121
Aluminosilicate composition, 61
Aluminum, 329–334
  ion, 331–333
  xylenol orange complex, 330
American Conference of Governmental Industrial Hygenists (ACGIH), 142
Ammonia
  solution, 184, 185
  bromide, 122
Amorphous systems, 223
Analytical
  grade, 74, 102, 187, 322
  purification, 321
Anionic
  dye, 138
  dyes, 121, 138
  surfactants, 60
  protection method, 182
ANOVA result, 44
Anthropogenic activities, 98
Anti-aflatoxigenic activity, 321
Antibacterial
  activity, 200
  agents, 201
  properties, 200
Antibiotics, 200, 201
Anti-corrosion

properties, 189, 195
coatings, 182
pigment, 183
Anti-inflammatory properties, 309
Anti-malarial, 309
Anti-microbial
activity, 200
properties, 321
Anti-oxidant, 283, 320
Anti-retroviral, 309
Aqueous
foam, 248, 249, 255
solution, 9, 36, 73, 84, 88, 94, 99, 101, 105, 109, 113, 122, 128, 131, 132, 145, 158, 184, 204, 207, 248, 250, 259, 261, 265, 322
Aromatic
nature, 183
phenolic groups, 224
rings, 194
Arrhenius equation, 313
Artificial
muscles, 162
oil spill source, 62
Asymmetric/symmetric vibration, 124
Autoclave, 293
Autofine sputter coater, 222
Autogenous pressure, 62
Automated analysis, 19, 20
Axial
dispersion coefficient, 21
mixing mechanism, 21

B

Bagasse fly ash, 120
Barium sulfate nanoparticles, 259
Batch
adsorption of Ni (II), 145
reactor, 214
sorption studies, 102, 106
Battery manufacture, 98
BDST model, 78, 79, 82
Bentonite, 36, 37, 39, 50
Benzene, 57, 59
Best-fit model, 78, 82
Betulinic acid, 308–314
Billion cubic meters (BCM), 72
Binary
mixture, 26
system, 26, 27, 28, 29
Biochemical activities, 321
Biodegradability, 59
Bio-functional nanoparticles, 202
Biological
environment, 183
fluids, 320
Biomass, 203, 204, 229–231, 235, 236, 242
Biomedical
applications, 202
engineering, 5
Biomoites, 205
Bio-separations, 20
Biot number (Bi), 21
Bohart Adams model, 85
Botanic structure, 218
Boundary condition, 10, 11
Breakthrough curve analysis, 33
British Petroleum, 59
Bromine, 287
Brunauer Emmet Teller (BET) surface area, 76, 222
Bulk
phase, 25, 248
fluid phase, 22
solution, 40, 105, 312
Butylated hydroxyanisole (BHA), 320

C

Cadmium, 98–100
chloride, 100
concentration, 104, 109, 113
ions, 99
oxide, 100
sulfate, 100
sulfide, 100
Calorific value, 231, 234–236
Capacity constant, 174, 313
Capture liquids, 60
Carbohydrates/vitamins, 308

Carbon black (CB) nanoparticles, 59
Carboxyl group, 147
Carcinogen, 57, 142
Carcinogenicity, 88, 120
Cassia auriculata flower, 201
Catalyst weight reaction, 293
Cation exchange capacity, 121, 123
Cationic dye adsorption, 121
Cavitation technique, 163
Cellulose aerogel, 59
Cellulose fibers, 59
Central composite design, 41
Central nervous system, 100
Centrifugation, 144, 146, 204
Charcoal treatment, 283, 284
Chemical
  activation, 218, 226
  analysis, 19
  exfoliation, 144, 159
  precipitation, 73, 98, 142, 162, 261
  Preparation, 74
  process industries, 297
  reaction, 78, 82, 259, 287, 290
  sorption, 157
  structure, 120
  synthesis, 201
Chiral flavonoid, 321
Chromatogram, 323, 324
Chromatographic
  analysis, 321
  methods, 19
  processes, 20
  separation, 20
Chromatography column, 20
Chrome plating, 101
Chromium, 73, 80, 83, 88–91, 94, 98–101, 108–110, 115, 116
  adsorption, 115
  Cr(VI), 73– 75, 77–99, 101, 103, 105, 107, 109, 111, 113, 115
Cis-unsaturated fats, 285
Citric acid, 74, 101
Citrus
  fruit, 320
  samples, 322
Class C oil, 57
Clay minerals, 121
Coal
  ash, 239
  consumption, 228, 229
Cold fluids, 299, 300
Cold-water flow rate, 300
Colloidal
  nanoparticles, 276
  ZnS nanoparticles, 261
Colorimetric method, 20
Complexometric methods, 330
Computational Fluid Dynamics (CFD), 262, 263
Congo red (CR), 36
  dye, 36, 122
  dye adsorption, 130–132
  uptake, 47
Conservative solution, 264
Consistency index, 9, 12, 16
Conventional
  extraction technique, 309
  method, 36, 137, 162, 242, 243, 259
Cooling engine systems, 298
Correlation coefficients, 135, 155
Corrosion
  analysis, 184
  inhibition, 182, 183, 188, 194, 195
  rate analysis, 193
  species, 182, 183
Cosmetic, 330–334
  products, 330, 334
  formulation, 248
Creamer/margarine, 289
Critical micelle concentration (CMC), 9, 12–16, 253, 254, 271
Cross cut adhesion, 189, 194
Cross-linked polymer structures, 162
Cross-linking agent, 163, 167
Crude oil spill cleaning, 59
Crystal structure, 191
Crystalline planes, 204
Crystallinity, 64, 240, 242
Crystallite size, 191, 205
Cu-Kα radiation, 222
Cumulative effects, 99
Cyanidine, 320

Cyclic corrosion, 189, 193, 194, 195
resistance, 194
Cyclohexane, 288
Cylindrical morphology, 125

D

Debye Scherrer's equation, 205
Deepwater Horizon Gulf oil spill, 59
Degradation, 36, 59, 162, 202
De-ionized water, 63, 101, 144, 145
Dense barrier coating, 182
Deoxygenation, 147
Design matrix, 41
Design of experiments (DOE) tools, 32
Design-expert software, 41, 45
Desorption
method, 222
studies, 67
Detergents, 248, 249, 254
Diffractogram, 242
Dioctahedral mineral, 121
Dioxane, 288
Diphenyl carbazide, 74
Dirt particles, 101
Discharge coefficient, 5, 11–13, 16
Dispovan syringes, 261
Dosage (Do), 40
Double beam spectrophotometer, 124
Drug delivery devices, 162
Dry MnO_2, 145, 146, 148
D-spacing values, 64, 240
Dye
adsorbents, 120
concentration, 124, 166, 171, 172
pollutant, 120
powder, 122
Dynamic
equilibrium, 155
light-scattering, 204
response, 78, 82
studies, 84

E

Eco-friendly solution, 243
Economic production, 243
Ecosystems, 59
Edible oils, 295
EDTA, 144, 330–334
Effect of
adsorbent dose, 103, 104, 107, 108, 112, 113
AgNO3 concentration, 210
contact time, 38, 102, 103, 106, 107, 111, 112
flow rate, 268
initial concentration, 104, 105, 108, 109, 112, 114
particle size, 106, 110, 111, 114
pH, 39, 105, 106, 109–114, 132, 150, 151, 186
temperature, 40, 46, 50, 150, 269, 293
Effluent concentration, 40
Electrical energy, 228, 238
Electro dialysis, 36
Electro static precipitator (ESP), 232
Electrochemical
action, 194
reduction, 73
treatment, 142, 162
Electrodialysis, 99
Electroflotation, 99
Electrolytic corrosion, 194
Electroplating, 73, 88, 100
Electrostatic force, 92, 93, 132
Elemental silver, 206
Enantiomeric forms, 320, 321
Endothermic, 40, 46, 50, 157, 159
Energy, 243, 288
capacity, 238
conservation, 263
Engineering applications, 4, 162
Environmental impacts, 229
Enzymatic oxidation, 283
Epoxy
polyamide, 194
resin, 183, 184
Equilibrium, 39, 40, 50, 78, 80, 82, 92, 94, 103, 104, 111–113, 123, 129, 130–135, 138, 151, 156, 162, 163, 166, 168, 171, 173, 286, 292

concentration, 123
state, 130
studies, 40
Ethanol, 145
Euphorbia prostrate, 201
Evaporation, 57, 67
Exfoliated nanosheet, 147
Exhibiting symptoms, 230
Exothermic
energy, 293
reaction, 286
Experimental
conditions, 99, 332
data, 41, 42, 73, 84, 90, 104, 113, 138, 174, 314
verification studies, 20
Extraction, 309–311, 313, 315
process, 66, 310, 312–314
rate, 310, 312, 313

F

Factorial methods, 41
Fats, 282, 285, 295
Fatty acids, 284, 285, 308
Ferulic acid, 320
Finite
difference method (FDM), 264
element method (FEM), 21, 264
volume method (FVM), 263
Fixed
average concentration, 24
bed chromatography, 33
bed column, 85
Flavonoids, 308, 320, 321, 326
Flow
behavior index, 9, 16
profile (upstream side), 8
Fluid
dynamics governing equations, 264
pressure, 5
properties, 298
Foamability, 248–251, 253, 255
Food, 283, 285, 287, 289, 291, 293, 295
industries, 297
processing, 5, 231, 248
Fossil fuel, 101, 229, 231
Fourier transform infrared (FTIR), 76, 124, 137, 146, 147, 185, 188–191, 219, 222, 224–226
FTIR spectra, 191, 219, 224, 226
FT-Raman Analysis, 147
Fractionation, 20, 289
Freundlich
constants, 134
isotherm, 134, 138, 176
Langmuir isotherm, 90
model, 174, 175
parameters, 174
Frontal
adsorption, 25
analysis, 25
F-test, 41
Fusion technique, 63

G

Galvanization, 142, 182
Gas molecules, 76
Gasoline, 56, 57
Gastrointestinal
disorders, 100
tract, 321
Gelation, 62
General disease syndromes, 100
Geometries, 8
Geometry, 5, 8, 64, 262, 263, 264
Gibbs energy, 157
Glacial acetic acid, 20
Glucoside, 320
Glycerol, 300, 304
GNS/δ-MnO_2, 141–147, 149–159
Gradual decay, 254
Graphene, 143, 145, 146, 148, 149, 159
Nanosheets (GNS), 143, 145
oxide powder, 145
Graphite oxide, 144
Green synthesis, 214
Group of Experts on Scientific Aspects of Marine Environmental Protection (GESAMP), 58

H

Halloysite, 121–123, 125, 127, 129, 131, 133, 135, 137, 138
- clays, 121
- nanotubes (HNTs), 121, 122, 137

Heat
- exchanger, 298–304
- transfer, 297, 298, 302–304
  - applications, 304
  - rate, 304

Heavy metal, 72, 73, 88–90, 98, 99, 108, 115, 142, 143, 162
- ions, 73, 90, 108

Herbal products, 321
Hesperetin, 320
Hesperidin (HD), 320–326
Hexane, 57, 66
Hexavalent chromium, 73
High adsorption capacity, 36, 73, 143
Homogeneous suspension, 144
Homogenous mixture, 164
Hot and cold channels, 299
Hot water flow rate, 300, 302, 304
HPLC
- analytical methods, 321
- grade, 322

Human activity, 56
Human health, 98, 325
Hummers-Offeman method, 144
Hydration sphere, 105
Hydraulic systems, 4
Hydrazine solution, 145
Hydrocarbons, 56, 183, 249
- chains

Hydrocinnamic acid, 320
Hydrogel, 163, 167, 168, 170, 176
Hydrogenation, 283, 285–287, 289–291, 293–295
- processes, 286
- reaction, 282, 293
- reactor, 286
- vegetable oil, 290

Hydrogenator, 293
Hydrolysis, 105
Hydrolytic rancidity, 282, 283
Hydrophobicity, 59
Hydroxyl acids, 282
Hydroxyl group, 183

I

Industrial
- application, 29, 73, 163, 183, 231
- industrialization, 72, 98
- operations, 88
- processes, 248
- wastewater, 73, 142, 162, 176
- wind turbines, 230

Infertility, 100
Infrared spectroscopy (IR), 66, 124
Inorganic
- materials, 293
- solids, 293

Insecticides, 100
Intrinsic velocity, 21
Ion exchanger, 243
Ion-exchange, 99
Ionization, 150, 185
Ions concentration, 109, 151
Irrigation, 98, 230
Isomerization, 287
Isotherm models, 90, 94
Isothermal, 21

K

Keta acids, 282
Kinetic
- behavior, 143, 155
- energy, 40
- model, 73, 78, 82, 135, 155, 313, 314
- parameters, 90
- rate equation, 136

L

Laboratory scale, 20, 73, 248
Lagergen model, 134
Laminar flow regime, 8
Langmuir
- adsorption constant, 151
- adsorption isotherm, 90, 138, 153, 175

constants, 24, 32
equation, 133
Freundlich adsorption isotherms, 133
isotherm, 25, 92, 133, 150, 151, 153, 154, 159, 173–175
model, 175
sorption isotherms model, 133
Large-scale chromatography, 20
Lead poisoning, 100
Leather tanning, 73, 88, 101
Lemon fruit peel, 324
Lignocellulose materials, 218
Linearized model, 79
Linolenic acid, 282, 283, 294
Liquid
chromatography, 20, 23, 322, 325
petroleum hydrocarbon, 56
temperatures, 9
Liver function, 58
Low-density polyethylene (LDPE), 261

M

Magnetic drive pump, 299
Magnetic Stirrer, 62
Malvern instruments, 188, 190, 204
Malvern Zetasizer instrument, 188, 190
Marine destruction, 61
Mass diffusion coefficient, 267
Mass transfer resistances, 20, 91
Material
panel, 9
properties, 10, 13, 16
Mathematical
equation, 78, 82
modeling, 20
Mechanical agitation, 248
Medical applications, 201, 259
Medicinal
applications, 200, 309
concoctions, 100
Membrane separation, 36, 99, 162
Mental retardation, 100
Mercaptopropyl trimethoxysilane, 61
Mesh
generation, 263
refinement, 9
Metabolic imbalance, 58
Metal
adsorption, 104, 108, 112
cat ions, 105
ion, 91, 99, 103–105, 107–109, 112–115, 121, 150, 152, 159, 200
nanocomposite, 143, 147
nanocomposites, 159
plating, 98
Metalloproteins, 283
Methanol, 310, 314
Methylene
blue, 176
groups, 124
Methyltrimethoxysilane (MTMS), 59
Microfluidic devices, 259
Micronaire (immature) cotton, 61
Micronaire cotton, 61
Micro-organisms, 283
Microporasil column, 20
Microreactor, 203, 210, 259–263, 266, 267, 270–276
Microscopic liquid circulation, 184
Mild steel, 184, 195
Millipore water, 37, 322
Mineral
fertilizers, 100
processing plants, 5
Mini Thermal Power Plants, 232
Mini Thermal Power Stations (MTPS), 231
Mining activities, 98
Modified bentonite, 36, 39, 51
Molar absorbtivity coefficient, 331, 332
Molecular
biology, 202
diffusion, 267
weight compounds, 67
Monoatomic graphene, 149
Monoclinic lamellar structure, 147
Monoenoic acid, 282
Monomer, 163, 166, 167
Monounsaturated, 282
Montmorillonite clay, 163
Morphological

characterization, 64
properties, 259
structure, 125
Multicomponent Langmuir isotherm, 23
Multi-component rate model, 22, 33
Multistage column system, 33
Multi-walled Carbon Nanotubes, 143
Mycaminose, 308
Myrica esculenta, 201

N

Nano
clay materials, 163
clay, 124, 167
composite, 163–166, 168–176, 194
particles, 182, 271
Nanoparticles, 200–203, 205, 207, 209, 211, 213, 259, 261, 263–267, 269, 271, 273, 275
Nanostructured hydrous titanium, 143
Nanotechnology, 200, 201
Nanotubes, 122, 124, 132, 138, 143
National Research Council (NRC), 58
Natural
antioxidants, 283–285, 320
clay, 121
log transformation, 42
seeps, 56
Navier-Stokes equations, 263
Nephritis, 98
Neuromuscular effects, 100
Neurotoxin, 329
Neutral salt spray, 193, 194
Newtonian fluid, 5
N-heptane, 20
Nickel, 142, 143, 145, 147, 149, 151–153, 155, 157–159, 282, 287, 288
Nitrogen adsorption, 222
Non-bio-degradability, 98
Non-conventional adsorbents, 73, 89
Non-ionic surfactants, 60
Non-Newtonian
character, 5
liquid, 9
Non-Petroleum oils, 57
Non-profit environmental organization, 60
Non-viscous fluid system, 304
Non-zero positive values, 42
Nuclear
power, 230
waste, 230
Nucleation coefficient, 265
Nuclei formation, 207, 212, 268, 270, 276
Nutrients, 315

O

Ocimum sanctum, 201
Octahedral
aluminum, 121
layers, 121
Octyl decyl Silane (ODS), 321, 323
Oil
adsorption capacity, 67
extraction, 248, 284
properties, 292
spill trajectory models (OSTM), 61
spillage, 57, 61, 62
Oleophilicity, 59
Optical devices, 259
Optimum process
conditions, 276
scheme, 20
Organic
intercalation, 121
salts, 121
solvent, 310
structure of graphene, 147
Orifice meter, 4, 6, 9, 11, 16
Orthogonal collocation method, 25
Outlet average concentration, 24, 30, 32
Overall heat transfer coefficient, 304
Oxidation, 36, 88, 142, 144, 162, 186, 201, 282, 283
Oxidative
rancidity, 282
stability, 284
Oxygen
content, 249

functionalities, 146
molecules, 121

P

PAA/MMT hydrogel, 165, 168, 172, 176
Paint
coatings, 195
manufacture, 98
pigments, 100
Paper industries, 59
Parallel flow, 298
Particle size, 37, 106, 115, 188–190, 201, 204, 206–213, 219, 223, 232, 233, 241, 243, 259, 265, 267–271, 309–311, 322, 323
analysis, 189, 219
Passivation technique, 182
Passive films, 195
Pathogenic bacteria, 201
Pe and Bi range, 24, 27
Peclet and Biot number, 21, 32
Peclet number, 21, 30, 267
Peleg model parameters, 313
Peleg rate constant, 313
Pentacyclic, 315
Percentage adsorption, 90, 91, 93, 104, 106, 108, 110, 112, 115
Pesticides, 98, 200
Petroleum refining, 98
Pharmaceutical companies, 201, 297
Phenols, 308
Phosphoric acid, 184
Photodetectors, 259
Photographic image, 240
Photography, 73
Physical
activation, 218
adsorption, 76
effects, 190, 206
properties, 20, 21, 143, 294, 309
terms, 11
Physico-chemical
characteristics, 219, 226
aspect, 92
Phytochemicals, 308
Phytoconstituents, 308
Pigment manufacture, 98
Pipe diameter, 5, 8, 10, 12–16
Planar graphene sheets, 144
Plate type heat exchanger, 298, 299, 302, 304
Point of zero charge (pHpzc), 221, 223, 225
Polyacrylic acid, 163
Polymer conversion, 167
Polymer processing, 5
Polymerization, 164–166, 175
Polymorphic phases, 143
Polymorphs, 121
Polyphenolic compounds, 320, 321
Polypropylene, 59, 61
Polyunsaturated, 282
Polyvinyl pyrrolidone (PVP), 260, 261
Pomegranate peel extract, 201
Pongamia pinnata, 218, 224–226
seed shell carbon, 224–226
Potassium
bromide (KBr), 222
dichromate, 74, 89, 102
phosphate, 184–186
Power-law fluid, 16
Precursor concentration, 260, 269
Pred R-squared, 44
Pressure loss coefficient, 5, 7, 8, 16
Process
parameter, 276
scale up, 33
Proximate analysis, 219, 221
Pseudo firstorder kinetics, 90, 156
Pseudomonas aeruginosa, 201
Pseudo-second order, 136–138, 155–157
kinetics, 137, 138, 156
reaction, 157
Pyrophilite clay, 62, 63

Q

Quadratic model, 41, 44, 50
polynomial model, 41
Quaternary ammonium salts, 166, 167

R

Raman spectra, 147
Rancidity, 283, 285, 287, 289, 291, 293, 295
R-and S-hesperidin, 320
Raney nickel catalyst, 288
Raphanus sativus, 202, 203, 204, 207, 209, 210, 212, 214
Red Gram, 89–91, 93, 94
Redox reaction, 145
Refrigeration, 298
Refrigerator dough, 289
Remediation, 57, 59, 61, 63, 65, 67, 68
Residence time distribution (RTD), 266
  analysis, 260, 266
Resistance to neutral salt spray, 189, 194
Reverse osmosis, 73, 89, 99, 162
Reynolds number, 4, 5, 8, 10–13, 16, 267, 275
Rhamnoglucoside, 320
Rice Husk, 99, 101, 235, 243
Rice Husk Ash (RHA), 231
Ross-Miles method, 248
Rubbery substance, 60
Rushton type six-blade impeller, 290
Rutinose sugar molecule, 321

S

Scanning electron microscope (SEM) , 64, 65, 75, 124–127, 137, 149, 185, 189, 192, 195, 204–206, 222, 225, 226, 240, 241, 288
  analysis, 149, 189
  images, 126, 127, 149, 204–206
Semi-implicit method, 11
Semi-solid fats, 294
Sherewood number, 290
Silica aerogel, 61
Silver
  cadmium batteries, 100
  ions, 206, 207, 210
  nanoparticles, 200–207, 209, 210, 212, 259
    analysis, 204
SIMPLE algorithm, 11
Simulation process, 24
Simulation strategy, 20
Skeletal vibrations, 147
Skimming vessels, 60
Small-scale plants, 242
Sodium aluminate, 233
Sodium dodecyl sulfate (SDS), 202, 203, 210, 212–214, 260, 261, 270–273, 276
Sodium hydroxide solution, 233
Sodium lauryl sulfate (SLS), 249
Sodium silicate, 61
  aluminate, 63
Solid
  liquid extraction techniques, 313
  liquid separation, 88
  loading, 309–312, 314
  matrix, 310
  phase, 23, 289, 292
Solidify spilled oil, 60
Solvent molecule, 310
Solving governing equations, 11
Sonication, 145, 164, 166, 185, 321
Sorbent weight, 67
Sorbents, 61, 104, 113
Sorbing species, 78, 82
Sorption capacities, 99, 104, 109, 113
Soxhlet
  apparatus, 309
  extraction, 65
  method, 61
Spectrophotometer, 146, 204, 222, 309, 310, 323, 330
Spectrophotometric, 330, 333, 334
  technique, 330
Spectrophotometrically, 77, 310
Spectrophotometry technique, 67
Stabilization and aggregation, 265
Steroids, 308
Stock solution, 37, 102, 122, 330
Stoichiometry, 232, 236, 238
Stoppard borosil glass bottles, 106
Suction devices, 60
Sulphuric acid, 99
Surface modified Zeolite-Y, 68
Surfactants, 248, 255

Sweetlime peel powder, 73, 75–77, 79, 81, 83–85
Synthetic dye stuff, 36
Syzygium cumini, 308

T

Tanneries, 98
Tannins, 308
Temperature (T), 40
Terpenoids, 308
Tertiary butylhydroquinone (TBHQ), 320
Tetrabutylammonium chloride (TBAC), 122
Tetrahedral (silicic), 121
Textile industries, 298
Theoretical
  discharge, 6, 7
  maximum adsorption capacity, 173
Therapeutic value, 308
Thermal
  applications, 230
  energy, 93
  length, 302–304
  oxidation, 283
  power plants, 228, 229, 232
  stability, 137, 286, 294
Thermochemical
  modifications, 121
  parameters, 157
  study, 159
Thermogravimetric analysis, 124
Toxic components, 57
Toxic effects, 36, 334
Toxicity level, 334
Toxicological effects, 283
Transmission electron microscopy (TEM), 124–126, 137, 167, 261, 271–273
Trapezoidal numerical integration, 26
Turbulence equations, 263
Two-dimensional, 8, 11, 21

U

Ultra-light, 59
Ultrasonic
  irradiation, 168, 176, 184, 190, 192
  waves, 310
Ultrasonicated assisted extraction technique, 314
Ultrasonication, 145, 314
Ultrasound, 163, 165, 167, 169, 171–173, 175, 176, 191, 309, 311, 313, 315
Underwater dispersants, 59
Uniform mass transport, 266
Univariate analysis, 40
Universal medical instruments, 261
Unoxidized
  acids, 284
  graphitic domains, 147
  oils, 287
Uranium, 230
Urinary 5-hydroxyindole acetic acid, 20
Uroporphyrins, 20
UV-detector, 321
UV-laser diodes, 259
UV-Vis spectrometer, 102–109, 111–113, 124, 203
UV-Vis spectrophotometer, 37, 77, 124, 261, 315

V

Vacuum filtration technique, 322
Vacuum oven, 145
Vacuum pump, 63, 102, 233
Vanaspati manufacturing units, 284
Variation of
  Cd with material properties, 12
  Cd with pipe diameter, 12
  Cd with β, 13
  kor with material properties, 16
  kor with pipe diameter, 13
  kor with β, 16
Vegetable oils, 200, 285, 286, 289
Velocities, 7, 298, 302
Velocity, 4, 5, 7, 8, 11, 21, 22, 156, 260, 267, 273–275, 290
Velocity profile, 260, 273–275
Vertigo, 230
Vinylpyrrolidone, 163

Viscosity, 9, 12, 59, 298, 300, 302, 304
Viscous
fluid system, 304
forces, 12, 267
Visual observations, 203
Vitamin C, 308, 320
Vitamins, 283, 308
Volatile matter, 233
Volatiles content, 220
Volume element, 263
Volume ratio, 145, 200, 210, 214, 259, 275
Volumetric flask, 75, 322
Volumetric flow rate, 7
Volumetric measurement, 222

W

Warrant equilibrium, 146
Waste heat recovery systems, 298
Wastewater, 88, 89, 91, 93, 98
treatment, 90, 120, 143, 162, 163
Water absorption property, 165
Water condensation, 189, 193–195
Water pollution control, 120
Water Sanitation and Hygiene (WSH), 142
Water swelling, 176
Water system, 300
Water uptake capacity, 169
Water-filled center, 284
Wavelength, 37, 102, 106, 111, 124, 146, 165, 207, 310, 323, 324, 331
photoelectronic, 259
Wax content, 61
Weather conditions, 57
Weight loss, 195
Wind power capacity, 229
Wind turbine syndrome, 230

X

X-rays, 64
(EDX) analysis, 64, 147, 190, 204, 219
diffraction, 148, 188, 204, 223, 273
diffractographs, 64, 147, 185, 188–190, 195, 204–207, 219, 222–225, 240, 261, 273
diffractometer, 189, 222, 240
Xylenol, 330, 331, 332
orange solution, 330, 331

Y

Y-joint, 263
microreactor, 263
Y-junction, 263
Yoon–Nelson model, 79, 85

Z

Zeolite, 61, 64, 121
Zeolite-A, 232, 233, 239–243
Zeolite-Y, 57, 59, 61–65, 67
Zeolite-Y resin, 63
Zinc, 100, 182–189, 191–195, 260–262, 268, 270–273, 276
Zinc chloride, 184–186, 260, 261
Zinc phosphate, 183
nanopigment, 186, 187, 189–193
synthesis, 184
Zinc rich coating, 194
Zinc sulfide (ZnS), 259, 264, 269–271, 276
nanoparticles, 259–261, 264–266, 268–271, 273
nanostructures, 259
particle size, 269
particles, 265, 270, 273
precipitate, 268
samples, 273
Zinc-rich paints, 194
ZP nanoparticles, 190–193
Zwitterionic, 248